In Memory of Stefan Kunz

In Memory of Stefan Kunz

Editors

Michael B.A. Oldstone
Juan De la Torre

MDPI • Basel • Beijing • Wuhan • Barcelona • Belgrade • Manchester • Tokyo • Cluj • Tianjin

Editors
Michael B.A. Oldstone
The Scripps Research Institute
USA

Juan De la Torre
The Scripps Research Institute
USA

Editorial Office
MDPI
St. Alban-Anlage 66
4052 Basel, Switzerland

This is a reprint of articles from the Special Issue published online in the open access journal *Viruses* (ISSN 1999-4915) (available at: https://www.mdpi.com/journal/viruses/special_issues/Stefan_Kunz).

For citation purposes, cite each article independently as indicated on the article page online and as indicated below:

LastName, A.A.; LastName, B.B.; LastName, C.C. Article Title. *Journal Name* **Year**, *Volume Number*, Page Range.

ISBN 978-3-0365-2947-9 (Hbk)
ISBN 978-3-0365-2946-2 (PDF)

Contents

 MDPI

Obituary

Tribute to Professor Stefan Kunz

Manuel Pascual [1,2]

1 Faculty of Biology and Medicine, University of Lausanne, 1011 Lausanne, Switzerland; Manuel.Pascual@unil.ch
2 Lausanne University Hospital, 1011 Lausanne, Switzerland

It has been a year since Stefan Kunz, a Full Professor since 2017 at the Faculty of Biology and Medicine of the University of Lausanne, Switzerland, passed away. His death was a shock and an enormous loss, not only for his family and friends, but also the entire Swiss and international scientific community, and especially our Faculty of Biology and Medicine and the University Hospital Center of Lausanne (CHUV). Professor Kunz always impressed us for being a deeply engaging, charismatic and gifted teacher, and for his benevolence, his energy and his discoveries.

We miss Professor Kunz, who specialized in the discipline of emerging viruses, focusing his efforts on understanding how viruses succeed in breaking species barriers to then hijack the human host cell machinery.

He started his career in the field of neurosciences, earning his PhD thesis at the University of Zurich, with a special interest in neural signaling. He then went to the Scripps Research Institute in California as a postdoctoral fellow, where he worked with one of the greatest experts in virology worldwide, Michael Oldstone. He worked in San Diego for almost 10 years, rising to the rank of Associate Professor at the Scripps Research Institute. Returning to Switzerland in 2008, he joined the Institute for Microbiology of the University of Lausanne.

His contributions are invaluable on every level: first and foremost, through his research on hemorrhagic fever viruses, dangerous pathogens for mankind, with his special focus on Lassa Virus. Furthermore, he produced excellent work on transmission and infection with emerging viruses such as Ebola.

As a mentor, he also created a veritable "school" of virology in Lausanne, involving dozens of researchers. He conceptualized new courses, conveying his own philosophy of teaching, his passion for medicine and his curiosity for research. Stefan Kunz was an extraordinary colleague and personality, and he will forever remain in our hearts and our memory. We would like to honor Professor Kunz and celebrate all that he has given us.

Funding: This research received no external funding.

Conflicts of Interest: The author declares no conflict of interest.

Citation: Pascual, M. Tribute to Professor Stefan Kunz. *Viruses* **2021**, *13*, 1862. https://doi.org/10.3390/v13091862

Received: 30 August 2021
Accepted: 31 August 2021
Published: 18 September 2021

Publisher's Note: MDPI stays neutral with regard to jurisdictional claims in published maps and institutional affiliations.

 viruses

MDPI

Obituary

In Tribute to Stefan Kunz

Mar Perez

Spanish Ministry of Defense, Paseo de la Castellana, 109, 28071 Madrid, Spain; mapema@gmail.com

We do not always remember the exact moment in which we first met our friends. However, I perfectly recall the day I talked to Stefan for the very first time. Or, to be precise, the first time that Stefan talked to me as I was desperately trying not to choke on a glass of water. It was Stefan's first day at TSRI and he had set out on a mission to introduce himself to every other person in the Viral-Immunology lab, where I had started my postdoc a couple of months before. There was no way that I could have known at the time, but I had just met a superb colleague, a remarkable scientist, a generous teacher, a charismatic mentor, and a great friend. But there he was, just standing right in front of a water cooler.

The first thing everyone noticed about Stefan is that he dressed in black almost every day or, if feeling particularly daring, in jeans and a white t-shirt. When asked about his limited choice of colours, he would hide behind the claim of being partially colour-blind. Nevertheless, it was quite evident that, for him, life was too short, and brimful of interesting projects as to waste even a second of time on petty decisions.

Dr. Stefan Kunz (August 1966–January 2020).

Citation: Perez, M. In Tribute to Stefan Kunz. *Viruses* **2021**, *13*, 1840. https://doi.org/10.3390/v13091840

Received: 12 August 2021
Accepted: 31 August 2021
Published: 15 September 2021

As a true Swiss, Stefan deeply enjoyed being outdoors, hiking, running, or skiing, but he also loved travelling, reading, and of course, cooking. He used to say that his Italian genes made him extremely fond of good food but had not taught him how to cook, so he had to fix that by taking cooking classes. Those fortunate enough to try his cooking would agree that the lessons really paid off. If you do not believe me, ask Nathalie, Kurt, Noemí, Elina, Hanna, or any of the many friends that he made while working in San Diego. At home, my husband and I still regularly prepare the recipes he shared with us.

Stefan had been gifted with a bright mind and boundless energy that he put to work for the benefit of science and society. Despite his love for research, he was a man of many interests, who could talk about the Bauhaus or a virus entry mechanism with the exact same ease and passion. He was always eager to learn, be it a new immunological technique, another language (he inadvertently insulted me a few times while practicing his Spanish), or even a step of salsa (although truth be told, he was a terrible dancer).

Over the years, we had endless chats about every single topic, heated scientific discussions at virology meetings, and many trips and meals. When the time came to marry the love of his life, Karin, he asked me to be his best man, notwithstanding the fact that I am a woman, which, in my opinion, is a great example of Stefan's personality. Conventionalism never got in his way!

In passing away, Stefan leaves an enormous void, both in the scientific community and in the lives of his friends and family. His contributions to the field of virology are well known, but what his friends and colleagues will remember is the integrity and human warmth of his personality.

At his funeral, his friend Leonard read one of Stefan's favourite poems, "The Palace" by Rudyard Kipling. This beautiful text perfectly reflects Stefan's vision on science, a worldwide collaborative effort in which every achievement, big or small, rests on the work of other scientists with whom we share the passion for discovery. Therefore, it just feels right to also share here its last line: "After me cometh a Builder. Tell him, I too have known!"

"Hasta siempre, suizo loco".

(So long, crazy Swiss).

Funding: This research received no external funding.

Conflicts of Interest: The author declares no conflict of interest.

Review

Lassa Fever Virus Binds Matriglycan—A Polymer of Alternating Xylose and Glucuronate—On α-Dystroglycan

Soumya Joseph and Kevin P. Campbell *

Howard Hughes Medical Institute, Senator Paul D. Wellstone Muscular Dystrophy Specialized Research Center, Department of Molecular Physiology and Biophysics and Department of Neurology, Roy J. and Lucille A. Carver College of Medicine, The University of Iowa, Iowa City, IA 52242, USA; soumya-joseph@uiowa.edu
* Correspondence: kevin-campbell@uiowa.edu

Abstract: Lassa fever virus (LASV) can cause life-threatening hemorrhagic fevers for which there are currently no vaccines or targeted treatments. The late Prof. Stefan Kunz, along with others, showed that the high-affinity host receptor for LASV, and other Old World and clade-C New World mammarenaviruses, is matriglycan—a linear repeating disaccharide of alternating xylose and glucuronic acid that is polymerized uniquely on α-dystroglycan by like-acetylglucosaminyltransferase-1 (LARGE1). Although α-dystroglycan is ubiquitously expressed, LASV preferentially infects vascular endothelia and professional phagocytic cells, which suggests that viral entry requires additional cell-specific factors. In this review, we highlight the work of Stefan Kunz detailing the molecular mechanism of LASV binding and discuss the requirements of receptors, such as tyrosine kinases, for internalization through apoptotic mimicry.

Keywords: Lassa fever virus; matriglycan; α-dystroglycan; lymphocytic choriomeningitis; Axl; Gas6; dystrophin-glycoprotein complex; LARGE1; laminin; apoptotic mimicry

Citation: Joseph, S.; Campbell, K.P. Lassa Fever Virus Binds Matriglycan—A Polymer of Alternating Xylose and Glucuronate—On α-Dystroglycan. *Viruses* **2021**, *13*, 1679. https://doi.org/10.3390/v13091679

Academic Editor: Luis Martinez-Sobrido

Received: 21 July 2021
Accepted: 20 August 2021
Published: 25 August 2021

1. Introduction

The surfaces of host cells and pathogens are covered with glycoproteins. Glycans are well-known targets for virus binding and entry [1]. We highlight the work of the late Stefan Kunz who showed the surface glycoprotein of Old World, including LASV, and New World clade-C mammarenaviruses, binds matriglycan [2–5]—a polysaccharide of alternating xylose and glucuronate—on the highly glycosylated mucin-like receptor, dystroglcyan on rodent and human hosts (Figure 1). On the other hand, the spike glycoproteins (GP) of New World mammarenavirus clades use transferrin as a receptor [6–8]. We limit the scope of our review to virus binding and entry. Other publications detailing events post cell-entry and broader aspects of arenaviruses have been published [9–12].

The high-affinty receptor for the trimeric glycoprotein-1 (GP1) of Old World and clade-C New World mammarenaviruses is matriglycan, which is polymerized on α-dystroglycan post-translationally by like-acetylglucosaminyltransferase (LARGE1) [13]. Dystroglycan consists of α-dystroglycan and β-dystroglycan [14] and is part of the plasma membrane-embedded dystrophin-glycoprotein-complex (DGC), which is ubiquitously expressed but whose composition varies between cell types [15–17]. Under normal circumstances, proteins that contain laminin-globular (LG) domains bind matriglycan and components of the DGC bind the actin cytoskeleton such that, at its simplest, the DGC anchors the cytoskeleton to the extracellular matrix (ECM) [18–22]. Genetic mutations that diminish or abolish matriglycan polymerization can cause muscular dystrophies, known as dystroglycanopathies, with or without brain and eye involvement [23,24]. Much like mutations that cause sickle-cell anemia can protect against malarial parasites, LARGE1 alleles have undergone recent positive selection in West African populations where LASV is endemic [25].

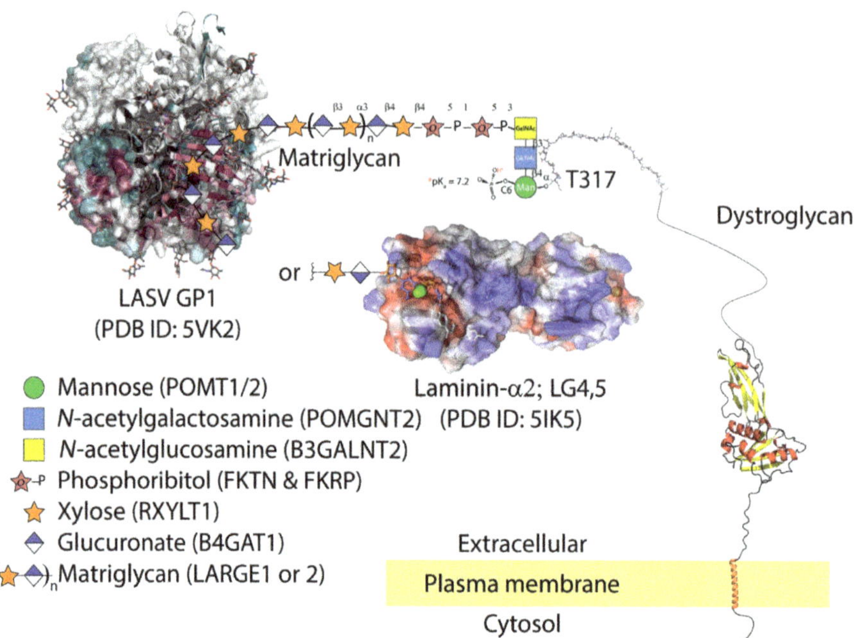

Figure 1. LASV GP1 is able to bind matriglycan (xylose and glucuronate), but only gains entry to cells that co-express apoptotic phagocytic machinery. The molecular details of LASV GP1 binding to matriglycan are unknown. Matriglycan is polymerized on a primer of extended phosphocore M3 on threonine-317 and possibly 379 of α-dystroglycan. The core M3 trisaccharide is phosphorylated by Protein O-Mannosyl Kinase (POMK); other glycosyltransferases are listed in parentheses next to their corresponding sugars. The conserved surface residues of LASV trimer from 5VK2 are shown as a gradient of magenta (conserved) to green (non-conserved); (accessed on 6 July 2021: https://consurf.tau.ac.il/). LASV GP1 binding displaces LG domains from matriglycan. The semi-transparent electrostatic surface of LG4-5 domains from laminin-2α is shown binding a unit of xylose-glucuronate via calcium (green sphere). Parts of the dystroglycan structure were downloaded from AlphaFold [26].

Despite ubiquitous expression of matriglycan on dystroglycan throughout human hosts, LASV preferentially infects antigen-presenting cells, including dendritic cells [27], but not muscle cells [28], which suggests that additional factors influence viral entry. The differential manifestations of both genetic and infectious diseases indicate that dystroglycan associates with dynamic, cell-type-specific complexes [29,30] on which pathologies depend. Many reports [31–33] have implicated auxiliary cell surface receptors as co-factors for viral entry: the tyrosine kinase family, Tyro3/Axl/MERTK (TAM) or hepatocyte growth factor receptor (HGFR) [34], T-cell immunoglobulin and mucin domain-containing proteins that bind phosphatidylserine, TIM1 and 4 [35], and the lectin family-4 members M and G (CLEC4G and M) that are popularly known as DC-SIGN and LSECtin [33].

LASV is a highly prevalent mammarenavirus in Western Africa, where it infects a few hundred thousand individuals annually [36]. Lassa fever (LF) is associated with high morbidity and a recent average case fatality rate of ~20% in Nigeria [37]. Currently, neither a vaccine nor specific treatments for LF exist. The majority of basic scientific knowledge on LASV entry was generated in the Oldstone laboratory (Scripps Reseach Institute) and much of the detailed analysis was carried out by the late Dr. Stefan Kunz, first as a post-doctoral fellow in the Oldstone laboratory and later as an independent investigator at the University of Lausanne.

Conveniently, lymphocytic choriomeningitis virus (LCMV) and LASV both use α-dystroglycan as a high-affinity receptor [2]. Results from virus overlay protein blot assays

(VOPBA), which are like Western blots but use whole virus particles instead of primary antibodies, suggested that an SDS-resistant moiety on α-dystroglycan is responsible for binding virus particles. Additionally, viral infectivity was restored in α-dystroglycan-deficient cells by adenoviral dystroglycan transductions. Purified soluble α-dystroglycan blocked cell entry by acting as a decoy receptor. These initial studies also showed that bacterially expressed α-dystroglycan could not bind virus particles, suggesting that a mammalian-specific post-translational modification, perhaps glycosylation, was crucial to binding [2]. This formed the basis of Dr. Kunz' invaluable work on mammarenaviral cell entry. Using a series of dystroglycan deletion mutants, he deduced that residues 169–408 of dystroglycan were essential for virus binding and that the viral glycoprotein bound the same post-translational modification as laminin [4]. This work culminated in the understanding that GP1 from LASV, LCMV, Mobala, and Oliveros mammarenavirus [5] bound a chain of alternating xylose and glucuronic acid that is polymerized by LARGE1 on α-dystroglycan residues threonine-317 [13,38]. His conclusions were bosltered by the fact that the overexpression of LARGE1 increased mammarenaviral binding [5,39]. Interestingly, a novel genetic screen using haploid cells confirmed these results [40]. Also, replacing the transmembrane and cytosolic domains of β-dystroglycan did not affect viral entry [41]. This, along with tissue tropism, implicates auxiliary signalling receptors in viral entry. This was the foundation of Dr. Kunz' independent research.

2. LASV Binds Matriglycan and Is Internalized in Cells Co-Expressing Gas6-Axl

Phosphorylation of intracellular auxiliary receptors is key to LASV cell entry [31,32,42,43]. The application of the generic tyrosine kinase inhibitor genistein prevented virus internalization but not binding [34,42], which suggests that virus binding and internalization are separable. For example, tyrosine-892 on β -dystroglycan is phosphorylated in response to virus binding. The Tyro3/Axl/MERTK (TAM) tyrosine kinases that are expressed on dendritic cells can promote the internalization of virus particles via receptor phosphorylation [17,42].

Axl bound to its co-factor growth arrest-specific protein-6 (Gas6) is expressed in cells that phagocytose apoptotic cells and is key to internalizing other viruses such as Zika, Dengue, and Ebola [32,44–46], which use apoptotic mimicry [47,48]. Although there are conflicting results regarding the role of Axl for mammarenavirus cell entry [31,49], Shimojima et al. showed that deletion of the first immunoglobulin-like domain of Axl abolished LASV cell entry [32]. Morizono et al. showed that the presence of Gas6 enhanced LASV infection [33]. A crystal structure of the LG domains from Gas6 in complex with Axl shows the interaction is mediated via an anti-parallel β-zipper and buries a total surface area of ~1100 Å^2 [50]. The LG domains of laminin-α2 can bind matriglycan through the chelation of calcium via two aspartate residues [51] (Figure 1). There is no evidence that the Gas6 LG domains bind matriglycan like those of laminin-α2 [52]. A more likely scenario is that LASV binds cells that express matriglycan but is only internalized in cells which co-express the Axl phagocytic machinery.

Functionally, deletion or substitution of the ATP binding site (K567M), or phosphotyrosine site (Y821F), of Axl cytoplasmic domains prevents LASV entry, suggesting that Axl tyrosine autophosphorylation is essential for LASV entry [35]. This research suggests that cells co-expressing α-dystroglycan modified with matriglycan, Gas6, and Axl or Tyro3 are highly susceptible to LASV entry, whereas non-phagocytic cells such as myocytes, although they express plenty of matriglycan and can bind virions, are incapable of engulfing LASV particles via apoptotic mimicry [34,47,48].

3. Subunit GP1 of the Viral Spike Glycoprotein Binds Matriglycan at the Interface Formed upon Trimerization

Unlike the calcium chelating mode in which LG domains bind matriglycan [4,51], GP1 from LASV and LCMV, which share ~50% sequence identity, likely bind matriglycan using a region formed at the central surface of the trimer. This deduction comes from the fact that the F260 [L/V/I] substitution in the LCMV GP1 increases the affinity for matriglycan by 2–2.5 orders of magnitude and promotes persistent infection in mice [3]. The residues surrounding leucine-260 are essential for viral fusion [53]. The equivalent leucine residue in the trimeric pre-fusion structure of LASV GP1 also maps to their central interface [54], which hints at the location of matriglycan binding (Figure 2).

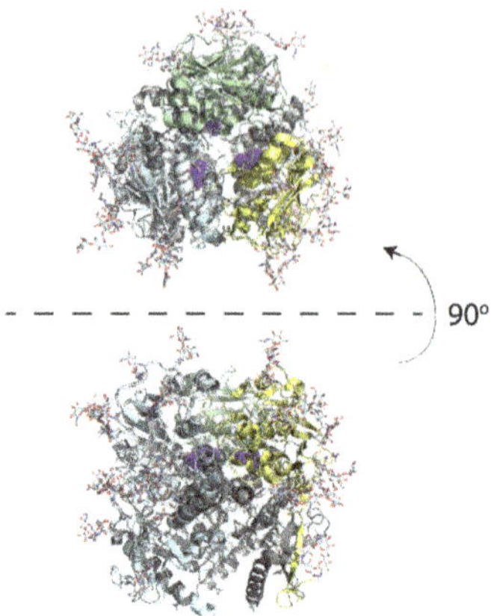

Figure 2. The crystal structure of LASV GP1 trimer is show in cartoon (PDB ID: 5vk2). Leucine-260 is shown in purple on the LCMV trimer. The upper panel shows the top view; the lower panel shows a side view.

4. Hypothetical Mechanism of Hemorrhage

LASV-infected cells appear to downregulate matriglycan polymerization [55]. The simplest mechanism might be that the spike protein encounters matriglycan that is nascently polymerized on α-dystroglycan in the Golgi as it traverses the secretory pathway. The incidental exposure of GP1 could disrupt and decrease matriglycan polymerization by LARGE1, which might explain the complex formation and self-limiting infection [55]. A decrease in matriglycan expression has been linked to membrane fragility and treatment of muscle cells with inactivated LCMV interferes with membrane integrity [56]. Patients with high viral load can succumb to hemorrhage, although the mechanism of hemorrhage remains unclear. A physical hypothesis is that compromised cell membranes of blood vessels contribute to hemorrhage. A biochemical hypothesis is that interference with vitamin K-dependent protein-S (ProS1), which is another TAM co-receptor, such as Gas6, acts in complex with ProtC to degrade blood clotting factors Va and VII. Interestingly, ProS1 shares domain architecture and 40% sequence identity with Gas6. Moreover, platelets, which express TAMs, are activated by Gas6 [57]. Perturbation of the ProS1-dependent blood coagulation mechanism eventually causes TAM triple-negative mice to hemorrhage [58]. Interference with platelet function may inhibit coagulation and can cause hemorrhage. Both of these physical and biochemical hypotheses remain to be tested experimentally.

5. Conclusions and Remarks

LASV binds host cells via its high-affinity interaction with matriglycan, a repeating disaccharide that is synthesized post-translationally on α-dystroglycan by LARGE1. The spike protein likely binds matriglycan in the central interface formed by trimerization of GP1 monomers. Although virus binding and entry appear to be independent, the former increases the probability of the latter. Viral entry is determined additionally by cell-type-specific factors [17] such as the tyrosine kinase Axl and cell surface lectins. Although studies conflict, many other viruses enter cells using Gas6-Axl mediated apoptotic mimicry [33,44,57,59,60], and this may also explain the cell specificity of the virus. The mechanism of hepatocyte infection and hemorrhage may have to do with parallel tyrosine kinase and LG-domain containing receptors in the liver, such as hepatocyte growth factor receptor (HGFR), and remains to be tested. A decrease in the polymerization of matriglycan may be due to GP1 interacting with the nascent matriglycan polymer on LARGE1-dystroglycan enzyme-substrate complex in the Golgi, which coincidentally prevents further virus entry. Hemorrhage might be the side-effect of conscripting ProS to bind viral glycoprotein resulting in poor blood clotting. We can honor Dr Kunz' legacy by building on his invaluable work to further explore these unanswered questions.

Author Contributions: S.J. and K.P.C. wrote the manuscript. All authors have read and agreed to the published version of the manuscript.

Funding: This work was supported in part by a Paul D. Wellstone Muscular Dystrophy Specialized Research Center grant (1U54NS053672 to K.P.C.). K.P.C. is an investigator of the Howard Hughes Medical Institute.

Institutional Review Board Statement: Not available.

Informed Consent Statement: Not available.

Data Availability Statement: All data generated or analyzed during this study are included in this published article.

Acknowledgments: We are grateful to Amber Mower for assistance with administrative support. We would also like to thank Erhard Hohenester, Juan Carlos de la Torre, Michael Oldstone, Lance Wells, and members of the Campbell lab for their helpful comments.

Conflicts of Interest: The authors declare no conflict of interest.

References

1. Lentz, T.L. The recognition event between virus and host cell receptor: A target for antiviral agents. *J. Gen. Virol.* **1990**, *71*, 751–766. [CrossRef]
2. Cao, W.; Henry, M.D.; Borrow, P.; Yamada, H.; Elder, J.H.; Ravkov, E.V.; Nichol, S.T.; Compans, R.W.; Campbell, K.P.; Oldstone, M.B.A. Identification of α-Dystroglycan as a Receptor for Lymphocytic Choriomeningitis Virus and Lassa Fever Virus. *Science* **1998**, *282*, 2079–2081. [CrossRef]
3. Sevilla, N.; Kunz, S.; Holz, A.; Lewicki, H.; Homann, D.; Yamada, H.; Campbell, K.P.; de la Torre, J.C.; Oldstone, M.B.A. Immunosuppression and Resultant Viral Persistence by Specific Viral Targeting of Dendritic Cells. *J. Exp. Med.* **2000**, *192*, 1249–1260. [CrossRef]
4. Kunz, S.; Sevilla, N.; McGavern, D.B.; Campbell, K.P.; Oldstone, M.B.A. Molecular analysis of the interaction of LCMV with its cellular receptor α-dystroglycan. *J. Cell Biol.* **2001**, *155*, 301–310. [CrossRef] [PubMed]
5. Kunz, S.; Rojek, J.M.; Kanagawa, M.; Spiropoulou, C.F.; Barresi, R.; Campbell, K.P.; Oldstone, M.B.A. Posttranslational Modification of Dystroglycan, the Cellular Receptor for Arenaviruses, by the Glycosyltransferase LARGE Is Critical for Virus Binding. *J. Virol.* **2005**, *79*, 14282–14296. [CrossRef] [PubMed]
6. Radoshitzky, S.R.; Abraham, J.; Spiropoulou, C.F.; Kuhn, J.H.; Nguyen, D.; Li, W.; Nagel, J.; Schmidt, P.J.; Nunberg, J.H.; Andrews, N.C.; et al. Transferrin receptor 1 is a cellular receptor for New World haemorrhagic fever arenaviruses. *Nature* **2007**, *446*, 92–96. [CrossRef] [PubMed]
7. Flanagan, M.L.; Oldenburg, J.; Reignier, T.; Holt, N.; Hamilton, G.A.; Martin, V.K.; Cannon, P.M. New World Clade B Arenaviruses Can Use Transferrin Receptor 1 (TfR1)-Dependent and -Independent Entry Pathways, and Glycoproteins from Human Pathogenic Strains Are Associated with the Use of TfR1. *J. Virol.* **2008**, *82*, 938–948. [CrossRef]
8. Abraham, J.; Corbett, K.D.; Farzan, M.; Choe, H.; Harrison, S.C. Structural basis for receptor recognition by New World hemorrhagic fever arenaviruses. *Nat. Struct. Mol. Biol.* **2010**, *17*, 438–444. [CrossRef] [PubMed]
9. Jae, L.T.; Raaben, M.; Herbert, A.S.; Kuehne, A.I.; Wirchnianski, A.S.; Soh, T.K.; Stubbs, S.H.; Janssen, H.; Damme, M.; Saftig, P.; et al. Lassa virus entry requires a trigger-induced receptor switch. *Science* **2014**, *344*, 1506–1510. [CrossRef]
10. Peng, R.; Xu, X.; Jing, J.; Wang, M.; Peng, Q.; Liu, S.; Wu, Y.; Bao, X.; Wang, P.; Qi, J.; et al. Structural insight into arenavirus replication machinery. *Nature* **2020**, *579*, 615–619. [CrossRef]
11. Pasquato, A.; Fernandez, A.H.; Kunz, S. *Studies of Lassa Virus Cell Entry BT—Hemorrhagic Fever Viruses: Methods and Protocols*; Salvato, M.S., Ed.; Springer: New York, NY, USA, 2018; pp. 135–155. ISBN 978-1-4939-6981-4.
12. Bhadelia, N. Understanding Lassa fever. *Science* **2019**, *363*, 30. [CrossRef]
13. Inamori, K.; Yoshida-Moriguchi, T.; Hara, Y.; Anderson, M.E.; Yu, L.; Campbell, K.P. Dystroglycan Function Requires Xylosyl- and Glucuronyltransferase Activities of LARGE. *Science* **2012**, *335*, 93–96. [CrossRef]
14. Ervasti, J.M.; Ohlendieck, K.; Kahl, S.D.; Gaver, M.; Campbell, K.P. Deficiency of a Glycoprotein Component of the Dystrophin Complex in Dystrophic Muscle. *Nature* **1990**, *345*, 315–319. [CrossRef] [PubMed]
15. Durbeej, M.; Henry, M.D.; Ferletta, M.; Campbell, K.P.; Ekblom, P. Distribution of Dystroglycan in Normal Adult Mouse Tissues. *J. Histochem. Cytochem.* **1998**, *46*, 449–457. [CrossRef]

16. Culligan, K.; Ohlendieck, K. Diversity of the Brain Dystrophin-Glycoprotein Complex. *J. Biomed. Biotechnol.* **2002**, *2*, 390232. [CrossRef]
17. Herrador, A.; Fedeli, C.; Radulovic, E.; Campbell, K.P.; Moreno, H.; Gerold, G.; Kunz, S.; Dermody, T.S. Dynamic Dystroglycan Complexes Mediate Cell Entry of Lassa Virus. *MBio* **2021**, *10*, e02869-18. [CrossRef] [PubMed]
18. Ibraghimov-Beskrovnaya, O.; Ervasti, J.M.; Leveille, C.J.; Slaughter, C.A.; Sernett, S.W.; Campbell, K.P. Primary Structure of Dystrophin-Associated Glycoproteins Linking Dystrophin to the Extracellular Matrix. *Nature* **1992**, *355*, 696–702. [CrossRef] [PubMed]
19. Ervasti, J.M.; Campbell, K.P. A Role for the Dystrophin-Glycoprotein Complex as a Transmembrane Linker Between Laminin and Actin. *J. Cell Biol.* **1993**, *122*, 809–823. [CrossRef] [PubMed]
20. Campbell, K.P. Three muscular dystrophies: Loss of cytoskeleton-extracellular matrix linkage. *Cell* **1995**, *80*, 675–679. [CrossRef]
21. Henry, M.D.; Campbell, K.P. Dystroglycan: An extracellular matrix receptor linked to the cytoskeleton. *Curr. Opin. Cell Biol.* **1996**, *8*, 625–631. [CrossRef]
22. Ohlendieck, K. Towards an understanding of the dystrophin-glycoprotein complex: Linkage between the extracellular matrix and the membrane cytoskeleton in muscle fibers. *Eur. J. Cell Biol.* **1996**, *69*, 1–10.
23. Michele, D.E.; Barresi, R.; Kanagawa, M.; Saito, F.; Cohn, R.D.; Satz, J.S.; Dollar, H.; Nishino, I.; Kelley, R.I.; Somer, H.; et al. Post-translational Disruption of Dystroglycan-Ligand Interactions in Congenital Muscular Dystrophies. *Nature* **2002**, *418*, 417–422. [CrossRef]
24. Yoshida-Moriguchi, T.; Campbell, K.P. Matriglycan: A novel polysaccharide that links dystroglycan to the basement membrane. *Glycobiology* **2015**, *25*, 702–713. [CrossRef]
25. Sabeti, P.C.; Varilly, P.; Fry, B.; Lohmueller, J.; Hostetter, E.; Cotsapas, C.; Xie, X.; Byrne, E.H.; McCarroll, S.A.; Gaudet, R.; et al. Genome-wide detection and characterization of positive selection in human populations. *Nature* **2007**, *449*, 913–918. [CrossRef] [PubMed]
26. Jumper, J.; Evans, R.; Pritzel, A.; Green, T.; Figurnov, M.; Ronneberger, O.; Tunyasuvunakool, K.; Bates, R.; ŽíDek, A.; Potapenko, A.; et al. Highly accurate protein structure prediction with AlphaFold. *Nature* **2021**. [CrossRef]
27. Baize, S.; Kaplon, J.; Faure, C.; Pannetier, D.; Georges-Courbot, M.-C.; Deubel, V. Lassa Virus Infection of Human Dendritic Cells and Macrophages Is Productive but Fails to Activate Cells. *J. Immunol.* **2004**, *172*, 2861–2869. [CrossRef] [PubMed]
28. Iwasaki, M.; Urata, S.; Cho, Y.; Ngo, N.; de la Torre, J.C. Cell entry of lymphocytic choriomeningitis virus is restricted in myotubes. *Virology* **2014**, *458–459*, 22–32. [CrossRef] [PubMed]
29. Barresi, R. Dystroglycan: From biosynthesis to pathogenesis of human disease. *J. Cell Sci.* **2006**, *119*, 199–207. [CrossRef] [PubMed]
30. Durbeej, M.; Campbell, K.P. Biochemical Characterization of the Epithelial Dystroglycan Complex*. *J. Biol. Chem.* **1999**, *274*, 26609–26616. [CrossRef]
31. Fedeli, C.; Torriani, G.; Galan-Navarro, C.; Moraz, M.-L.; Moreno, H.; Gerold, G.; Kunz, S.; Dermody, T.S. Axl Can Serve as Entry Factor for Lassa Virus Depending on the Functional Glycosylation of Dystroglycan. *J. Virol.* **2021**, *92*, e01613-17. [CrossRef] [PubMed]
32. Shimojima, M.; Ströher, U.; Ebihara, H.; Feldmann, H.; Kawaoka, Y. Identification of Cell Surface Molecules Involved in Dystroglycan-Independent Lassa Virus Cell Entry. *J. Virol.* **2012**, *86*, 2067–2078. [CrossRef]
33. Morizono, K.; Xie, Y.; Olafsen, T.; Lee, B.; Dasgupta, A.; Wu, A.M.; Chen, I.S.Y. The Soluble Serum Protein Gas6 Bridges Virion Envelope Phosphatidylserine to the TAM Receptor Tyrosine Kinase Axl to Mediate Viral Entry. *Cell Host Microbe* **2011**, *9*, 286–298. [CrossRef] [PubMed]
34. Oppliger, J.; Torriani, G.; Herrador, A.; Kunz, S.; Dermody, T.S. Lassa Virus Cell Entry via Dystroglycan Involves an Unusual Pathway of Macropinocytosis. *J. Virol.* **2021**, *90*, 6412–6429. [CrossRef]
35. Brouillette, R.B.; Phillips, E.K.; Patel, R.; Mahauad-Fernandez, W.; Moller-Tank, S.; Rogers, K.J.; Dillard, J.A.; Cooney, A.L.; Martinez-Sobrido, L.; Chioma, O.; et al. TIM-1 Mediates Dystroglycan-Independent Entry of Lassa Virus. *J. Virol.* **2021**, *92*, e00093-18. [CrossRef]
36. Ogbu, O.; Ajuluchukwu, E.; Uneke, C.J. Lassa fever in West African sub-region: An overview. *J. Vector Borne Dis.* **2007**, *44*, 1.
37. Yaro, C.A.; Kogi, E.; Opara, K.N.; Batiha, G.E.-S.; Baty, R.S.; Albrakati, A.; Altalbawy, F.M.A.; Etuh, I.U.; Oni, J.P. Infection pattern, case fatality rate and spread of Lassa virus in Nigeria. *BMC Infect. Dis.* **2021**, *21*, 149. [CrossRef]
38. Hara, Y.; Kanagawa, M.; Kunz, S.; Yoshida-moriguchi, T.; Satz, J.S. modification of dystroglycan at Thr-317/319 is required for laminin binding and arenavirus infection. *Proc. Natl. Acad. Sci. USA* **2011**, *108*, 17426–17432. [CrossRef] [PubMed]
39. Rojek, J.M.; Spiropoulou, C.F.; Campbell, K.P.; Kunz, S. Old World and Clade C New World Arenaviruses Mimic the Molecular Mechanism of Receptor Recognition Used by α-Dystroglycan's Host-Derived Ligands. *J. Virol.* **2007**, *81*, 5685–5695. [CrossRef]
40. Jae, L.T.; Raaben, M.; Riemersma, M.; van Beusekom, E.; Blomen, V.A.; Velds, A.; Kerkhoven, R.M.; Carette, J.E.; Topaloglu, H.; Meinecke, P.; et al. Deciphering the Glycosylome of Dystroglycanopathies Using Haploid Screens for Lassa Virus Entry. *Science* **2013**, *340*, 479–483. [CrossRef]
41. Kunz, S.; Campbell, K.P.; Oldstone, M.B.A. α-Dystroglycan can mediate arenavirus infection in the absence of β-dystroglycan. *Virology* **2003**, *316*, 213–220. [CrossRef]
42. Moraz, M.-L.; Pythoud, C.; Turk, R.; Rothenberger, S.; Pasquato, A.; Campbell, K.P.; Kunz, S. Cell entry of Lassa virus induces tyrosine phosphorylation of dystroglycan. *Cell. Microbiol.* **2013**, *15*, 689–700. [CrossRef] [PubMed]

43. Rojek, J.M.; Moraz, M.-L.; Pythoud, C.; Rothenberger, S.; Van der Goot, F.G.; Campbell, K.P.; Kunz, S. Binding of Lassa virus perturbs extracellular matrix-induced signal transduction via dystroglycan. *Cell. Microbiol.* **2012**, *14*, 1122–1134. [CrossRef]
44. Meertens, L.; Labeau, A.; Dejarnac, O.; Cipriani, S.; Sinigaglia, L.; Bonnet-Madin, L.; Le Charpentier, T.; Hafirassou, M.L.; Zamborlini, A.; Cao-Lormeau, V.-M.; et al. Axl Mediates ZIKA Virus Entry in Human Glial Cells and Modulates Innate Immune Responses. *Cell Rep.* **2017**, *18*, 324–333. [CrossRef]
45. Richard, A.S.; Shim, B.-S.; Kwon, Y.-C.; Zhang, R.; Otsuka, Y.; Schmitt, K.; Berri, F.; Diamond, M.S.; Choe, H. AXL-dependent infection of human fetal endothelial cells distinguishes Zika virus from other pathogenic flaviviruses. *Proc. Natl. Acad. Sci. USA* **2017**, *114*, 2024–2029. [CrossRef]
46. Chen, J.; Yang, Y.; Yang, Y.; Zou, P.; Chen, J.; He, Y.; Shui, S.; Cui, Y.; Bai, R.; Liang, Y.; et al. AXL promotes Zika virus infection in astrocytes by antagonizing type I interferon signalling. *Nat. Microbiol.* **2018**, *3*, 302–309. [CrossRef]
47. Lemke, G.; Rothlin, C.V. Immunobiology of the TAM receptors. *Nat. Rev. Immunol.* **2008**, *8*, 327–336. [CrossRef] [PubMed]
48. Amara, A.; Mercer, J. Viral apoptotic mimicry. *Nat. Rev. Microbiol.* **2015**, *13*, 461–469. [CrossRef]
49. Sullivan, B.M.; Welch, M.J.; Lemke, G.; Oldstone, M.B.A. Is the TAM Receptor Axl a Receptor for Lymphocytic Choriomeningitis Virus? *J. Virol.* **2013**, *87*, 4071–4074. [CrossRef] [PubMed]
50. Sasaki, T.; Knyazev, P.G.; Clout, N.J.; Cheburkin, Y.; Göhring, W.; Ullrich, A.; Timpl, R.; Hohenester, E. Structural basis for Gas6–Axl signalling. *EMBO J.* **2006**, *25*, 80–87. [CrossRef] [PubMed]
51. Briggs, D.C.; Yoshida-Moriguchi, T.; Zheng, T.; Venzke, D.; Anderson, M.E.; Strazzulli, A.; Moracci, M.; Yu, L.; Hohenester, E.; Campbell, K.P. Structural basis of laminin binding to the LARGE glycans on dystroglycan. *Nat. Chem. Biol.* **2016**, *12*, 810–814. [CrossRef] [PubMed]
52. Sasaki, T.; Knyazev, P.G.; Cheburkin, Y.; Göhring, W.; Tisi, D.; Ullrich, A.; Timpl, R.; Hohenester, E. Crystal Structure of a C-terminal Fragment of Growth Arrest-specific Protein Gas6: Receptor tyrosine kinase activation by laminin G-like domains. *J. Biol. Chem.* **2002**, *277*, 44164–44170. [CrossRef] [PubMed]
53. Hastie, K.M.; Igonet, S.; Sullivan, B.M.; Legrand, P.; Zandonatti, M.A.; Robinson, J.E.; Garry, R.F.; Rey, F.A.; Oldstone, M.B.; Saphire, E.O. Crystal structure of the prefusion surface glycoprotein of the prototypic arenavirus LCMV. *Nat. Struct. Mol. Biol.* **2016**, *23*, 513–521. [CrossRef] [PubMed]
54. Hastie, K.M.; Zandonatti, M.A.; Kleinfelter, L.M.; Heinrich, M.L.; Rowland, M.M.; Chandran, K.; Branco, L.M.; Robinson, J.E.; Garry, R.F.; Saphire, E.O. Structural basis for antibody-mediated neutralization of Lassa virus. *Science* **2017**, *356*, 923–928. [CrossRef] [PubMed]
55. Rojek, J.M.; Campbell, K.P.; Oldstone, M.B.A.; Kunz, S. Old World Arenavirus Infection Interferes with the Expression of Functional α-Dystroglycan in the Host Cell. *Mol. Biol. Cell* **2007**, *18*, 4493–4507. [CrossRef]
56. Han, R.; Kanagawa, M.; Yoshida-Moriguchi, T.; Rader, E.P.; Ng, R.A.; Michele, D.E.; Muirhead, D.E.; Kunz, S.; Moore, S.A.; Iannaccone, S.T.; et al. Basal lamina strengthens cell membrane integrity via the laminin G domain-binding motif of α-dystroglycan. *Proc. Natl. Acad. Sci. USA* **2009**, *106*, 12573–12579. [CrossRef]
57. Gould, W.R.; Baxi, S.M.; Schroeder, R.; Peng, Y.W.; Leadley, R.J.; Peterson, J.T.; Perrin, L.A. Gas6 receptors Axl, Sky and Mer enhance platelet activation and regulate thrombotic responses. *J. Thromb. Haemost.* **2005**, *3*, 733–741. [CrossRef]
58. Lu, Q.; Lemke, G. Homeostatic Regulation of the Immune System by Receptor Tyrosine Kinases of the Tyro 3 Family. *Science* **2001**, *293*, 306–311. [CrossRef]
59. Masayuki, S.; Ayato, T.; Hideki, E.; Gabriele, N.; Kouki, F.; Tatsuro, I.; Steven, J.; Heinz, F.; Yoshihiro, K. Tyro3 Family-Mediated Cell Entry of Ebola and Marburg Viruses. *J. Virol.* **2006**, *80*, 10109–10116. [CrossRef]
60. Shimojima, M.; Ikeda, Y.; Kawaoka, Y. The Mechanism of Axl-Mediated Ebola Virus Infection. *J. Infect. Dis.* **2007**, *196*, S259–S263. [CrossRef]

 viruses 

Review

How Do Enveloped Viruses Exploit the Secretory Proprotein Convertases to Regulate Infectivity and Spread?

Nabil G. Seidah [1,*], Antonella Pasquato [2] and Ursula Andréo [1]

1 Laboratory of Biochemical Neuroendocrinology Montreal Clinical Research Institute, University of Montreal, Montreal, QC H2W1R7, Canada; Ursula.Andreo@ircm.qc.ca
2 Antonella Pasquato, Department of Industrial Engineering, University of Padova, Via Marzolo 9, 35131 Padova, Italy; Antonella.pasquato@unipd.it
* Correspondence: seidahn@ircm.qc.ca; Tel.: +1-514-987-5609

Abstract: Inhibition of the binding of enveloped viruses surface glycoproteins to host cell receptor(s) is a major target of vaccines and constitutes an efficient strategy to block viral entry and infection of various host cells and tissues. Cellular entry usually requires the fusion of the viral envelope with host plasma membranes. Such entry mechanism is often preceded by "priming" and/or "activation" steps requiring limited proteolysis of the viral surface glycoprotein to expose a fusogenic domain for efficient membrane juxtapositions. The 9-membered family of Proprotein Convertases related to Subtilisin/Kexin (PCSK) serine proteases (PC1, PC2, Furin, PC4, PC5, PACE4, PC7, SKI-1/S1P, and PCSK9) participate in post-translational cleavages and/or regulation of multiple secretory proteins. The type-I membrane-bound Furin and SKI-1/S1P are the major convertases responsible for the processing of surface glycoproteins of enveloped viruses. **Stefan Kunz** has considerably contributed to define the role of SKI-1/S1P in the activation of arenaviruses causing hemorrhagic fever. Furin was recently implicated in the activation of the spike S-protein of SARS-CoV-2 and Furin-inhibitors are being tested as antivirals in COVID-19. Other members of the PCSK-family are also implicated in some viral infections, such as PCSK9 in Dengue. Herein, we summarize the various functions of the PCSKs and present arguments whereby their inhibition could represent a powerful arsenal to limit viral infections causing the present and future pandemics.

Keywords: enveloped virus; proprotein convertases; Furin; SKI-1/S1P; PCSK9; SARS-CoV-2; COVID-19; pandemic

Citation: Seidah, N.G.; Pasquato, A.; Andréo, U. How Do Enveloped Viruses Exploit the Secretory Proprotein Convertases to Regulate Infectivity and Spread?. *Viruses* **2021**, *13*, 1229. https://doi.org/10.3390/v13071229

Academic Editors: Michael B. A. Oldstone and Juan C. De la Torre

Received: 16 April 2021
Accepted: 18 June 2021
Published: 25 June 2021

1. Introduction

Infectious diseases that threatened the life of humans since at least the Neolithic times, are believed to have started around 12,000 years ago, when roaming human hunter-gatherers became sedentary and settled into small camps and villages to domesticate animals and cultivate crops [1,2]. The domestication of animals and their proximity to humans favored the transmission of animal-human (zoonotic) diseases. The latter implicated various infectious microorganisms, such as bacteria, fungi, parasites, and viruses. Some of these pathogens caused widespread disease worldwide with "pandemic proportions" resulting in the death of a significant fraction of the human population.

The earliest recorded pandemic disease, called Antonine Plague (A.D. 165–180), resulted from infectious smallpox or measles virus and led to ~5 million (M) deaths throughout the Roman empire (~10% of the empire's population). The next three big pandemics were caused by deadly bacterial (Yersinia pestis) infections. The first was the Justinian's Plague (A.D. 541–542), which resulted in 5–10 M fatalities, was traced to China and northeast India. The bacterial infection was transmitted via land and sea trade routes to Egypt where it entered the Byzantine Empire through Mediterranean ports. The second was also caused by the same gram-negative bacteria transmitted from rats via infected fleas to humans and was at the origin of the bubonic/pneumonic plague (black death;

A.D. 1347–1351), which resulted in the highest number of fatalities ever (200 M). The third was a major bubonic plague pandemic that began in Yunnan, China, in 1855, spread to all inhabited continents leading to 12–15 M+ deaths in India and China, with about 10 M killed in India alone. The Smallpox virus (Variola, A.D. 1520) killed more than 56 M people over the years resulting in the decimation of 90% of Native Americans and 400,000 death/year in Europe in the 19th century. This virus must have circulated for a long time since signs of smallpox have been found in Egyptian mummies, including Ramses V, who died in 1157 B.C. The deadliest pandemic virus in the 20th century (A.D. 1918–1919) was traced to an H1N1 variant of the flu virus (Spanish Flu) and resulted in 50 M+ death worldwide close to the end of the first world war. This virus spread between the soldiers in the war trenches in Europe and was transmitted to the general population when they came back home. Such a deadly pandemic never went away and is still causing problems today. Finally, another persisting pandemic that has hit the world starting in 1981 is due to HIV retroviral infections causing AIDS, leading to more than 35 M+ death worldwide. The latest ongoing viral-related pandemic that started in December 2019 causes a coronavirus-associated disease (called COVID-19) due to SARS-CoV-2 infections [3], which so far have resulted in more than 2.8 M+ death worldwide and against which we now have access to various efficient vaccine formulations [4–6].

From the above summary, it is apparent that a substantial proportion of the deadly newly emerging and re-emerging diseases in the past century have been of viral origin [6]. While not all viruses have envelopes, e.g., enteric RNA viruses [7], the majority (RNA & DNA viruses and retroviruses) have a viral envelope that protects the genetic material in their life cycle when traveling between host cells. The envelopes are typically derived from portions of the host cell membranes (phospholipids and proteins) but include some viral glycoproteins. Glycoproteins on the surface of the envelope serve to identify and bind to receptor sites on the host's membrane. The viral envelope then fuses with the host's cell membrane [8], allowing the capsid and viral genome to enter the host. Another mechanism of virus spread requiring fusion is the formation of syncytia through cell-to-cell fusion. An infected cell harboring the glycoprotein on the plasma membrane can fuse with an adjacent cell forming polynucleated infected cells called syncytia. This mechanism allows the virus to spread while limiting the release of the virus in the extracellular milieu.

The proprotein convertases (PCs; genes *PCSKs*) constitute a family of nine secretory serine proteases that regulate various processes in both health and disease states [9]. Seven basic amino acid (aa)-specific convertases (PC1, PC2, Furin, PC4, PC5, PACE4, and PC7) cleave precursor proteins at single or paired basic amino acids after the general motif (K/R)-X_n-(K/R)↓, where Xn = 0, 2, 4 or 6 spacer X residues [9,10]. The 8th member (SKI-1/S1P) cleaves precursor proteins at R-X-Aliphatic-Z↓, where X is any residue except Pro and Cys, and Z is any aa except Val, Pro, Cys, or Glu [11–15]. The last member, PCSK9, cleaves itself once, but its inhibitory prodomain remains non-covalently attached rendering it inactive as a protease *in trans*, but rather it can act as a chaperone escorting some surface receptors (e.g., LDLR and MHC-class I receptor) toward lysosomes for degradation [16–19].

Through proteolysis, PCs are responsible for the activation and/or inactivation of many secretory precursor proteins, including viral surface glycoproteins [9,10,20] (Table 1).

Because of their critical functions, PCs, especially the ubiquitously expressed Furin [22] and SKI-1/S1P [11] are implicated in many viral infections via specific cleavages of envelope glycoproteins, a condition that allows not only the fusion of the viral lipid envelope with host cell membranes [9] but also for cell-to-cell fusion forming syncytia of certain viruses leading to important cytopathogenic effects [23,24].

Table 1. Families of pathogenic viruses that depend on the basic aa-specific proprotein convertases for host cell entry. RT = reverse transcriptase. Modified from [21].

Family	Virus	Capsid	Genome
Retroviridae	HIV, Leukemia viruses	Enveloped	Linear ssRNA(−), RT
Flaviridae	HCV, Dengue, Zika, West Nile	Enveloped	Linear ssRNA(+)
Togaviridae	Chikungunya	Enveloped	Linear ssRNA(+)
Coronaviridae	SARS-CoV-1,2, MERS	Enveloped	Linear ssRNA(+)
Filoviridae	Ebola, Marburg	Enveloped	Linear ssRNA(−)
Orthomyxoviridae	Avian Influenza H5N1	Enveloped	Linear ssRNA(−)
Paramixoviridae	Measle, RSV, Nipah, MPV	Enveloped	Linear ssRNA(−)
Hepadnaviridae	Hepatitis B	Enveloped	Linear ssDNA(−), RT
Herpesviridae	Herpes, CMV, Varicella-Zoster	Enveloped	Linear dsDNA
Papillomaviridae	HPV	Naked	Circular dsDNA

In this review, we will summarize our knowledge of the role of enzymatic cleavage by proprotein convertases and type-II transmembrane serine proteases required to facilitate virus entry through fusion as well as an innate antiviral strategy that the host deploys to counteract this step. A number of excellent reviews have appeared on the importance of cleavage of the surface glycoproteins of enveloped viruses in productive infections [21,25]. Herein, we will mostly concentrate on the roles of the proprotein convertases [9] in this process and emphasize the seminal contributions of the group of **Stefan Kunz** in the analysis of the function of SKI-1/S1P in the activation of arenaviruses causing hemorrhagic fever.

2. Proprotein Convertases and Enveloped Viruses

2.1. *Furin in Viral Infections and Pathogenicity*

Furin is the third and best characterized member of the PCSK family of secretory convertases. It cleaves at basic amino acid motifs and recognizes a prototypical sequence R-X-(K/R)-R↓ [10]. The presence of this Furin(-like) motif in viral envelope glycoproteins constitutes a way to activate the fusion-dependant entry of several viruses such as HIV gp160, influenza hemagglutinin (Table 2), and some coronaviruses spike proteins [26,27].

Table 2. Various enveloped viruses and the sequences of surrounding their surface glycoprotein cleavage sites (designated by an arrow) by Furin(-like) proprotein convertases. The bold and underlined residues at positions P8, P6, P4, P2, P1 and P2′ emphasize the importance of these amino acids for protease recognition.

Virus	Glycoprotein	P8		P6		P4		P2		↓	P2′
HIV	gp160			V	Q	**R**	E	**K**	**R**	A	**V**
H7N1 A/FPV/Rostock/34	HA			**K**	K	**R**	E	**K**	**R**	G	**L**
Avian H5N8 TKY/IRE	HA			**R**	K	**R**	K	**K**	**R**	G	**L**
Avian H5N1 A/HK/97	HA	**R**	E	**R**	R	**R**	K	**K**	**R**	G	**L**
Avian H5N1 TKY/ENG	HA	N	T	P	Q	**R**	K	**K**	**R**	G	**L**
Human CMV	gB			**H**	N	**R**	T	**K**	**R**	S	T
Human MPV	F Protein			N	P	**R**	Q	S	**R**	F	**V**
Human RSV	F Protein			**K**	K	**R**	K	**R**	**R**	F	**L**
Dengue Virus (DENG2)	PrM			**H**	R	**R**	E	**K**	**R**	S	**V**
Ebola Virus	gp160			G	R	**R**	T	**R**	**R**	E	A
Chikungunya (CHIKV)	E3E2			P	R	**R**	Q	**R**	**R**	S	**I**
Zika Virus	PrM			A	R	**R**	S	**R**	**R**	A	**V**
SARS-CoV-2	S			S	P	**R**	R	A	**R**	S	**V**

In 1992, Furin was identified to be the cellular protease cleaving hemagglutinin (HA) of fowl plague/influenza virus (FPV; H7N1 A/FPV/Rostock/34), as well as the HIV glycoprotein gp160 into gp120 and gp41 [28–30]. The presence of a Furin cleavage sequence can shift the pathogenicity of certain viruses from non-virulent to virulent. For instance, Newcastle disease virus (NDV) fusion protein contains only a monobasic cleavage site, while the virulent NDV fusion protein possesses a polybasic sequence of amino acids that can be cleaved by Furin [31]. Similarly, the acquisition of a Furin cleavage site has been linked to the high pathogenicity of avian influenza [32], especially for the Hong Kong variant (H5N1 A/HK/97) [33,34] that has an insertion of RERR between the NTPQ and the Furin-site RKKR↓GL in HA (Table 2), resulting in a very efficient cleavage by more than one convertase, such as Furin, PC5 and PC7 [35]. Other pathogenic influenza virus do not possesses a Furin cleavage site and rely on trypsin-like proteases, e.g., HAT and TMPRSS2 [36]. The expression of these enzymes is restricted to the upper respiratory tract [37], while Furin is ubiquitously expressed, and the acquisition of a Furin(-like) site enhances the cellular tropism of influenza. However, the presence of a polybasic Furin(-like) cleavage site is not the only determining factor for cleavage by Furin, as the amino-acids adjacent to the Furin(-like) sequence are important too, as described in Table 2. Furthermore, the presence of oligosaccharides near the cleavage site can prevent processing [38], unless it is compensated by an increased number of basic residues as described in the non-pathogenic variant of influenza hemagglutinin A of H5N2 [39].

Aside from glycoproteins from NDV and influenza hemagglutinin, HIV glycoprotein, as well as other retrovirus glycoproteins such as Rous sarcoma virus and murine leukemia virus are cleaved by Furin. Interestingly, these glycoproteins form trimers, and some viruses assemble with un-cleaved glycoproteins that remain either inactive or can be cleaved at the plasma membrane. The Furin-specific cleavage of gp160 in HIV-1 has been described as dispensable as other proprotein convertases [40] could substitute for Furin in some tissues, and plasmin could activate the gp160 at the plasma membrane. However, these hypotheses remain to be validated in vivo in humans.

Human metapneumovirus (hMPV) is a paramyxovirus responsible for acute respiratory tract infections which can result in hospitalization of both children and adults. We showed that blocking the activity of the protease-activated receptor 1 (PAR1), a G-coupled receptor, induced by inflammation, protects against hMPV infections [41]. Furthermore, Furin activates the envelope F-protein of hMPV by cleavage at NP**R**QS**R**$_{102}$↓FV (Table 2, where bold residues emphasise the critical P1 and P4 positions) [41]. Unexpectedly, PAR1 itself potently inhibits cellular Furin activity [42]. Indeed, PAR1 exhibits in its second luminal loop a Furin(-like) motif (P**R**SFLL**R**$_{46}$-NP) with a P1 and P6 Arg and is cleaved by PC5A and PACE4 but not by Furin. The presence of an Asn$_{47}$ at the P1′ site rather made it a potent Furin-inhibitor that binds its catalytic subunit and sequesters the [PAR1-Furin] complex in the *trans*-Golgi network (TGN), thereby inactivating Furin and preventing PAR1 from reaching the cell surface [42]. The overexpression of PAR1 has been described in the brain upon neuroinflammation and in patients with neurocognitive disorders associated with HIV infection (HAND) [42]. Interestingly, while PAR1 binds and inhibits Furin, PAR2 is cleaved by Furin at the motif **R**SSKG**R**$_{36}$↓SL, which also exhibits a P1 and P6 Arg but a favorable Ser at P1′ and Leu at P2′ (Table 2) [38]. Recently, another strategy of viral host-cell defense has been identified. Guanylate-binding proteins (GBPs) are interferon stimulated genes. Upon HIV infection, the cytosolic GBP2 and 5 expressions are induced, and these proteins can bind the cytosolic tail of Furin and prevent its trafficking beyond the *cis/medial* Golgi, thus keeping it in an inactive state bound to its inhibitory prodomain. This effectively inhibits the Furin processing of gp160 into gp120 and gp41 [25,43]. Thus, while PAR1 is the first secretory endogenous natural inhibitor of Furin that blocks its enzymatic activity and prevents its exit from the TGN, the cytosolic GBP2,5 also inhibits Furin activity by blocking it in the *cis/medial* Golgi in an inactive state.

Another family of viruses requiring Furin is the *filoviridae*. Marburg and Ebola are enveloped single negative strand RNA viruses of the *filoviridae* family. These viruses are highly pathogenic in humans causing hemorrhagic fever with a very high mortality rate. The glycoprotein of Marburg and Ebola (GPs) are cleaved by Furin into GP1 containing the receptor binding domain (RBD) and membrane-bound GP2 [44]. Pathogenic viruses of the family possess a Furin(-like) cleavage site, but experiments design to establish the role of Furin in pathogenicity indicated that Furin cleavage is not required for replication in cell culture [45] and disease severity in non-human primates [46]. Recent work identified a new Furin site in Ebola GP that requires N-glycosylation to be processed [47]. Interestingly, MARCH8 a member of the Membrane-associated RING-CH-type 8 (MARCH8) that has been described to have broad antiviral activity against several viruses, can inhibit Furin by forming a complex with the GP and Furin thereby sequestering it in the Golgi, and preventing the maturation of the viral particles [48]. This is the third example of endogenous Furin-inhibitors that sequester the enzyme in a subcellular Golgi compartment where it remains inactive.

The Flaviviruses envelope also requires Furin to get activated. The first evidence was reported for the Tick Borne Encephalitis virus. Flavivirus are enveloped positive strand RNA viruses. Their surface protein is composed of a heterodimer formed by prM and E [49]. The immature prM is cleaved by Furin at GS**R**T**RR**$_{205}$↓SV before the viral particles are released from the cell. The cleavage allows fusion and is required for infectivity. An acidic pH is required to promote a change of conformation of the heterodimer to facilitate Furin cleavage. While, the cleavage of prM into M is inefficient especially for Dengue virus (**H**R**R**E**KR**$_{205}$↓SV) (Table 2), likely due to the presence of a Glu at P3 [21], Furin has been confirmed to be required for infectivity of Dengue virus in vitro by using LoVo cells that are Furin deficient [50]. Additionally, the role of Furin in the antibody-dependent-enhancement (ADE) of Dengue virus infection is intriguing. ADE is the mechanism by which the immunisation against Dengue could exacerbate the pathogenicity of Dengue in the context of a second infection. This has been explained by the fact that the binding of antibodies to an immature particle could enhance its infectivity. Interestingly, while Furin inhibition has been described to block anti-Envelope dependant enhancement [51], antibodies against prM (uncleaved by Furin) allow tissue entry of immature particles [52]. Recently, broadly neutralizing monoclonal antibodies were reported to protect against multiple tick-borne flaviviruses (TBFVs) [53], opening the way to the design of vaccines and antibody therapeutics against clinically relevant TBFVs.

Chikungunya virus (CHIKV) is a mosquito-transmitted α-virus that causes in humans an acute infection characterized by polyarthralgia, fever, myalgia, and headache. Since 2005 this virus has been responsible for an epidemic outbreak of unprecedented magnitude. By analogy with other α-viruses, it is thought that cellular proteases can process the viral precursor protein E3E2 to produce the receptor binding E2 protein that associates as a heterodimer with E1. Destabilization of the heterodimer by exposure to low pH allows viral fusion and infection. We demonstrated that membrane-bound Furin but also PC5B can process E3E2 from African CHIKV strains at the H**R**Q**RR**$_{642}$↓ST site, whereas a CHIKV strain of Asian origin is cleaved at R**R**Q**RR**$_{642}$↓SI (Table 2) by membranous and soluble Furin, PC5A, PC5B, and PACE4 but not by PC7 or SKI-1/S1P [54]. This cleavage was also observed in CHIKV-infected cells and could be blocked by Furin inhibitor decanoyl-RVKR-chloromethyl ketone. This inhibitor was compared with chloroquine for its ability to inhibit CHIKV spreading in myoblast cell cultures, a cell-type previously described as a natural target of this virus [54]. We observed an additive effect of dec-RVKR-cmk and chloroquine, supporting the concept that these two drugs act by essentially distinct mechanisms. The combinatory action of chloroquine and dec-RVKR-cmk led to almost total suppression of viral spread and yield [54].

Furin cleavage of viral glycoproteins of the *Herpesviridae* family have been reported for Herpes simplex virus 1 and 2 and Varicella Zoster (Table 1), contributing to the pathogenesis of the later in vivo [55,56]. Finally, while Furin cleaves many glycoproteins, its activity is required for other viral proteins such as for Hepatitis B virus (HBV) core antigen. For reviews see [21,25].

HIV-1 encodes four accessory gene products—Vpr, Vif, Vpu, and Nef—which are thought to collectively manipulate host cell biology to promote viral replication, persistence, and immune escape [57]. Increasing evidence suggests that extracellular Vpr could contribute to HIV pathogenesis through its effect on bystander cells. Soluble forms of Vpr have been detected in the sera and cerebrospinal fluids of HIV-1-infected patients, and in vitro studies have implicated extracellular Vpr as an effector of cellular responses, including G2 arrest, apoptosis, and induction of cytokines and chemokines production, presumably through its ability to transduce into multiple cell types. Thus, Vpr is a true virulence factor and is a potential and promising target in different strategies aiming to fight infected cells including latently HIV-infected cells [58]. While Furin is clearly implicated in the processing activation of multiple viral surface glycoproteins, including gp160 of HIV-1, other cell-surface associated convertases such as PC5A and PACE4 [9,59] seem to negatively regulate the activity of the HIV-1 by cleavage-inactivation of secreted Vpr at **RQRR**$_{88}$↓ [60]. PC-mediated processing of extracellular Vpr results in a truncated Vpr product that was defective for the induction of cell cycle arrest and apoptosis when expressed in human cells [60]. Thus, inhibitors of PC5/PACE4 may enhance HIV-1 virulence, and Furin-specific inhibitors would restrict HIV-1 infectivity, without affecting Vpr inactivation by PC5A/PACE4.

2.2. *Coronavirus Infections, Including SARS-CoV-2*

There are seven known human coronaviruses (CoV), which are enveloped positive-stranded RNA viruses belonging to the order *Nidovirales* and are mostly responsible for upper respiratory tract infections. All these coronaviruses exhibit a "crown-like" structure composed of a trimeric spike (S) protein. The S-protein is a ~180–200 kDa type I transmembrane protein, with the N-terminus facing the extracellular space and anchored to the viral membrane via its transmembrane domain followed by a C-terminal short tail facing the cytosol [61]. During infection, the trimetric S-protein is first processed at the "priming site" S1/S2 by host cell proteases, separating the N-terminal S1 domain that binds a specific host cell receptor(s) via its receptor binding domain (RBD), and the C-terminal membrane-bound S2 domain involved in viral entry through fusion at the plasma membrane after cleavage. S1 and S2 have been described to remain non-covalently bound therefore protecting the fusion peptide from being exposed too early before the membranes can fuse. Four coronaviruses are now endemic in the human population and cause mild disease, including the α-coronaviruses HCoV-229E and HCoV-NL63 and β-coronaviruses HCoV-OC43 and HCoV-HKU1 [24]. Interestingly, the insertion of a more favourable S1/S2 Furin(-like) site of the neuro-invasive HCoV-OC43 (R**R**S**RR**$_{758}$↓A**I**) resulted in the establishment of less pathogenic but more persistent viral infections in the brain [26].

In the last 18 years three new highly pathogenic human β-coronaviruses appeared, SARS-CoV-1 in 2002–2003, MERS-CoV in 2013, and SARS-CoV-2 (Table 2) at the end of 2019. The first two were associated with severe human pathologies, such as pneumonia and bronchiolitis, and even meningitis in more vulnerable populations [62]. However, SARS-CoV-1 was rapidly contained and MERS-CoV only spread from dromedary to human and no human to human transmissions were reported. The RBDs of the S1-subunits of SARS-CoV-1 and MERS-CoV recognise angiotensin-converting enzyme 2 (ACE2) [63] and dipeptidylpeptidyl 4 (CD26/DPP4) [64] as their entry receptors, respectively. While the single basic Arg at the S1/S2 site of SARS-CoV-1 (TVSLL**R**$_{667}$-ST) does not contain a Furin(-like) site, that of MERS-CoV does (TP**R**SV**R**$_{751}$↓S**V**). Indeed, the priming of SARS-CoV-1 in the lung is thought to be performed by elastase [65,66], whereas MERS-CoV processing at S1/S2 is thought to be mostly performed by Furin and possibly by TMPRSS2, a type-II membrane-bound serine protease [67,68]. The S2-product generated following S1/S2 cleavage contains a second proteolytic site (called S2′ cleavage site), which when cleaved would generate an S2′-fragment that starts with a fusion peptide (FP) followed by two heptad-repeat domains preceding the transmembrane domain (TM) and cytosolic tail. Cleavage at the S2′ site triggers membrane fusion and is essential for efficient viral infection. The S2′ cleavage sites of SARS-CoV-1 (PT**KR**$_{797}$↓SF) and MERS-CoV (**R**SA**R**$_{887}$↓SA) suggest that both could be cleaved by a Furin(-like) enzyme, but the single Arg-specific enzyme TMPRSS2 has solely been proposed to cleave both spike glycoproteins at the S2′ site [69].

The Furin(-like) S2′ cleavage site at **KR**$_{797}$↓S**F** with P1 and P2 basic residues and a P2′ hydrophobic Phe [9], is identical between the SARS-CoV-1 and SARS-CoV-2 (Figure 1).

In the MERS-CoV and OC43-CoV it is replaced by the less favourable Furin(-like) site **R**XX**R**↓SA, with P1 and P4 basic residues, and an Ala (not hydrophobic) at P2′. However, in other less pathogenic circulating human coronaviruses, the S2′ cleavage site only exhibits a Furin-unfavourable monobasic **R**↓S sequence [70] with no basic residues at either P2 and/or P4. Even though processing at the S2′ site in the spike glycoprotein of SARS-CoV-2 is thought to be key in the activation of the S-protein, leading to cell-fusion and entry, multiple protease(s) might be involved in S-cleavage at different sites and subcellular compartments [71]. The ability of the Arg/Lys-specific TMPRSS2 to directly cleave at the S2′ site was inferred from the viral entry blockade by the relatively non-specific TMPRSS2 inhibitor Camostat [72,73]. We recently demonstrated that Furin is capable of performing both S1/S2 (P**R**RA**R**$_{685}$↓S**V**) (Table 2) and S2′ (KPS**KR**$_{815}$↓S**F**) cleavages generating an N-terminal S1 subunit (with an RBD domain) and a C-terminal membrane bound fusogenic S2′-fragment in the presence of the Spike-glycoprotein receptor ACE2 [74]. Interestingly, binding of the RBD of the S-protein of SARS-CoV-2 to ACE2 was demonstrated to exert a conformational change that allosterically enhances the exposure of the S1/S2 site to Furin cleavage [75]. Whether it also enhances the exposure of the S2′ site is not clear, but our results clearly show that Furin cleavage at the S2′ site is enhanced in the presence of ACE2 [74]. Our data also showed that non-peptide small molecule BOS-inhibitors of Furin(-like) enzymes block the processing of the S-protein at both S1/S2 and S2′ sites and result in a significant reduction of cell-to-cell fusion. In the presence of overexpressed TMPRSS2 together with the S-protein, the latter was cleaved into an apparently lower sized fragment than S1 (called S1′) [74] released in the cell culture media, which turned out to be an authentic S1 product lacking O-glycosylation as it is generated in the endoplasmic reticulum (ER) and secreted by an unconventional pathway (*submitted*). Furthermore, overexpressed TMPRSS2 generated two C-terminal membrane bound fragments S2a and S2b, which are S2 and S2′ products lacking O-glycosylation. These products may be artificially produced in the ER upon overexpression of TMPRSS2 with the S-glycoprotein. However, in co-culture conditions in which ACE2 and TMPRSS2 are expressed in acceptor cells and S-protein in donor cells, we concluded that both Furin (in the TGN) and TMPRSS2 (at the cell surface) cleave S-protein into S1/S2 and S2′ in the presence of ACE2 (*submitted*), but that in addition TMPRSS2 can also process and shed ACE2 into a soluble form thereby modulating viral entry and infection. In agreement, an inhibitor cocktail that combines a Furin inhibitor (BOS) with a TMPRSS2 inhibitor (Camostat) can reduce by 99% SARS-CoV-2 infectivity of lung-derived Calu-3 cells [74]. It should be noted that coronaviruses may enter the cells through fusion at the plasma membrane or the following endocytosis into endosomes [76]. The versatility of strategies to penetrate the cell allow coronaviruses to depend less on the availability of the different proteases in specific cell type and in cellular compartments. Furthermore, the activation process of SARS-CoV-2 may be more complex, as other proteases may also participate in processing the S-protein at multiple sites in various tissues, including cathepsins [71], HAT [77], TMPRSS11D, and TMPRSS13 [78].

Interestingly, we previously showed that chloroquine, initially used as an antimalarial, can reduce viral infection and spread of both the Chikungunya virus [54] and SARS-CoV-1 [79]. In the latter case, chloroquine was also shown to affect the N-glycosylation pattern of ACE2, the receptor of SARS-CoV-1 [79] and SARS-CoV-2 [27]. Interfering with terminal glycosylation of ACE2 may negatively influence the virus-receptor binding and abrogate infection, with further ramifications by the elevation of TGN/vesicular pH, resulting in the inhibition of infection and spread of SARS CoV-1 at clinically admissible concentrations. This may also be applicable for SARS-CoV-2, but the protective role of the less toxic hydroxychloroquine together with the antibiotic azithromycin against COVID-19 at the early infection stages remains controversial and begs for unbiased clinical trials [80,81].

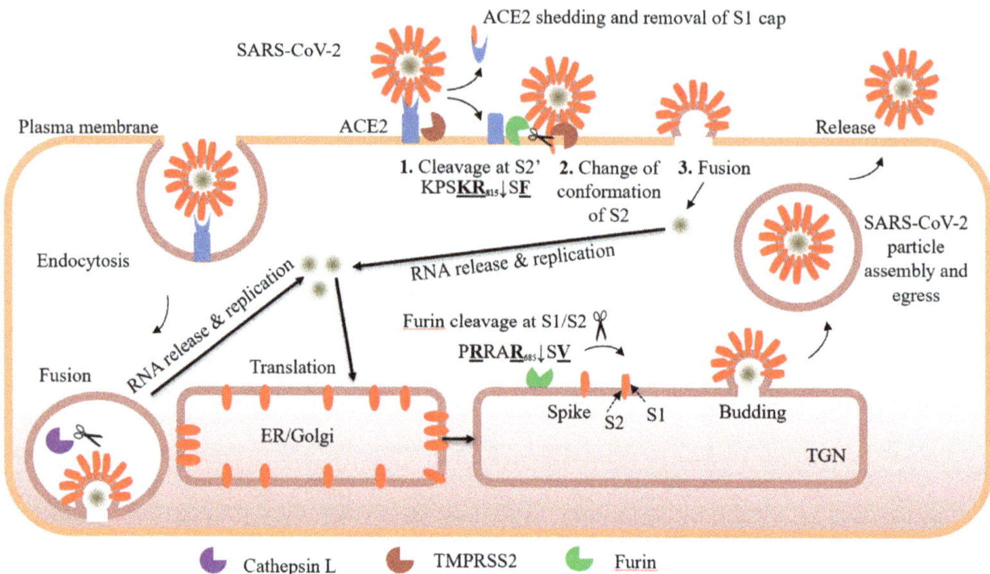

Figure 1. Role of protease processing in SARS-CoV-2 entry. SARS-CoV-2 can enter through fusion at the plasma membrane or by endocytosis. The processing of the spike-protein at the plasma membrane by Furin at the S1/S2 site is favored while Furin & TMPRSS2 are involved in the processing at S2′, thereby inducing fusion. Furin cleavage at S1/S2 is not required to promote entry through endocytosis and Cathepsin L may cleave at S1/S2 and S2′ in endosomes.

Worldwide distribution of various SARS-CoV-2 vaccines conceived to mount an immune response against the S protein, therefore, blocking accessibility of the S-protein to ACE2 (https://www.raps.org/news-and-articles/news-articles/2020/3/covid-19-vaccine-tracker, accessed on 17 June 2021) represents an efficient strategy to prevent SARS-CoV-2 infections end to eliminate the burden of the pandemic on a global scale. Time is needed to determine if these vaccines will confer long-lasting protection. Furthermore, since April 2020 new variants of SARS-CoV-2 emerged, some of which are becoming prominent in Europe as well as North and South America. The emergence of such variants indicated an enhanced SARS-CoV-2 fitness and are thought to have increased transmissibility. Variants in the United Kingdom (B.1.1.7; α-variant) [82], South Africa (B.1.351; β-variant) [83], Brazil, and Japan (B.1.1.248; γ-variant) and India (B.1.167; δ-variant) [84] present multiple substitutions in the spike protein, some of which are common among different variants. These include critical mutations in the ACE2 binding site (e.g., N501Y and E484K) within the RBD (Table 3), resulting in enhanced binding affinity of S-protein to ACE2 by 2.5-fold for N501Y and 13.5-fold for [N501Y + E484K] [85]. Interestingly, the α- and δ-variant exhibit mutations in the P5 position of the Furin cleavage site P681H and P681R, respectively, which may increase processing efficacy by Furin(-like) enzymes at the S1/S2 site and enhance viral entry.

Table 3. Amino acid changes in the S-protein of SARS-CoV-2 variants of worldwide concern covering the period of September 2020–June 2021. The **P681H** and P681R emphasize their P5 position in the Furin(-like) site (P/H/R)RRAR$_{685}$↓SV and variants in red are expected to affect the binding of S-protein to ACE2.

Variant	First Identification	S-Protein Mutations
B.1.1.7 α-variant	UK September 2020	del69-70 HV, del144Y, N501Y, A570D, D614G, **P681H**, T761I, S982A, D1118H
B.1.351 β-variant	South Africa October 2020	K417N, E484K, N501Y, D614G, A701V
B.1.1.248 γ-variant	Brazil, Japan January 2021	L18F, T20N, P26S, D138Y, R190S, K417T, E484K, N501Y, H655Y, T1027I
B.1.167 δ-variant	India December 2020	T95I, G142D, E154K, K417N, L452R, E484Q, D614G, **P681R**

Accordingly, it is not surprising that these variants are associated with a reduced efficacy of certain vaccines against SARS-CoV-2 infection but fortunately remain protective against severe disease and death [86,87]. Amazingly, screening various artificial mutants in vitro showed that some mutant-combinations could result in a very worrisome 600-fold higher binding affinity of S-protein to ACE2 [85]. Thus, B.1.351 and similar spike mutations present new challenges for antibody therapy and threaten the protective efficacy of current vaccines.

While the protective effect of the vaccination on the whole world population remains incomplete, additional effective antiviral drugs that block viral entry in target cells (Figure 1) are still needed and could help in the early treatment of the disease. Ultimately, in the case of new emerging coronavirus pandemics [6,88], the availability of such treatments would constitute a powerful antiviral arsenal.

2.3. PCSK9 and Viral Infections

The secreted protein PCSK9 is the 9th member of the PCSKs family that autocatalytically cleaves its prodomain in the ER allowing its exit from this compartment and its secretion from cells [16], as a protease-inactive [prodomain-PCSK9] complex [89,90]. Its high expression in the liver and the presence of its gene on chromosome 1p32 [16], led to the genetic association of gain-of-function (GOF) variants of *PCSK9* with an autosomal dominant form of hypercholesterolemia [91]. This strongly suggested that PCSK9 may play a critical role in low density lipoprotein- cholesterol (LDLc) regulation. Indeed, while GOF variants were associated with high levels LDLc [91], the reverse was true for loss-of-function (LOF) variants [92]. Mechanistically, circulating PCSK9 was shown to reduce the levels of liver LDL-receptor (LDLR) protein by sorting the PCSK9-LDLR complex to lysosomes for degradation, implicating an interaction of the PCSK9 catalytic domain with the EGF-A domain of the LDLR [89,93,94].

In a first study, we showed that PCSK9 can reduce HCV infection of HuH7 cells using the cell culture-derived HCV clone JFH1 (genotype 2a). We demonstrated that the effect was due to the PCSK9-induced decrease of the LDLR protein levels on the surface of HuH7 cells. Also, HCV upregulates the expression of LDLR in vitro and in the chronically infected liver [95], likely due to an upregulation of the LDLR promotor activity through SREBPs and a decreased expression of PCSK9 [96]. Befittingly, Alirocumab, a therapeutic human monoclonal antibody (mAb) to PCSK9 used to treat hypercholesterolemia through an increase of cell surface LDLR failed to reduce HCV entry and infectivity in hepatocyte in vitro [97].

Additionally, the role of PCSK9 during natural HCV infection is not fully understood. We showed that HCV genotype 3 (G3) infected patients with high viral load have low levels of PCSK9 and LDLc compared to HCV genotype 1 (G1) infected patients, suggesting that the particular interplay between HCV and lipid metabolism and PCSK9 might be genotype specific [98]. In contrast, in a retrospective study, PCSK9 levels were found to be elevated especially in chronic genotype 2 (G2) HCV patients with or without hepatocellular carcinoma (HCC), but the increased PCSK9 levels were not significantly associated with changes in LDLc [99]. Overall, the data reported illustrate the complexity of the relation between HCV and lipid metabolism and more work would be required to better understand the role of PCSK9 in HCV infection.

Dengue virus (DENV), a single positive-stranded RNA virus of the family Flaviviridae, is transmitted to humans by the urban-adapted *Aedes* mosquitoes. It is estimated that 400 M+ individuals are infected with 1 of the 4 types of DENV [100]. The DENV surface glycoprotein prM-E complex needs cleavage by Furin(-like) enzymes to remove the prodomain of the envelope glycoprotein prM and allow the formation of the activated M-E complex and acquisition of infectivity. However, unlike cleavage of the prM of other flaviviruses, cleavage of DENV prM is incomplete in many cell lines, likely due to the negative influence of the aspartic acid (D) at the P3 position of the processing site HR**R**D**KR**↓**SV** at the pr-M junction (Table 2) [101]. Therefore, it is not surprising that inhibitors of Furin(-like) enzymes, while significantly reducing viral entry, did not completely block viral infections [102] (*unpublished results*).

Very recently, we showed that DENV infection reduces the antiviral response of the host hepatocytes. Thus, DENV infection induces expression of PCSK9, thereby reducing cell surface levels of LDLR and LDLc uptake resulting in enhanced *de novo* cholesterol synthesis and its enrichment in the ER. In turn, high levels of ER cholesterol suppressed the phosphorylation and activation of the ER-resident stimulator of interferon (IFN) gene (STING), leading to reduction of type I interferon (IFN) signaling through antiviral IFN-stimulated genes (ISGs) [103]. This was supported by the detection of elevated plasma PCSK9 levels in patients infected with DENV resulting in high viremia and increased severity of plasma leakage. This unexpected role of PCSK9 in dengue pathogenesis, led us to test the effect of inhibition of PCSK9 function by the mAb Alirocumab. Befittingly, this treatment resulted in higher LDLR levels and lower viremia. Our data suggested that PCSK9 inhibitors could be a suitable host-directed treatment for patients with dengue [103], possibly in combination with Furin(-like) inhibitors.

2.4. Implications of SKI-1/S1P in Viral Infections

The 8th member of the proprotein convertase family was discovered in 1998/1999 as a type-I membrane-bound protease called site-1 protease (S1P) [104] or alternatively subtilisin-kexin isozyme-1 (SKI-1) [11]. Since then, it is commonly named SKI-1/S1P, and its gene is given the name Membrane-Bound Transcription-factor Site-1 Protease (*MBTPS1*) [9]. Extensive studies of its cleavage specificity revealed that it recognizes the general motif **R**-X-**Aliphatic**-Z↓, where X is any residue except Pro and Cys, and Z is any aa (best Leu) except Val, Pro, Cys, or Glu [11–15] (Table 4).

Different from the basic-aa specific convertases this enzyme undergoes in the ER two autocatalytic zymogen processing events in the middle of the prodomain **R**K**V**F$_{133}$↓**R**S**L**K$_{137}$↓YA (sites B and B′; Table 4) [13,105], allowing it to exit the ER and reach its final destination, the *cis/medial* Golgi, where it is further cleaved at the C-terminus of the prodomain at **R**R**A**S$_{166}$↓LS and **R**R**L**L$_{186}$↓RA (sites C′ and C; Table 4) [11,12]. This results in the maximal activation of the protease, which would then act *in trans* to cleave its substrates usually in the *cis/medial* Golgi where it is expected to be fully active.

SKI-1/S1P was first identified in our group because of its ability to cleave in the *cis/medial* Golgi brain derived neurotrophic factor (BDNF) at **R**G**L**T$_{57}$↓SL, thereby preceding the Furin cleavage in the TGN at **R**V**RR**$_{128}$↓HS to release mature BDNF [11]. The enzyme was then linked to the activation of several transcription factors. Among others, the sterol regulatory element binding proteins 1 and 2 (SREBP1,2; implicated in sterol and fatty acid synthesis), the type-II membrane-bound ER-stress factor ATF6, and other type-II membrane-bound substrates including at least six CREB-like basic leucine zipper transcription factors (Table 4). Other substrates identified revealed a role of SKI-1/S1P in the regulation of the kinase FAM20C that phosphorylates luminal serine residues in the secretory pathway [106], and in the activation of the α/β-GlcNAc-1-phospho-transferase that phosphorylates mannose residues in glycoproteins sorted to lysosomes (Table 4) [107]. The last identified substrate is the (pro)renin receptor which is shed by SKI-1/S1P in the bloodstream where it plays a major role in hypertension [108]. In humans, *MBTPS1* mutations are linked to the Silver–Russell syndrome (SRS) in patients affected by defective inter-organelle protein trafficking and skeletal malformations [109,110].

Table 4. SKI-1/S1P processing sites of various cellular substrates and those of enveloped viruses with emphasis on the sequences surrounding their surface glycoprotein cleavage sites, designated by an arrow. The bold and underlined residues at positions P4 and P2 emphasize the importance of these amino acids for protease recognition.

	Substrate	P8		P6		P4		P2		↓	P2′		P4′
CELLULAR	h Pro-SKI-1 site B	R	K	V	F	**R**	S	**L**	K	Y	A	E	S
	h Pro-SKI-1 site B′	V	T	P	Q	**R**	K	**V**	F	R	S	L	K
	h Pro-SKI-1 site C	R	H	S	S	**R**	R	**L**	L	R	A	I	P
	h SREBP2	S	G	S	G	**R**	S	**V**	L	S	F	E	S
	h SREBP1	H	S	P	G	**R**	N	**V**	L	G	T	E	S
	h ATF6	A	N	Q	R	**R**	H	**L**	L	G	F	S	A
	h Luman	G	V	L	S	**R**	Q	**L**	R	A	L	P	S
	m OASIS (CREB3L1)	Q	M	P	S	**R**	S	**L**	L	F	Y	D	D
	h CREB-H	R	V	F	S	**R**	T	**L**	H	N	D	A	A
	h pro-BDNF	K	A	G	S	**R**	G	**L**	T	S	L	A	D
	h α/β-GlcNAc-1-pTr	K	N	T	G	**R**	Q	**L**	K	D	T	F	A
	h FAM20C	K	H	T	L	**R**	I	**L**	Q	D	F	S	S
	h pro-Renin receptor	I	R	K	T	**R**	T	**I**	L	E	A	K	Q
VIRAL	Lassa Virus (LASV) GP-C	I	Y	I	S	**R**	R	**L**	L	G	T	F	T
	CCHFV PreGn	S	S	G	S	**R**	R	**L**	L	S	E	E	S
	LCMV GP-C	K	F	L	T	**R**	R	**L**	A	G	T	F	T
	Junin Virus (JUNV) GP-C	Q	L	P	R	**R**	S	**L**	K	A	F	F	S

The unique cleavage specificity of SKI-1/S1P puts it in a class of its own, distinct from the basic-aa specific convertases. This led to the search for possible viral glycoprotein substrates that are activated via a non-canonical Furin(-like) cleavage, possibly by SKI-1/S1P or through a combination of both processing enzymes. Thus, SKI-1/S1P was shown to play a major role in the processing of surface glycoproteins precursors (GP-C) of infectious arenaviruses such as Lassa virus (LASV) at **R**R**L**L↓GT (Table 4) [111–113], which is endemic in West Africa and is estimated to affect some 500,000 people annually resulting in several thousand deaths each year [114].

Interestingly, while SKI-1/S1P is predicted to be fully mature in the *cis/medial* Golgi, it can cleave the GP-C of LASV in the ER/*cis* Golgi indicating a functional activity in early secretory compartments. We speculate that forms of SKI-1/S1P that are only cleaved at the B′/B sites can already be partially active. This may be due to the fact that the optimal **R**R**L**L↓ cleavage motif [14] found in site C of the SKI-1/S1P prodomain (Table 4) is identical to the processing site of LASV GP-C [111,112], representing an ideal substrate [14]. In line with this hypothesis, the Stefan Kunz group found that Golgi-cleavable substrates can be processed in the ER when engineered to contain **R**R**L**L↓, regardless of the rest of the backbone protein that carries the motif [112]. The N-terminal B′/B fragment of the prodomain represents an autonomous structural and functional unit that is necessary and sufficient for folding and partial activation of SKI-1/S1P. In contrast, the C-terminal B-C fragment lacks a defined structure but is crucial for autocatalytic processing and full enzymatic activity [115]. Therefore, it is not surprising that the partial processing of the prodomain in the ER at the B′/B site is enough to confer some enzymatic activity on specific viruses with very favorable glycoprotein processing sites, e.g., LASV.

Attempts to generate vaccines against the GP-C of LASV revealed that major histocompatibility class I (MHC-I) receptors play a critical role in the T-cells response to LASV infections [116], suggesting that conditions that favor higher levels of MHC-I receptors may enhance the inflammatory response by activated macrophages following LASV infection contributing to its pathogenesis. In this context, the recent observation that PCSK9 inhibition/silencing in vivo enhances the levels of MHC-I via its PCSK9-induced degradation in endosomes/lysosomes [19] would suggest that this clinically safe treatment, though very effective in reducing hypercholesterolemia [117] and in cancer/metastasis treatments [19], may actually also enhance the contributions of the immune response to viral disease severity by exacerbating the LASV-induced inflammatory response, which represents a major hurdle in vaccine development against viral infections [118]. Thus, administration of PCSK9 inhibitors [117] would not be recommended in LASV-infected and HCV co-infected patients, where MHC-I and LDLR, respectively, play important roles in the pathogenesis.

Another GP-C activated by SKI-1/S1P includes the prototypic arenavirus lymphocytic choriomeningitis virus (LCMV) processed at **R**RLA↓GT (Table 4) [112,119–121]. In contrast to LASV, SKI-1/S1P cleaves the GP-C of LCMV in the *trans* Golgi. Of note, the **R**RLA↓ motif is sufficient to direct the maturation of the glycoprotein within this late compartment. Indeed, a single point mutation of LASV GP-C engineered to replace Leu at P1 with Ala results in a glycoprotein resistant to the cleavage in the ER but sensitive to SKI-1/S1P activity much later in the secretory pathway [112,122] (Figure 1). These key findings by Stefan Kunz are important to further classify SKI-1/S1P as a unique enzyme, different from the rest of the basic aa-specific PC-members. Using the Arenaviruses as a tool, his team was able to confirm not only that SKI-1/S1P possesses a distinctive *consensus* motif but also to show that the nature of the single amino acids around the cleavage site is critical to allow the enzymatic activity within a specific sub-cellular compartment. Such behavior in the panorama of proteases is unique and probably reflects the complex maturation mechanism of SKI-1/S1P that Stefan Kunz unraveled, resulting in a plethora of forms retaining prodomain fragments of this enzyme.

LASV and LCMV were the first Arenaviruses found to be SKI-1/S1P dependent [111]. Since then, other members of the Old World (OW) Arenaviruses showed this distinctive feature. The GP-C of Lujo Virus (LUJV) recently isolated in a cluster of lethal hemorrhagic fever cases in Africa, is an example [123]. Using a very elegant approach, Stefan Kunz showed that SKI-1/S1P processes the LUJV GP-C by engineering a luciferase sensor carrying the glycoprotein specific cleavage site **R**KLM↓ [124]. Like Old World counterparts, the New World (NW) arenaviruses—which are endemic in the Americas—strictly depend on SKI-1/S1P for their maturation.

The JUNV transmitted by rodents is a well characterized member of the NW arenaviruses with a mortality rate of 20–30% among infected populations in Argentina [125]. The JUNV surface glycoprotein GP-C is cleaved by SKI-1/S1P at **R**SLK↓AF. The processing site was confirmed by Stefan Kunz using the luciferase sensor platform [124] (Table 4). Interestingly, the JUNV maturation site closely resembles the C-site auto-cleavage motif of SKI-1/S1P [11] (Table 4). As expected, it was suggested that the JUNV envelope glycoprotein is activated in a late cellular Golgi compartment [126] where also the last steps of SKI-1/S1P maturation take place. In general, the present data suggest that arenavirus GP-Cs evolved to mimic SKI-1/S1P auto-processing sites [127,128], likely ensuring effective cleavage and avoiding competition with cellular substrates of SKI-1/S1P [129]. Old World and clade C New World GP-Cs further possess a distinctive signature that consists in displaying an aromatic residue at position P7, which is far away from the actual cleavage site. Such aromatic residues are implicated in the molecular recognition of the counterpart Tyr_{285} located on the molecular surface of the enzyme. This finding suggests that during co-evolution with their mammalian hosts, arenavirus GPCs expanded the molecular contacts with SKI-1/S1P beyond the classical four-amino-acid recognition sequences and currently occupy an extended binding pocket [113]. The specificity of the interaction between Tyr_{285} of SKI-1/S1P and aromatic P7 residues found in viral but not cellular substrates makes this interaction a promising target for the development of specific antiviral drugs which should not interfere with the catalytic triad directly, thus preserving full cellular activities of SKI-1/S1P. Overall, Stefan Kunz work has brought to light the outstanding ability of Arenaviruses to exploit SKI-1/S1P in a way to optimize the molecular contacts, outmatching the enzyme natural substrates. His results should be of great value for further identification of antivirals.

The crucial role of SKI-1/S1P in the virus life cycle and therefore in the establishment of infection has been beautifully shown by a recent work by Manning and colleagues. Candid #1 (Can), derived from the pathogenic Junin virus, is a vaccine against JUNV whose use is limited to Argentina. One single mutation of the envelope glycoprotein of Candid #1 is primarily responsible for the virus attenuation by impairing an effective GP-C glycosylation that in turn drastically limits the envelope glycoprotein processing [130]. The above example relies on a variation that perturbs the recognition of the viral substrate by the cellular enzyme. In a similar fashion, we can imagine that misfunctions of SKI-1/S1P can lead to similar outcomes. Indeed, studies on mice carrying a hypomorphic mutation in the *MBTPS1* gene (woodrat) showed that arenaviruses are incapable of establishing persistent infections in vivo [131]. Besides natural (Candid #1) and engineered (woodrat) variants, pharmacological treatments targeting SKI-1/S1P were also shown to be effective against arenavirus spread. Different from other enveloped viruses, only the cleaved glycoprotein is incorporated into budding virions. Naked viral particles fully lacking spikes are not infectious and cannot be primed by other proteases, e.g., during entry. Thus, SKI-1/S1P mediated maturation of arenavirus glycoproteins represents a key event and a druggable target, prompting several groups to test inhibitors against various arenaviruses [132].

SKI-1/S1P inhibitors can be grouped into three major classes:

1. Proteins: We first showed that overexpression of the prodomain alone of SKI-1/S1P, as well as variants of α1-antitrypsin, can inhibit the function of this convertase [13]. The effect is striking against LASV multicycle replication [133].
2. Peptides: These include among others, tetrapeptide-chloromethylketones (CMKs) developed in our group [14] and the enediynyl peptides [134]. CMKs potently block LCMV infection [135] but their use is limited to research purposes since they are highly toxic in vivo.
3. Non-peptide small molecules: Because of their properties and stability, this class of inhibitors are generally preferred over others for in vivo use. A small molecule SKI-1/S1P inhibitor PF-429242 was developed by Pfizer [136,137] and tested as an antiviral targeting GP-C processing and productive infection of arenaviruses. SKI-1/S1P inhibition by PF-429242 suppresses viral replication in cells infected with LASV, LCMV [138], and New World arenaviruses [139]. Interruption of drug treatment did not result in re-emergence of infection, indicating that PF-429242 treatment leads to virus extinction. Of note, Stefan Kunz found that the drug is capable of clearing LCMV from chronically infected cells with no emergence of escape variants [139]. This finding is intriguing since an LCMV mutant engineered to carry the RR**RR**↓ mutation is viable and fit [135]. The replacement of the wild type RRLA↓ motif with RR**RR**↓ does switch LCMV dependence from SKI-1/S1P to Furin. Therefore, it seems that arenaviruses are not prompted to use other members of the PCs family. The reason(s) for this selectivity of SKI-1/S1P has not yet been elucidated.

SKI-1/S1P also cleaves the glycoprotein PreGn of the Crimean Congo Hemorrhagic Fever Virus (CCHFV), a member of the *Bunyaviridae* family, at **R**R**L**L↓SE [140] (Table 4) to release an N-terminal mucin-like domain (MLD) fused to a non-structural glycoprotein GP38 and a C-terminal fusogenic Gn-containing domain. This virus is unique since after the SKI-1/S1P processing at **R**R**L**L↓, the N-terminal product is further processed at **R**S**KR**↓ by Furin(-like) enzymes separating the MLD from GP38 [141,142]. Functional data further revealed that cleavage at the Furin(-like) **R**S**KR**↓ site is not essential for CCHFV production or cell-to-cell spread, but that Furin cleavage enhances virion production [142]. In contrast, processing at **R**R**L**L↓ is critical for virus infectivity. Cells deficient in SKI-1/S1P produce no infectious virus, although PreGn accumulates normally in the Golgi apparatus where the virus assembles [141].

Aside from direct glycoprotein cleavage by SKI-1/S1P, indirect roles in viral infection are now being documented. In a first example, while SKI-1/S1P is not implicated in the direct processing of the surface glycoprotein GP of the Ebola virus (EBOV), the SKI-1/S1P activation of α/β-GlcNAc-1-phospho-transferase is required to allow cathepsins sorting to lysosomes and their promotion of EBOV assembly [143]. Thus, active cathepsins B and L in lysosomes cleave EBOV GP to remove the glycan cap and the mucin-like domains, thereby favoring binding of GP to the NPC1 receptor and priming it for fusion [143].

In a second example, loss of SKI-1/S1P activity is expected to reduce the processing of SREBP1,2 and hence reduce the levels of intracellular lipids that in turn drastically impact viral infections and replications. In line with this vision, the SREBP-dependent lipidomic reprogramming represents an attractive antiviral target [144]. However, since lipids are at the crossroad of various cellular pathways/signaling, deciphering the complex mechanism(s) behind the SREBPs-mediated antiviral effects may plead different explanations. Accordingly, robust inhibitory effects on SARS-CoV-2 entry and replication are observed following *MBTPS1* silencing or inhibition of SKI-1/S1P by PF-429242 treatment (IC_50 = 300 nM) in cells [145]. The present evidence excludes any interference with the maturation of the envelope S-glycoprotein, which is indeed cleaved by another Proprotein Convertase, i.e., Furin [74]. Rather, this member of the *Coronaviridae* family depends on healthy cellular lipid homeostasis more than other coronaviruses through a mechanism that has not yet been fully described. Lipid droplets formation were further confirmed to be essential in SARS-CoV-2 replication and pathogenesis [146]. On the other hand, it was suggested that inhibition of SREBP2 activation (implicating SKI-1/S1P cleavage) may represent a successful strategy to avoid a possible cytokine storm event in COVID-19 patients [147,148]. Flaviviruses represent another group of pathogens whose ability to infect is intimately associated with cholesterol and therefore SKI-1/S1P. Along this line, flaviviruses rearrange intracellular membranes and orchestrate a profound reorganization of the host cell lipid metabolism to create a favorable environment for viral replication and assembly. Major members of the *Flaviviridae* family are the Dengue virus (DENV), hepatitis C virus (HCV), West Nile virus (WNV), and Zika virus (ZIKV). DENV is highly sensitive to SKI-1/S1P inhibitors at the late stage of the infection indicating that PF-429242 does not target virus surface glycoprotein prM processing directly, but rather regulates host cellular factors needed to synthesize lipids which are essential for suitable assembly platforms [149]. The inhibitory mechanism may be associated with a dramatic reduction in the abundance of neutral lipids and their markers which drastically impairs virus propagation. Similarly, we [150] and others [151] showed that mis-regulation of the cellular SKI-1/S1P activity is a successful strategy to eradicate HCV infection.

WNV and ZIKV were likewise shown to be sensitive to SKI/S1P inhibitors, along with structurally unrelated drugs targeting the SREBPs pathway [152]. Interestingly, we showed that SKI-1/S1P regulates HCV viral replication early in the viral lifecycle with no change in LDLR or CD81 receptor levels [150], closely resembling the effect of SKI-1/S1P inhibition in infections by SARS-CoV-2 [148]. In this scenario, indirect effects induced by SKI-1/S1P inhibition may contribute to reducing virus infection [153].

It is plausible that in the presence of PF-429242, lower cholesterol and fatty acids syntheses may upregulate the antiviral interferon (IFN) response of cells, as it was observed by PCSK9 inhibitors that would lower ER-cholesterol availability and hence active SREBP-2 levels, resulting—for example—in reduced DENV assembly [103]. Flaviviruses are somehow expected to be affected by SKI-1/S1P manipulation due to their lipid dependence during virion assembly and/or viral protein trafficking. In contrast, there are pathogens that are mainly affected in their entry step upon disrupting the normal cellular lipid distribution. This is the case of Hantaviruses, members of the family *Bunyaviridae* associated with hemorrhagic fever with renal and cardiopulmonary/pulmonary syndromes in humans. Both depletion of *MBTPS1* gene and SKI-1/S1P inactivation efficiently inhibit Hantavirus membrane fusion at or near the lipid mixing step and block cytoplasmic delivery of viral matrix protein entry. As a matter of fact, Hantavirus membrane fusion directly depends on the cholesterol abundance of the target cells [154].

It is becoming more and more evident that SKI-1/S1P is a key regulator of multiple types of viral infection. Thus, targeting such a host component represents a powerful strategy to contain the multiplication and spread of viruses. In turn, this would provide the patient's immune system a window of opportunity to develop an appropriate antiviral immune response to clear the virus off. Unfortunately, the only small inhibitor available PF-429242 failed to reach clinical applications due to its poor pharmacokinetic properties [137], thus begging for a potent and safe SKI-1/S1P inhibitor as an antiviral to fight such deadly infections [132,155–157].

3. Discussion

From the above considerations, it became clear that most enveloped viruses require the processing of their surface glycoproteins for productive infectivity. Enhanced pathogenicity has been associated with the apparition of a new cleavage site in certain viruses. While clearly favoring Furin cleavage [74,158], the exact role of the acquisition of a Furin site in SARS-CoV-2 [70] and its role in the spread of the virus and/or its pathogenicity have not been adequately established. Experimental evidence is not easy to obtain. One would need to generate a virus mutant to be tested in an animal model exhibiting the multiple pathogeneses that mimic the human clinical manifestation of COVID-19. A clear advantage of SARS-CoV-2 over SARS-CoV-1 has been its ability to spread asymptomatically and with great efficiency in a large portion of the population. One can speculate that the acquisition of a Furin cleavage site has contributed to confer to SARS-CoV-2 such an advantage and to enhance the tropism of this virus, as Furin is widely expressed in most tissues, whereas TMPRSS2 implicated in SARS-CoV-1 activation [24] is more limited in its tissue expression.

A clear distinction was made between the choice of the two ubiquitously expressed convertases: (i) the basic aa-specific Furin *versus* (ii) the non-basic aa-specific SKI-1/S1P. The mechanistic rationale may be, in part, related to the subcellular localization of the active forms of these convertases, which are initially synthesized as zymogens in the ER that are autocatalytically activated via two or more cleavages of their prodomains. Thus, whereas SKI-1/S1P autocatalytically activates itself early in the ER/*cis* Golgi (B′/B cleavages), becoming maximally active in the *medial* Golgi (C/C′ cleavages), Furin is first autocatalytically processed at the C-terminus of its prodomain in the ER, but still remains inactive as it is non-covalently bound to its prodomain. It is only in the TGN that Furin completely separates from its inhibitory prodomain, whereupon it processes various substrates *in trans* in the TGN, cell surface and/or endosomes [9,20,159,160]. This allows the processing of a wide spectrum of substrates that are preferentially cleaved by Furin at either the cell surface neutral pH (Anthrax toxin and PCSK9) or acidic pH of the TGN (e.g., SARS-CoV-2 S-protein, Ebola Zaire pro-GP) and endosomes (e.g., Diphtheria and Shiga toxins). The same also seems to apply for SKI-1/S1P that can be partially activated at neutral pH in the ER, or much more active in the slightly acidic pH environment of the *cis/medial* Golgi (Figure 2). An additional selectivity is provided by the substrate itself via its protein folding and subcellular trafficking, whereupon some SKI-1/S1P substrates are activated in the ER (e.g., LASV GP-C), and the majority of others in the *cis/medial* Golgi (SREBPs, and multiple viral glycoproteins). Finally, one single pathogen is known to exploit both pathways; CCHFV uses both SKI-1/S1P for viral activation and Furin to enhance viral production [142].

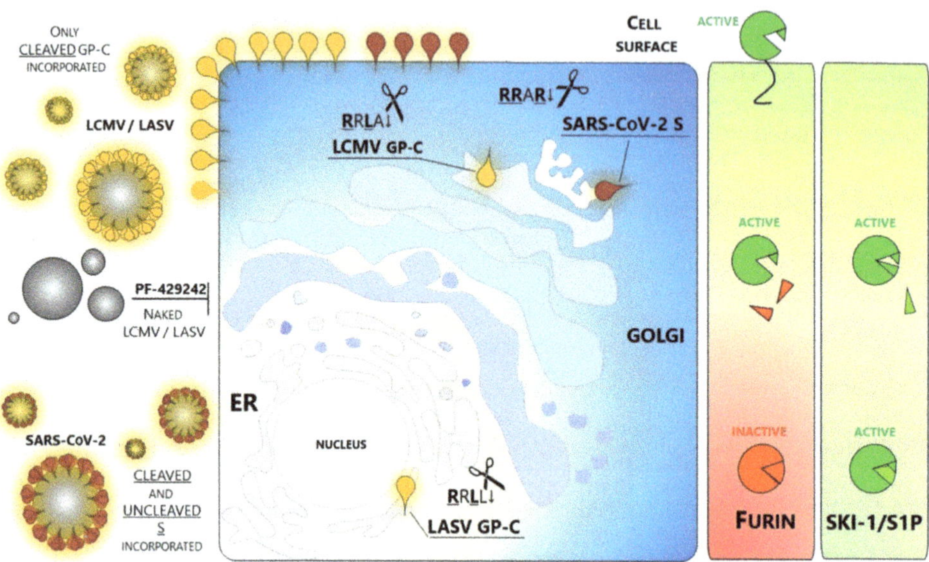

Figure 2. Schematic representation of the cellular compartments and where Proprotein Convertases are active against viral substrates. Furin processes the coronavirus SARS-CoV-2 spike S in the TGN; SKI-1/S1P matures the arenavirus LASV and LCMV GP-Cs in the ER and late Golgi, respectively. The cleaved viral glycoproteins reach the cell surface. Arenaviruses do incorporate only cleaved GP-C into viral budding particles since treatment with the SKI-1/S1P inhibitor PF-429242 blocks GP-C processing and its incorporation into budding particles; SARS-CoV-2 can include both cleaved and uncleaved envelope glycoprotein spike S.

In all examples described above, the link between protease(s) and the pathogen is clear cut, being the surface glycoprotein matured into active forms (e.g., acquisition of fusogenic properties) to attain full infectivity. This event is visible at the cell surface and therefore easily traceable by standard biochemical techniques. In contrast, how pathogens interact with the host as well as hijack, avoid and re-programme pre-existing pathways is still a matter of study. We may identify some common strategies, keeping in mind—though—that each infectious agent leaves its own fingerprints behind. For instance, upregulation of the SKI-1/S1P-dependent SREBP1,2 pathway is a well-documented example of host manipulation by Flaviviruses, since the increase of ER/Golgi lipids is critical for virion assembly, as was recently shown for the Dengue virus [103]. Other viruses, e.g., SARS-CoV-2 and Hantaviruses, require cholesterol for effective entry, thereby affecting early stages of infection. Along the same line, we are beginning to appreciate the master role of PCSK9, as a key regulator of cholesterol content. Some other pathways are carefully avoided by the pathogen allowing them to "silently" colonize the host without raising any red flag. This is the case of LCMV chronic infections that do not trigger any cellular unfolded protein response (UPR) pathway, the latter known to be activated by SKI-1/S1P mediated cleavage of ATF6 [161]. Overall, however, our current knowledge just gives us a glimpse over a very complex scenario where actors play roles in a plot not yet fully unraveled. More and more evidence, for example, start unveiling the contribution of PCs to the immune response: the same pathways that regulate lipid homeostasis—thought to be manipulated to favor the establishment of successful infections—are now shown to be a "wake-up call" for immunity [19,162,163]. In view of this situation, more efforts are required to fill the gaps. The current SARS-CoV-2 pandemic further highlights that the scientific community is due for a shake-up. We need to invest more energies and resources to better understand pathogens and, of course, to develop effective drug treatments. In this context, PCs represent ideal drug targets. First, by being proteases, these molecules possess defined substrate-binding pockets, which can be exploited as binding sites for pharmaceutical inhibitors [164]. In addition, the interference with PCs activity can be quantitatively measured by wellestablished assays that allow rational lead optimization. Second, PCs are at the crossroad of multiple types of infections. Identification of powerful and safe PCs inhibitors may come in handy for current and newly arising pathogens, including SARS-CoV-2 and the emergence of a Furin-dependent influenza virus. The latter has been a matter of discussion during the last decade as a causative agent of a potential novel global pandemic [165,166]. Likewise, targeting PCs would provide a powerful tool to modulate lipids, which are a "common denominator" of many enveloped viruses, e.g., flaviviruses, α-viruses, coronaviruses, filoviruses, retroviruses and more (Table 1).

Thus, PCs-targeted antivirals provide great potential as general pandemic-preparedness tools. The advantage of targeting the host—i.e., PCs—consists in delivering a drug to a patient early in the infection and before the pandemic occurs and without knowing any details on the nature of the pathogen other than its dependence on these enzymes. Moreover, PCs fall into the "usual suspects" category, meaning they are popular and very often hijacked proteases and their involvement in future diseases is highly likely. In answer to those who claim toxic effects by PC-inhibition of critical host factors, we highlight that the drug administration should cover only a limited amount of time (few days), sufficiently to allow an appropriate immune response to the pathogen. Also, pharmaceutical companies manage to find drugs and dosing regimens against host proteins that people can tolerate. So, why should PCs antivirals be any different?

The present review clearly emphasized the seminal contribution of **Stefan Kunz** to our better understanding of the complex interplay between hemorrhagic fever viral surface glycoproteins folding and processing activation by SKI-1/S1P. The work of his team led the way towards a rational understanding of the structure-function of SKI-1/S1P, its zymogen activation steps, and their effects on the processing of various arenaviruses that rely on this enzyme for their activation. The combination of his contributions to the monumental literature on the activation of enveloped viruses provided a distinct framework for future studies on the development of specific inhibitors or therapies against a multitude of infectious viruses. This is a solid base to build upon for our preparedness to counter the onslaught of the present (e.g., COVID-19) and future pandemics that will surely arise because of the unrelenting interference and manipulation of wildlife and fauna by man and the ensuing loss of biodiversity.

Finally, two of us (AP and NGS) have had the privilege to work closely with Stefan Kunz, allowing us to appreciate his scientific intellect and curiosity. He will be fondly missed and remembered, and this review is a tribute and dedication to his memory.

Funding: This work was funded thanks to grants to NGS including a CIHR Foundation Scheme grant (# 148363), a Canada Chair in Precursor Proteolysis (# 950-231335) and a CHAMPS team grant (# HAL 157986) and to AP a EU Research Framework Programme H2020/Marie Skłodowska-Curie Actions grant #101024974.

Institutional Review Board Statement: Not applicable.

Informed Consent Statement: Not applicable.

Data Availability Statement: Not applicable.

Acknowledgments: The authors would like to acknowledge the expert secretarial help of Habiba Oueslati in the preparation of this manuscript.

Conflicts of Interest: The authors declare no conflict of interest.

List of Abbreviations

ACE2, angiotensin-converting enzyme 2; ADE, antibody-dependent-enhancement; CCHFV, crimean congo hemorrhagic fever virus; CHIKV, chikungunya virus; CMK, chloromethylketone; COVID-19, coronavirus disease 2019; DENV, dengue virus; EBOV, ebola virus; FPV, fowl plague virus; HAT, human airway trypsin-like protease; HA, hemagglutinin; HAND, HIV-associated neurocognitive disease; HBV, hepatitis B virus; HCV, hepatitis C virus; HIV, human immunodeficiency virus; hMPV, human metapneumovirus (hMPV); IFN, interferon; LASV, lassa virus; LCMV, lymphocytic choriomeningitis virus; LDLR, low density lipoprotein receptor; LUJV, Lujo Virus; MARCH8, membrane-associated RING-CH-type 8; MBTPS1, membrane-bound transcription-factor site-1 protease (*MBTPS1*); MERS-CoV, middle east respiratory syndrome coronavirus; MHC-I, major histocompatibility class I receptor; NDV, newcastle disease virus; NW, new world; PAR1/2, protease activated receptors 1/2; PCs, proprotein convertases; PCSK, proprotein convertase subtilisin-kexin; PCSK9, proprotein convertase subtilisin-kexin 9; SARS-CoV, severe acute respiratory syndrome coronavirus; RBD, receptor binding domain; SKI-1/S1P; subtilisin/kexin isozyme-1/site-1 protease; SREBP, sterol regulatory element binding protein; TBFV, tick-borne flavivirus; TMPRSS2, transmembrane protease-serine 2; WNV, west nile virus; ZIKV, zika virus.

References

1. Dobson, A.P.; Carper, E.R. Infectious Diseases and Human Population History: Throughout history the establishment of disease has been a side effect of the growth of civilization. *Bioscience* **1996**, *46*, 115–126. [CrossRef]
2. Morens, D.M.; Folkers, G.K.; Fauci, A.S. Emerging infections: A perpetual challenge. *Lancet Infect. Dis.* **2008**, *8*, 710–719. [CrossRef]
3. Cyranoski, D. Profile of a killer: The complex biology powering the coronavirus pandemic. *Nature* **2020**, *581*, 22–26. [CrossRef]
4. Singh, R.; Kang, A.; Luo, X.; Jeyanathan, M.; Gillgrass, A.; Afkhami, S.; Xing, Z. COVID-19: Current knowledge in clinical features, immunological responses, and vaccine development. *FASEB J.* **2021**, *35*, e21409. [CrossRef] [PubMed]
5. Funk, C.D.; Laferrière, C.; Ardakani, A. Target Product Profile Analysis of COVID-19 Vaccines in Phase III Clinical Trials and Beyond: An Early 2021 Perspective. *Viruses* **2021**, *13*, 418. [CrossRef]
6. Morens, D.M.; Fauci, A.S. Emerging Pandemic Diseases: How We Got to COVID-19. *Cell* **2020**, *182*, 1077–1092. [CrossRef] [PubMed]
7. Owusu, I.A.; Quaye, O.; Passalacqua, K.D.; Wobus, C.E. Egress of non-enveloped enteric RNA viruses. *J. Gen. Virol.* **2021**, *102*, 001557. [CrossRef]
8. Leroy, H.; Han, M.; Woottum, M.; Bracq, L.; Bouchet, J.; Xie, M.; Benichou, S. Virus-Mediated Cell-Cell Fusion. *Int. J. Mol. Sci.* **2020**, *21*, 9644. [CrossRef]
9. Seidah, N.G.; Prat, A. The biology and therapeutic targeting of the proprotein convertases. *Nat. Rev. Drug Discov.* **2012**, *11*, 367–383. [CrossRef] [PubMed]
10. Seidah, N.G.; Chretien, M. Proprotein and prohormone convertases: A family of subtilases generating diverse bioactive polypeptides. *Brain Res.* **1999**, *848*, 45–62. [CrossRef]
11. Seidah, N.G.; Mowla, S.J.; Hamelin, J.; Mamarbachi, A.M.; Benjannet, S.; Toure, B.B.; Basak, A.; Munzer, J.S.; Marcinkiewicz, J.; Zhong, M.; et al. Mammalian subtilisin/kexin isozyme SKI-1: A widely expressed proprotein convertase with a unique cleavage specificity and cellular localization. *Proc. Natl. Acad. Sci. USA* **1999**, *96*, 1321–1326. [CrossRef] [PubMed]
12. Espenshade, P.J.; Cheng, D.; Goldstein, J.L.; Brown, M.S. Autocatalytic processing of site-1 protease removes propeptide and permits cleavage of sterol regulatory element-binding proteins. *J. Biol. Chem.* **1999**, *274*, 22795–22804. [CrossRef]
13. Pullikotil, P.; Vincent, M.; Nichol, S.T.; Seidah, N.G. Development of protein-based inhibitors of the proprotein of convertase SKI-1/S1P: Processing of SREBP-2, ATF6, and a viral glycoprotein. *J. Biol. Chem.* **2004**, *279*, 17338–17347. [CrossRef] [PubMed]
14. Pasquato, A.; Pullikotil, P.; Asselin, M.C.; Vacatello, M.; Paolillo, L.; Ghezzo, F.; Basso, F.; Di Bello, C.; Dettin, M.; Seidah, N.G. The Proprotein Convertase SKI-1/S1P: In vitro analysis of lassa virus glycoprotein-derived substrates and ex vivo validation of irreversible peptide inhibitors. *J. Biol. Chem.* **2006**, *281*, 23471–23481. [CrossRef] [PubMed]
15. Seidah, N.G. Proprotein Convertases SKI-1/S1P and PCSK9. In *Handbook of the Biologically Active Peptides*; Minamino, N., Kastin, A., Eds.; Academic Press: Cambridge, MA, USA, 2012; pp. 1–8.
16. Seidah, N.G.; Benjannet, S.; Wickham, L.; Marcinkiewicz, J.; Jasmin, S.B.; Stifani, S.; Basak, A.; Prat, A.; Chretien, M. The secretory proprotein convertase neural apoptosis-regulated convertase 1 (NARC-1): Liver regeneration and neuronal differentiation. *Proc. Natl. Acad. Sci. USA* **2003**, *100*, 928–933. [CrossRef]
17. Seidah, N.G.; Awan, Z.; Chretien, M.; Mbikay, M. PCSK9: A Key Modulator of Cardiovascular Health. *Circ. Res.* **2014**, *114*, 1022–1036. [CrossRef]
18. Seidah, N.G.; Abifadel, M.; Prost, S.; Boileau, C.; Prat, A. The Proprotein Convertases in Hypercholesterolemia and Cardiovascular Diseases: Emphasis on Proprotein Convertase Subtilisin/Kexin 9. *Pharmacol. Rev.* **2017**, *69*, 33–52. [CrossRef]
19. Liu, X.; Bao, X.; Hu, M.; Chang, H.; Jiao, M.; Cheng, J.; Xie, L.; Huang, Q.; Li, F.; Li, C.Y. Inhibition of PCSK9 potentiates immune checkpoint therapy for cancer. *Nature* **2020**, *588*, 693–698. [CrossRef]
20. Thomas, G. Furin at the cutting edge: From protein traffic to embryogenesis and disease. *Nat. Rev. Mol. Cell Biol.* **2002**, *3*, 753–766. [CrossRef]
21. Izaguirre, G. The Proteolytic Regulation of Virus Cell Entry by Furin and Other Proprotein Convertases. *Viruses* **2019**, *11*, 837. [CrossRef]

22. Van de Ven, W.J.; Creemers, J.W.; Roebroek, A.J. Furin: The prototype mammalian subtilisin-like proprotein-processing enzyme. Endoproteolytic cleavage at paired basic residues of proproteins of the eukaryotic secretory pathway. *Enzyme* **1991**, *45*, 257–270. [CrossRef] [PubMed]
23. Moulard, M.; Decroly, E. Maturation of HIV envelope glycoprotein precursors by cellular endoproteases. *Biochim. Biophys. Acta* **2000**, *1469*, 121–132. [CrossRef]
24. Millet, J.K.; Whittaker, G.R. Host cell proteases: Critical determinants of coronavirus tropism and pathogenesis. *Virus Res.* **2015**, *202*, 120–134. [CrossRef]
25. Braun, E.; Sauter, D. Furin-mediated protein processing in infectious diseases and cancer. *Clin. Transl. Immunol.* **2019**, *8*, e1073. [CrossRef]
26. Le Coupanec, A.; Desforges, M.; Meessen-Pinard, M.; Dube, M.; Day, R.; Seidah, N.G.; Talbot, P.J. Cleavage of a Neuroinvasive Human Respiratory Virus Spike Glycoprotein by Proprotein Convertases Modulates Neurovirulence and Virus Spread within the Central Nervous System. *PLoS Pathog.* **2015**, *11*, e1005261. [CrossRef]
27. Tang, T.; Bidon, M.; Jaimes, J.A.; Whittaker, G.R.; Daniel, S. Coronavirus membrane fusion mechanism offers as a potential target for antiviral development. *Antivir. Res.* **2020**, *178*, 104792. [CrossRef]
28. Seidah, N.G. The proprotein convertases, 20 years later. *Methods Mol. Biol.* **2011**, *768*, 23–57.
29. Hallenberger, S.; Bosch, V.; Angliker, H.; Shaw, E.; Klenk, H.D.; Garten, W. Inhibition of furin-mediated cleavage activation of HIV-1 glycoprotein gp160. *Nature* **1992**, *360*, 358–361. [CrossRef] [PubMed]
30. Stieneke-Grober, A.; Vey, M.; Angliker, H.; Shaw, E.; Thomas, G.; Roberts, C.; Klenk, H.D.; Garten, W. Influenza virus hemagglutinin with multibasic cleavage site is activated by furin, a subtilisin-like endoprotease. *EMBO J.* **1992**, *11*, 2407–2414. [CrossRef]
31. Li, Z.; Sergel, T.; Razvi, E.; Morrison, T. Effect of cleavage mutants on syncytium formation directed by the wild-type fusion protein of Newcastle disease virus. *J. Virol.* **1998**, *72*, 3789–3795. [CrossRef] [PubMed]
32. Walker, J.A.; Molloy, S.S.; Thomas, G.; Sakaguchi, T.; Yoshida, T.; Chambers, T.M.; Kawaoka, Y. Sequence specificity of furin, a proprotein-processing endoprotease, for the hemagglutinin of a virulent avian influenza virus. *J. Virol.* **1994**, *68*, 1213–1218. [CrossRef]
33. Gao, P.; Watanabe, S.; Ito, T.; Goto, H.; Wells, K.; McGregor, M.; Cooley, A.J.; Kawaoka, Y. Biological heterogeneity, including systemic replication in mice, of H5N1 influenza A virus isolates from humans in Hong Kong. *J. Virol.* **1999**, *73*, 3184–3189. [CrossRef]
34. Claas, E.C.; Osterhaus, A.D.; van Beek, R.; de Jong, J.C.; Rimmelzwaan, G.F.; Senne, D.A.; Krauss, S.; Shortridge, K.F.; Webster, R.G. Human influenza A H5N1 virus related to a highly pathogenic avian influenza virus. *Lancet* **1998**, *351*, 472–477. [CrossRef]
35. Basak, A.; Zhong, M.; Munzer, J.S.; Chretien, M.; Seidah, N.G. Implication of the proprotein convertases furin, PC5 and PC7 in the cleavage of surface glycoproteins of Hong Kong, Ebola and respiratory syncytial viruses: A comparative analysis with fluorogenic peptides. *Biochem. J.* **2001**, *353*, 537–545. [CrossRef] [PubMed]
36. Böttcher, E.; Matrosovich, T.; Beyerle, M.; Klenk, H.D.; Garten, W.; Matrosovich, M. Proteolytic activation of influenza viruses by serine proteases TMPRSS2 and HAT from human airway epithelium. *J. Virol.* **2006**, *80*, 9896–9898. [CrossRef] [PubMed]
37. Murza, A.; Dion, S.P.; Boudreault, P.L.; Désilets, A.; Leduc, R.; Marsault, É. Inhibitors of type II transmembrane serine proteases in the treatment of diseases of the respiratory tract—A review of patent literature. *Expert Opin. Ther. Pat.* **2020**, *30*, 807–824. [CrossRef] [PubMed]
38. Sachan, V.; Lodge, R.; Mihara, K.; Hamelin, J.; Power, C.; Gelman, B.B.; Hollenberg, M.D.; Cohen, É.A.; Seidah, N.G. HIV-induced neuroinflammation: Impact of PAR1 and PAR2 processing by Furin. *Cell Death Differ.* **2019**, *26*, 1942–1954. [CrossRef] [PubMed]
39. Ohuchi, M.; Orlich, M.; Ohuchi, R.; Simpson, B.E.; Garten, W.; Klenk, H.D.; Rott, R. Mutations at the cleavage site of the hemagglutinin after the pathogenicity of influenza virus A/chick/Penn/83 (H5N2). *Virology* **1989**, *168*, 274–280. [CrossRef]
40. Gu, M.; Rappaport, J.; Leppla, S.H. Furin is important but not essential for the proteolytic maturation of gp160 of HIV-1. *FEBS Lett.* **1995**, *365*, 95–97. [CrossRef]
41. Aerts, L.; Hamelin, M.E.; Rhaume, C.; Lavigne, S.; Couture, C.; Kim, W.; Susan-Resiga, D.; Prat, A.; Seidah, N.G.; Vergnolle, N.; et al. Modulation of protease activated receptor 1 influences human metapneumovirus disease severity in a mouse model. *PLoS ONE* **2013**, *8*, e72529. [CrossRef]
42. Kim, W.; Zekas, E.; Lodge, R.; Susan-Resiga, D.; Marcinkiewicz, E.; Essalmani, R.; Mihara, K.; Ramachandran, R.; Asahchop, E.; Gelman, B.; et al. Neuroinflammation-Induced Interactions between Protease-Activated Receptor 1 and Proprotein Convertases in HIV-Associated Neurocognitive Disorder. *Mol. Cell Biol.* **2015**, *35*, 3684–3700. [CrossRef]
43. Braun, E.; Hotter, D.; Koepke, L.; Zech, F.; Groß, R.; Sparrer, K.M.J.; Müller, J.A.; Pfaller, C.K.; Heusinger, E.; Wombacher, R.; et al. Guanylate-Binding Proteins 2 and 5 Exert Broad Antiviral Activity by Inhibiting Furin-Mediated Processing of Viral Envelope Proteins. *Cell Rep.* **2019**, *27*, 2092–2104.e10. [CrossRef]
44. Volchkov, V.E.; Feldmann, H.; Volchkova, V.A.; Klenk, H.D. Processing of the Ebola virus glycoprotein by the proprotein convertase furin. *Proc. Natl. Acad. Sci. USA* **1998**, *95*, 5762–5767. [CrossRef] [PubMed]
45. Neumann, G.; Feldmann, H.; Watanabe, S.; Lukashevich, I.; Kawaoka, Y. Reverse genetics demonstrates that proteolytic processing of the Ebola virus glycoprotein is not essential for replication in cell culture. *J. Virol.* **2002**, *76*, 406–410. [CrossRef] [PubMed]
46. Neumann, G.; Geisbert, T.W.; Ebihara, H.; Geisbert, J.B.; Daddario-DiCaprio, K.M.; Feldmann, H.; Kawaoka, Y. Proteolytic processing of the Ebola virus glycoprotein is not critical for Ebola virus replication in nonhuman primates. *J. Virol.* **2007**, *81*, 2995–2998. [CrossRef] [PubMed]

47. Wang, B.; Wang, Y.; Frabutt, D.A.; Zhang, X.; Yao, X.; Hu, D.; Zhang, Z.; Liu, C.; Zheng, S.; Xiang, S.H.; et al. Mechanistic understanding of N-glycosylation in Ebola virus glycoprotein maturation and function. *J. Biol. Chem.* **2017**, *292*, 5860–5870. [CrossRef]
48. Yu, C.; Li, S.; Zhang, X.; Khan, I.; Ahmad, I.; Zhou, Y.; Li, S.; Shi, J.; Wang, Y.; Zheng, Y.H. MARCH8 Inhibits Ebola Virus Glycoprotein, Human Immunodeficiency Virus Type 1 Envelope Glycoprotein, and Avian Influenza Virus H5N1 Hemagglutinin Maturation. *mBio* **2020**, *11*, e01882-20. [CrossRef]
49. Stadler, K.; Allison, S.L.; Schalich, J.; Heinz, F.X. Proteolytic activation of tick-borne encephalitis virus by furin. *J. Virol.* **1997**, *71*, 8475–8481. [CrossRef]
50. Zybert, I.A.; van der Ende-Metselaar, H.; Wilschut, J.; Smit, J.M. Functional importance of dengue virus maturation: Infectious properties of immature virions. *J. Gen. Virol.* **2008**, *89*, 3047–3051. [CrossRef] [PubMed]
51. Da Silva Voorham, J.M.; Rodenhuis-Zybert, I.A.; Ayala Nuñez, N.V.; Colpitts, T.M.; van der Ende-Metselaar, H.; Fikrig, E.; Diamond, M.S.; Wilschut, J.; Smit, J.M. Antibodies against the envelope glycoprotein promote infectivity of immature dengue virus serotype 2. *PLoS ONE* **2012**, *7*, e29957. [CrossRef]
52. Rodenhuis-Zybert, I.A.; van der Schaar, H.M.; da Silva Voorham, J.M.; van der Ende-Metselaar, H.; Lei, H.Y.; Wilschut, J.; Smit, J.M. Immature dengue virus: A veiled pathogen? *PLoS Pathog.* **2010**, *6*, e1000718. [CrossRef]
53. VanBlargan, L.A.; Errico, J.M.; Kafai, N.M.; Burgomaster, K.E.; Jethva, P.N.; Broeckel, R.M.; Meade-White, K.; Nelson, C.A.; Himansu, S.; Wang, D.; et al. Broadly neutralizing monoclonal antibodies protect against multiple tick-borne flaviviruses. *J. Exp. Med.* **2021**, *218*, e20210174. [CrossRef] [PubMed]
54. Ozden, S.; Lucas-Hourani, M.; Ceccaldi, P.E.; Basak, A.; Valentine, M.; Benjannet, S.; Hamelin, J.; Jacob, Y.; Mamchaoui, K.; Mouly, V.; et al. Inhibition of Chikungunya Virus Infection in Cultured Human Muscle Cells by Furin Inhibitors: Impairment of the maturation of the E2 surface glycoprotein. *J. Biol. Chem.* **2008**, *283*, 21899–21908. [CrossRef] [PubMed]
55. Kropp, K.A.; Srivaratharajan, S.; Ritter, B.; Yu, P.; Krooss, S.; Polten, F.; Pich, A.; Alcami, A.; Viejo-Borbolla, A. Identification of the Cleavage Domain within Glycoprotein G of Herpes Simplex Virus Type 2. *Viruses* **2020**, *12*, 1428. [CrossRef] [PubMed]
56. Oliver, S.L.; Sommer, M.; Zerboni, L.; Rajamani, J.; Grose, C.; Arvin, A.M. Mutagenesis of varicella-zoster virus glycoprotein B: Putative fusion loop residues are essential for viral replication, and the furin cleavage motif contributes to pathogenesis in skin tissue in vivo. *J. Virol.* **2009**, *83*, 7495–7506. [CrossRef]
57. Emerman, M.; Malim, M.H. HIV-1 regulatory/accessory genes: Keys to unraveling viral and host cell biology. *Science* **1998**, *280*, 1880–1884. [CrossRef]
58. Wallet, C.; Rohr, O.; Schwartz, C. Evolution of a concept: From accessory protein to key virulence factor, the case of HIV-1 Vpr. *Biochem. Pharmacol.* **2020**, *180*, 114128. [CrossRef]
59. Mayer, G.; Hamelin, J.; Asselin, M.C.; Pasquato, A.; Marcinkiewicz, E.; Tang, M.; Tabibzadeh, S.; Seidah, N.G. The regulated cell surface zymogen activation of the proprotein convertase PC5A directs the processing of its secretory substrates. *J. Biol. Chem.* **2008**, *283*, 2373–2384. [CrossRef]
60. Xiao, Y.; Chen, G.; Richard, J.; Rougeau, N.; Li, H.; Seidah, N.G.; Cohen, E.A. Cell-surface processing of extracellular human immunodeficiency virus type 1 Vpr by proprotein convertases. *Virology* **2008**, *372*, 384–397. [CrossRef]
61. Bosch, B.J.; van der, Z.R.; de Haan, C.A.; Rottier, P.J. The coronavirus spike protein is a class I virus fusion protein: Structural and functional characterization of the fusion core complex. *J. Virol.* **2003**, *77*, 8801–8811. [CrossRef] [PubMed]
62. De Wit, E.; van Doremalen, N.; Falzarano, D.; Munster, V.J. SARS and MERS: Recent insights into emerging coronaviruses. *Nat. Rev. Microbiol.* **2016**, *14*, 523–534. [CrossRef] [PubMed]
63. Li, W.; Moore, M.J.; Vasilieva, N.; Sui, J.; Wong, S.K.; Berne, M.A.; Somasundaran, M.; Sullivan, J.L.; Luzuriaga, K.; Greenough, T.C.; et al. Angiotensin-converting enzyme 2 is a functional receptor for the SARS coronavirus. *Nature* **2003**, *426*, 450–454. [CrossRef] [PubMed]
64. Lu, G.; Hu, Y.; Wang, Q.; Qi, J.; Gao, F.; Li, Y.; Zhang, Y.; Zhang, W.; Yuan, Y.; Bao, J.; et al. Molecular basis of binding between novel human coronavirus MERS-CoV and its receptor CD26. *Nature* **2013**, *500*, 227–231. [CrossRef]
65. Belouzard, S.; Madu, I.; Whittaker, G.R. Elastase-mediated activation of the severe acute respiratory syndrome coronavirus spike protein at discrete sites within the S2 domain. *J. Biol. Chem.* **2010**, *285*, 22758–22763. [CrossRef]
66. Belouzard, S.; Millet, J.K.; Licitra, B.N.; Whittaker, G.R. Mechanisms of coronavirus cell entry mediated by the viral spike protein. *Viruses* **2012**, *4*, 1011–1033. [CrossRef] [PubMed]
67. Shirato, K.; Kawase, M.; Matsuyama, S. Middle East respiratory syndrome coronavirus infection mediated by the transmembrane serine protease TMPRSS2. *J. Virol.* **2013**, *87*, 12552–12561. [CrossRef]
68. Kleine-Weber, H.; Elzayat, M.T.; Hoffmann, M.; Pöhlmann, S. Functional analysis of potential cleavage sites in the MERS-coronavirus spike protein. *Sci. Rep.* **2018**, *8*, 16597. [CrossRef] [PubMed]
69. Hörnich, B.F.; Großkopf, A.K.; Schlagowski, S.; Tenbusch, M.; Kleine-Weber, H.; Neipel, F.; Stahl-Hennig, C.; Hahn, A.S. SARS-CoV-2 and SARS-CoV spike-mediated cell-cell fusion differ in the requirements for receptor expression and proteolytic activation. *J. Virol.* **2021**, *95*, e00002-21. [CrossRef] [PubMed]
70. Coutard, B.; Valle, C.; de Lamballerie, X.; Canard, B.; Seidah, N.G.; Decroly, E. The spike glycoprotein of the new coronavirus 2019-nCoV contains a furin-like cleavage site absent in CoV of the same clade. *Antivir. Res.* **2020**, *176*, 104742. [CrossRef]
71. Bollavaram, K.; Leeman, T.H.; Lee, M.W.; Kulkarni, A.; Upshaw, S.G.; Yang, J.; Song, H.; Platt, M.O. Multiple Sites on SARS-CoV-2 Spike Protein are Susceptible to Proteolysis by Cathepsins B, K, L, S, and V. *Protein Sci.* **2021**. [CrossRef] [PubMed]

72. Kawase, M.; Shirato, K.; van der Hoek, L.; Taguchi, F.; Matsuyama, S. Simultaneous treatment of human bronchial epithelial cells with serine and cysteine protease inhibitors prevents severe acute respiratory syndrome coronavirus entry. *J. Virol.* **2012**, *86*, 6537–6545. [CrossRef] [PubMed]
73. Nimishakavi, S.; Raymond, W.W.; Gruenert, D.C.; Caughey, G.H. Divergent Inhibitor Susceptibility among Airway Lumen-Accessible Tryptic Proteases. *PLoS ONE* **2015**, *10*, e0141169. [CrossRef] [PubMed]
74. Essalmani, R.; Jain, J.; Susan-Resiga, D.; Andréo, U.; Evagelidis, A.; Derbali, R.M.; Huynh, D.N.; Dallaire, F.; Laporte, M.; Delpal, A.; et al. Furin cleaves SARS-CoV-2 spike-glycoprotein at S1/S2 and S2′ for viral fusion/entry: Indirect role of TMPRSS2. *bioRxiv* **2020**. [CrossRef]
75. Raghuvamsi, P.V.; Tulsian, N.K.; Samsudin, F.; Qian, X.; Purushotorman, K.; Yue, G.; Kozma, M.M.; Hwa, W.Y.; Lescar, J.; Bond, P.J.; et al. SARS-CoV-2 S protein:ACE2 interaction reveals novel allosteric targets. *eLife* **2021**, *10*, e63646. [CrossRef]
76. Tang, T.; Jaimes, J.A.; Bidon, M.K.; Straus, M.R.; Daniel, S.; Whittaker, G.R. Proteolytic Activation of SARS-CoV-2 Spike at the S1/S2 Boundary: Potential Role of Proteases beyond Furin. *ACS Infect Dis.* **2021**, *7*, 264–272. [CrossRef]
77. Heurich, A.; Hofmann-Winkler, H.; Gierer, S.; Liepold, T.; Jahn, O.; Pöhlmann, S. TMPRSS2 and ADAM17 cleave ACE2 differentially and only proteolysis by TMPRSS2 augments entry driven by the severe acute respiratory syndrome coronavirus spike protein. *J. Virol.* **2014**, *88*, 1293–1307. [CrossRef]
78. Kishimoto, M.; Uemura, K.; Sanaki, T.; Sato, A.; Hall, W.W.; Kariwa, H.; Orba, Y.; Sawa, H.; Sasaki, M. TMPRSS11D and TMPRSS13 Activate the SARS-CoV-2 Spike Protein. *Viruses* **2021**, *13*, 384. [CrossRef]
79. Vincent, M.J.; Bergeron, E.; Benjannet, S.; Erickson, B.R.; Rollin, P.E.; Ksiazek, T.G.; Seidah, N.G.; Nichol, S.T. Chloroquine is a potent inhibitor of SARS coronavirus infection and spread. *Virol. J.* **2005**, *2*, 69. [CrossRef]
80. Hache, G.; Rolain, J.M.; Gautret, P.; Deharo, J.C.; Brouqui, P.; Raoult, D.; Honoré, S. Combination of Hydroxychloroquine Plus Azithromycin As Potential Treatment for COVID-19 Patients: Safety Profile, Drug Interactions, and Management of Toxicity. *Microb. Drug Resist.* **2021**, *27*, 281–290. [CrossRef]
81. Maisonnasse, P.; Guedj, J.; Contreras, V.; Behillil, S.; Solas, C.; Marlin, R.; Naninck, T.; Pizzorno, A.; Lemaitre, J.; Gonçalves, A.; et al. Hydroxychloroquine use against SARS-CoV-2 infection in non-human primates. *Nature* **2020**, *585*, 584–587. [CrossRef]
82. Leung, K.; Shum, M.H.; Leung, G.M.; Lam, T.T.; Wu, J.T. Early transmissibility assessment of the N501Y mutant strains of SARS-CoV-2 in the United Kingdom, October to November 2020. *Eurosurveillance* **2021**, *26*, 2002106. [CrossRef]
83. Tegally, H.; Wilkinson, E.; Giovanetti, M.; Iranzadeh, A.; Fonseca, V.; Giandhari, J.; Doolabh, D.; Pillay, S.; San, E.J.; Msomi, N.; et al. Emergence and rapid spread of a new severe acute respiratory syndrome-related coronavirus 2 (SARS-CoV-2) lineage with multiple spike mutations in South Africa. *medRxiv* **2020**. [CrossRef]
84. Fujino, T.; Nomoto, H.; Kutsuna, S.; Ujiie, M.; Suzuki, T.; Sato, R.; Fujimoto, T.; Kuroda, M.; Wakita, T.; Ohmagari, N. Novel SARS-CoV-2 Variant Identified in Travelers from Brazil to Japan. *Emerg. Infect. Dis.* **2021**, *27*, 1243. [CrossRef]
85. Zahradník, J.; Marciano, S.; Shemesh, M.; Zoler, E.; Chiaravalli, J.; Meyer, B.; Dym, O.; Elad, N.; Schreiber, G. SARS-CoV-2 RBD in vitro evolution follows contagious mutation spread, yet generates an able infection inhibitor. *bioRxiv* **2021**. [CrossRef]
86. Wang, P.; Nair, M.S.; Liu, L.; Iketani, S.; Luo, Y.; Guo, Y.; Wang, M.; Yu, J.; Zhang, B.; Kwong, P.D.; et al. Antibody Resistance of SARS-CoV-2 Variants B.1.351 and B.1.1.7. *Nature* **2021**, *593*, 130–135. [CrossRef] [PubMed]
87. Chen, R.E.; Zhang, X.; Case, J.B.; Winkler, E.S.; Liu, Y.; VanBlargan, L.A.; Liu, J.; Errico, J.M.; Xie, X.; Suryadevara, N.; et al. Resistance of SARS-CoV-2 variants to neutralization by monoclonal and serum-derived polyclonal antibodies. *Nat. Med.* **2021**, *27*, 717–726. [CrossRef]
88. Yang, C.; Huang, Y.; Liu, S. Therapeutic Development in COVID-19. *Adv. Exp. Med. Biol.* **2021**, *1318*, 435–448. [PubMed]
89. Benjannet, S.; Rhainds, D.; Essalmani, R.; Mayne, J.; Wickham, L.; Jin, W.; Asselin, M.C.; Hamelin, J.; Varret, M.; Allard, D.; et al. NARC-1/PCSK9 and its natural mutants: Zymogen cleavage and effects on the low density lipoprotein (LDL) receptor and LDL cholesterol. *J. Biol. Chem.* **2004**, *279*, 48865–48875. [CrossRef]
90. Li, J.; Tumanut, C.; Gavigan, J.A.; Huang, W.J.; Hampton, E.N.; Tumanut, R.; Suen, K.F.; Trauger, J.W.; Spraggon, G.; Lesley, S.A.; et al. Secreted PCSK9 promotes LDL receptor degradation independently of proteolytic activity. *Biochem. J.* **2007**, *406*, 203–207. [CrossRef] [PubMed]
91. Abifadel, M.; Varret, M.; Rabes, J.P.; Allard, D.; Ouguerram, K.; Devillers, M.; Cruaud, C.; Benjannet, S.; Wickham, L.; Erlich, D.; et al. Mutations in PCSK9 cause autosomal dominant hypercholesterolemia. *Nat. Genet.* **2003**, *34*, 154–156. [CrossRef]
92. Cohen, J.; Pertsemlidis, A.; Kotowski, I.K.; Graham, R.; Garcia, C.K.; Hobbs, H.H. Low LDL cholesterol in individuals of African descent resulting from frequent nonsense mutations in PCSK9. *Nat. Genet.* **2005**, *37*, 161. [CrossRef]
93. Maxwell, K.N.; Breslow, J.L. Adenoviral-mediated expression of Pcsk9 in mice results in a low-density lipoprotein receptor knockout phenotype. *Proc. Natl. Acad. Sci. USA* **2004**, *101*, 7100–7105. [CrossRef]
94. Park, S.W.; Moon, Y.A.; Horton, J.D. Post-transcriptional regulation of low density lipoprotein receptor protein by proprotein convertase subtilisin/kexin type 9a in mouse liver. *J. Biol. Chem.* **2004**, *279*, 50630–50638. [CrossRef] [PubMed]
95. Syed, G.H.; Tang, H.; Khan, M.; Hassanein, T.; Liu, J.; Siddiqui, A. Hepatitis C virus stimulates low-density lipoprotein receptor expression to facilitate viral propagation. *J. Virol.* **2014**, *88*, 2519–2529. [CrossRef] [PubMed]
96. Li, Z.; Liu, Q. Hepatitis C virus regulates proprotein convertase subtilisin/kexin type 9 promoter activity. *Biochem. Biophys. Res. Commun.* **2018**, *496*, 1229–1235. [CrossRef]

97. Ramanathan, A.; Gusarova, V.; Stahl, N.; Gurnett-Bander, A.; Kyratsous, C.A. Alirocumab, a Therapeutic Human Antibody to PCSK9, Does Not Affect CD81 Levels or Hepatitis C Virus Entry and Replication into Hepatocytes. *PLoS ONE* **2016**, *11*, e0154498. [CrossRef] [PubMed]
98. Bridge, S.H.; Sheridan, D.A.; Felmlee, D.J.; Crossey, M.M.; Fenwick, F.I.; Lanyon, C.V.; Dubuc, G.; Seidah, N.G.; Davignon, J.; Thomas, H.C.; et al. PCSK9, apolipoprotein E and lipoviral particles in chronic hepatitis C genotype 3: Evidence for genotype-specific regulation of lipoprotein metabolism. *J. Hepatol.* **2014**, *62*, 763–770. [CrossRef]
99. Fasolato, S.; Pigozzo, S.; Pontisso, P.; Angeli, P.; Ruscica, M.; Savarino, E.; De Martin, S.; Lupo, M.G.; Ferri, N. PCSK9 Levels Are Raised in Chronic HCV Patients with Hepatocellular Carcinoma. *J. Clin. Med.* **2020**, *9*, 3134. [CrossRef] [PubMed]
100. Bhatt, S.; Gething, P.W.; Brady, O.J.; Messina, J.P.; Farlow, A.W.; Moyes, C.L.; Drake, J.M.; Brownstein, J.S.; Hoen, A.G.; Sankoh, O.; et al. The global distribution and burden of dengue. *Nature* **2013**, *496*, 504–507. [CrossRef] [PubMed]
101. Junjhon, J.; Lausumpao, M.; Supasa, S.; Noisakran, S.; Songjaeng, A.; Saraithong, P.; Chaichoun, K.; Utaipat, U.; Keelapang, P.; Kanjanahaluethai, A.; et al. Differential modulation of prM cleavage, extracellular particle distribution, and virus infectivity by conserved residues at nonfurin consensus positions of the dengue virus pr-M junction. *J. Virol.* **2008**, *82*, 10776–10791. [CrossRef]
102. Rana, J.; Slon Campos, J.L.; Poggianella, M.; Burrone, O.R. Dengue virus capsid anchor modulates the efficiency of polyprotein processing and assembly of viral particles. *J. Gen. Virol.* **2019**, *100*, 1663–1673. [CrossRef]
103. Gan, E.S.; Tan, H.C.; Le, D.H.T.; Huynh, T.T.; Wills, B.; Seidah, N.G.; Ooi, E.E.; Yacoub, S. Dengue virus induces PCSK9 expression to alter antiviral responses and disease outcomes. *J. Clin. Investig.* **2020**, *130*, 5223–5234. [CrossRef] [PubMed]
104. Sakai, J.; Rawson, R.B.; Espenshade, P.J.; Cheng, D.; Seegmiller, A.C.; Goldstein, J.L.; Brown, M.S. Molecular identification of the sterol-regulated luminal protease that cleaves SREBPs and controls lipid composition of animal cells. *Mol. Cell* **1998**, *2*, 505–514. [CrossRef]
105. Da Palma, J.R.; Burri, D.J.; Oppliger, J.; Salamina, M.; Cendron, L.; de Laureto, P.P.; Seidah, N.G.; Kunz, S.; Pasquato, A. Zymogen activation and subcellular activity of subtilisin kexin isozyme 1/site 1 protease. *J. Biol. Chem.* **2014**, *289*, 35743–35756. [CrossRef] [PubMed]
106. Tagliabracci, V.S.; Wiley, S.E.; Guo, X.; Kinch, L.N.; Durrant, E.; Wen, J.; Xiao, J.; Cui, J.; Nguyen, K.B.; Engel, J.L.; et al. A Single Kinase Generates the Majority of the Secreted Phosphoproteome. *Cell* **2015**, *161*, 1619–1632. [CrossRef] [PubMed]
107. Marschner, K.; Kollmann, K.; Schweizer, M.; Braulke, T.; Pohl, S. A key enzyme in the biogenesis of lysosomes is a protease that regulates cholesterol metabolism. *Science* **2011**, *333*, 87–90. [CrossRef] [PubMed]
108. Nakagawa, T.; Suzuki-Nakagawa, C.; Watanabe, A.; Asami, E.; Matsumoto, M.; Nakano, M.; Ebihara, A.; Uddin, M.N.; Suzuki, F. Site-1 protease is required for the generation of soluble (pro)renin receptor. *J. Biochem.* **2017**, *161*, 369–379. [CrossRef]
109. Kondo, Y.; Fu, J.; Wang, H.; Hoover, C.; McDaniel, J.M.; Steet, R.; Patra, D.; Song, J.; Pollard, L.; Cathey, S.; et al. Site-1 protease deficiency causes human skeletal dysplasia due to defective inter-organelle protein trafficking. *JCI Insight* **2018**, *3*, e121596. [CrossRef]
110. Meyer, R.; Elbracht, M.; Opladen, T.; Eggermann, T. Patient with an autosomal-recessive MBTPS1-linked phenotype and clinical features of Silver-Russell syndrome. *Am. J. Med. Genet. A* **2020**, *182*, 2727–2730. [CrossRef]
111. Lenz, O.; ter Meulen, J.; Klenk, H.D.; Seidah, N.G.; Garten, W. The Lassa virus glycoprotein precursor GP-C is proteolytically processed by subtilase SKI-1/S1P. *Proc. Natl. Acad. Sci. USA* **2001**, *98*, 12701–12705. [CrossRef]
112. Burri, D.J.; Pasqual, G.; Rochat, C.; Seidah, N.G.; Pasquato, A.; Kunz, S. Molecular characterization of the processing of arenavirus envelope glycoprotein precursors by subtilisin kexin isozyme-1/site-1 protease. *J. Virol.* **2012**, *86*, 4935–4946. [CrossRef] [PubMed]
113. Burri, D.J.; da Palma, J.R.; Seidah, N.G.; Zanotti, G.; Cendron, L.; Pasquato, A.; Kunz, S. Differential recognition of Old World and New World arenavirus envelope glycoproteins by subtilisin kexin isozyme 1 (SKI-1)/site 1 protease (S1P). *J. Virol.* **2013**, *87*, 6406–6414. [CrossRef] [PubMed]
114. McCormick, J.B.; Webb, P.A.; Krebs, J.W.; Johnson, K.M.; Smith, E.S. A prospective study of the epidemiology and ecology of Lassa fever. *J. Infect. Dis.* **1987**, *155*, 437–444. [CrossRef] [PubMed]
115. Da Palma, J.R.; Cendron, L.; Seidah, N.G.; Pasquato, A.; Kunz, S. Mechanism of Folding and Activation of Subtilisin Kexin Isozyme-1 (SKI-1)/Site-1 Protease (S1P). *J. Biol. Chem.* **2016**, *291*, 2055–2066. [CrossRef] [PubMed]
116. Flatz, L.; Rieger, T.; Merkler, D.; Bergthaler, A.; Regen, T.; Schedensack, M.; Bestmann, L.; Verschoor, A.; Kreutzfeldt, M.; Brück, W.; et al. T cell-dependence of Lassa fever pathogenesis. *PLoS Pathog.* **2010**, *6*, e1000836. [CrossRef]
117. Seidah, N.G.; Prat, A.; Pirillo, A.; Catapano, A.L.; Norata, G.D. Novel strategies to target proprotein convertase subtilisin kexin 9: Beyond monoclonal antibodies. *Cardiovasc. Res.* **2019**, *115*, 510–518. [CrossRef] [PubMed]
118. Hatch, S.; Mathew, A.; Rothman, A. Dengue vaccine: Opportunities and challenges. *IDrugs* **2008**, *11*, 42–45.
119. Beyer, W.R.; Popplau, D.; Garten, W.; Von Laer, D.; Lenz, O. Endoproteolytic Processing of the Lymphocytic Choriomeningitis Virus Glycoprotein by the Subtilase SKI-1/S1P. *J. Virol.* **2003**, *77*, 2866–2872. [CrossRef]
120. Bederka, L.H.; Bonhomme, C.J.; Ling, E.L.; Buchmeier, M.J. Arenavirus stable signal peptide is the keystone subunit for glycoprotein complex organization. *MBio* **2014**, *5*, e02063. [CrossRef] [PubMed]
121. Kunz, S.; Edelmann, K.H.; de la Torre, J.C.; Gorney, R.; Oldstone, M.B. Mechanisms for lymphocytic choriomeningitis virus glycoprotein cleavage, transport, and incorporation into virions. *Virology* **2003**, *314*, 168–178. [CrossRef]
122. Wright, K.E.; Spiro, R.C.; Burns, J.W.; Buchmeier, M.J. Post-translational processing of the glycoproteins of lymphocytic choriomeningitis virus. *Virology* **1990**, *177*, 175–183. [CrossRef]

123. Briese, T.; Paweska, J.T.; McMullan, L.K.; Hutchison, S.K.; Street, C.; Palacios, G.; Khristova, M.L.; Weyer, J.; Swanepoel, R.; Egholm, M.; et al. Genetic detection and characterization of Lujo virus, a new hemorrhagic fever-associated arenavirus from southern Africa. *PLoS Pathog.* **2009**, *5*, e1000455. [CrossRef]
124. Oppliger, J.; da Palma, J.R.; Burri, D.J.; Bergeron, E.; Khatib, A.M.; Spiropoulou, C.F.; Pasquato, A.; Kunz, S. A Molecular Sensor to Characterize Arenavirus Envelope Glycoprotein Cleavage by Subtilisin Kexin Isozyme 1/Site 1 Protease. *J. Virol.* **2016**, *90*, 705–714. [CrossRef]
125. Grant, A.; Seregin, A.; Huang, C.; Kolokoltsova, O.; Brasier, A.; Peters, C.; Paessler, S. Junín virus pathogenesis and virus replication. *Viruses* **2012**, *4*, 2317–2339. [CrossRef] [PubMed]
126. Candurra, N.A.; Damonte, E.B. Effect of inhibitors of the intracellular exocytic pathway on glycoprotein processing and maturation of Junin virus. *Arch. Virol.* **1997**, *142*, 2179–2193. [CrossRef] [PubMed]
127. Toure, B.B.; Munzer, J.S.; Basak, A.; Benjannet, S.; Rochemont, J.; Lazure, C.; Chretien, M.; Seidah, N.G. Biosynthesis and enzymatic characterization of human SKI-1/S1P and the processing of its inhibitory prosegment. *J. Biol. Chem.* **2000**, *275*, 2349–2358. [CrossRef]
128. Pasquato, A.; Burri, D.J.; Traba, E.G.; Hanna-El-Daher, L.; Seidah, N.G.; Kunz, S. Arenavirus envelope glycoproteins mimic autoprocessing sites of the cellular proprotein convertase subtilisin kexin isozyme-1/site-1 protease. *Virology* **2011**, *417*, 18–26. [CrossRef]
129. Pasqual, G.; Burri, D.J.; Pasquato, A.; de la Torre, J.C.; Kunz, S. Role of the host cell's unfolded protein response in arenavirus infection. *J. Virol.* **2011**, *85*, 1662–1670. [CrossRef]
130. Manning, J.T.; Yun, N.E.; Seregin, A.V.; Koma, T.; Sattler, R.A.; Ezeomah, C.; Huang, C.; de la Torre, J.C.; Paessler, S. The Glycoprotein of the Live-Attenuated Junin Virus Vaccine Strain Induces Endoplasmic Reticulum Stress and Forms Aggregates prior to Degradation in the Lysosome. *J. Virol.* **2020**, *94*, e01693-19. [CrossRef] [PubMed]
131. Popkin, D.L.; Teijaro, J.R.; Sullivan, B.M.; Urata, S.; Rutschmann, S.; de la Torre, J.C.; Kunz, S.; Beutler, B.; Oldstone, M. Hypomorphic mutation in the site-1 protease Mbtps1 endows resistance to persistent viral infection in a cell-specific manner. *Cell Host Microbe* **2011**, *9*, 212–222. [CrossRef] [PubMed]
132. Burri, D.J.; da Palma, J.R.; Kunz, S.; Pasquato, A. Envelope glycoprotein of arenaviruses. *Viruses* **2012**, *4*, 2162–2181. [CrossRef] [PubMed]
133. Maisa, A.; Stroher, U.; Klenk, H.D.; Garten, W.; Strecker, T. Inhibition of Lassa virus glycoprotein cleavage and multicycle replication by site 1 protease-adapted α 1-antitrypsin variants. *PLoS. Negl. Trop. Dis.* **2009**, *3*, e446. [CrossRef] [PubMed]
134. Basak, A.; Goswami, M.; Rajkumar, A.; Mitra, T.; Majumdar, S.; O'Reilly, P.; Bdour, H.M.; Trudeau, V.L.; Basak, A. Enediynyl peptides and iso-coumarinyl methyl sulfones as inhibitors of proprotein convertases PCSK8/SKI-1/S1P and PCSK4/PC4: Design, synthesis and biological evaluations. *Bioorg. Med. Chem. Lett.* **2015**, *25*, 2225–2237. [CrossRef]
135. Rojek, J.M.; Pasqual, G.; Sanchez, A.B.; Nguyen, N.T.; de la Torre, J.C.; Kunz, S. Targeting the proteolytic processing of the viral glycoprotein precursor is a promising novel antiviral strategy against arenaviruses. *J. Virol.* **2010**, *84*, 573–584. [CrossRef]
136. Hay, B.A.; Abrams, B.; Zumbrunn, A.Y.; Valentine, J.J.; Warren, L.C.; Petras, S.F.; Shelly, L.D.; Xia, A.; Varghese, A.H.; Hawkins, J.L.; et al. Aminopyrrolidineamide inhibitors of site-1 protease. *Bioorg. Med. Chem. Lett.* **2007**, *17*, 4411–4414. [CrossRef] [PubMed]
137. Hawkins, J.L.; Robbins, M.D.; Warren, L.C.; Xia, D.; Petras, S.F.; Valentine, J.J.; Varghese, A.H.; Wang, I.K.; Subashi, T.A.; Shelly, L.D.; et al. Pharmacologic inhibition of site 1 protease activity inhibits sterol regulatory element-binding protein processing and reduces lipogenic enzyme gene expression and lipid synthesis in cultured cells and experimental animals. *J. Pharmacol. Exp. Ther.* **2008**, *326*, 801–808. [CrossRef]
138. Urata, S.; Yun, N.; Pasquato, A.; Paessler, S.; Kunz, S.; de la Torre, J.C. Antiviral activity of a small-molecule inhibitor of arenavirus glycoprotein processing by the cellular site 1 protease. *J. Virol.* **2011**, *85*, 795–803. [CrossRef]
139. Pasquato, A.; Rochat, C.; Burri, D.J.; Pasqual, G.; la Torre, J.C.; Kunz, S. Evaluation of the anti-arenaviral activity of the subtilisin kexin isozyme-1/site-1 protease inhibitor PF-429242. *Virology* **2012**, *423*, 14–22. [CrossRef]
140. Vincent, M.J.; Sanchez, A.J.; Erickson, B.R.; Basak, A.; Chretien, M.; Seidah, N.G.; Nichol, S.T. Crimean-Congo hemorrhagic fever virus glycoprotein proteolytic processing by subtilase SKI-1. *J. Virol.* **2003**, *77*, 8640–8649. [CrossRef]
141. Bergeron, E.; Vincent, M.J.; Nichol, S.T. Crimean-Congo hemorrhagic fever virus glycoprotein processing by the endoprotease SKI-1/S1P is critical for virus infectivity. *J. Virol.* **2007**, *81*, 13271–13276. [CrossRef]
142. Bergeron, É.; Zivcec, M.; Chakrabarti, A.K.; Nichol, S.T.; Albariño, C.G.; Spiropoulou, C.F. Recovery of Recombinant Crimean Congo Hemorrhagic Fever Virus Reveals a Function for Non-structural Glycoproteins Cleavage by Furin. *PLoS Pathog.* **2015**, *11*, e1004879. [CrossRef]
143. Flint, M.; Chatterjee, P.; Lin, D.L.; McMullan, L.K.; Shrivastava-Ranjan, P.; Bergeron, É.; Lo, M.K.; Welch, S.R.; Nichol, S.T.; Tai, A.W.; et al. A genome-wide CRISPR screen identifies N-acetylglucosamine-1-phosphate transferase as a potential antiviral target for Ebola virus. *Nat. Commun.* **2019**, *10*, 285. [CrossRef]
144. Yuan, S.; Chu, H.; Chan, J.F.; Ye, Z.W.; Wen, L.; Yan, B.; Lai, P.M.; Tee, K.M.; Huang, J.; Chen, D.; et al. SREBP-dependent lipidomic reprogramming as a broad-spectrum antiviral target. *Nat. Commun.* **2019**, *10*, 120. [CrossRef]
145. Wang, R.; Simoneau, C.R.; Kulsuptrakul, J.; Bouhaddou, M.; Travisano, K.A.; Hayashi, J.M.; Carlson-Stevermer, J.; Zengel, J.R.; Richards, C.M.; Fozouni, P.; et al. Genetic Screens Identify Host Factors for SARS-CoV-2 and Common Cold Coronaviruses. *Cell* **2021**, *184*, 106–119.e14. [CrossRef]

146. Dias, S.S.G.; Soares, V.C.; Ferreira, A.C.; Sacramento, C.Q.; Fintelman-Rodrigues, N.; Temerozo, J.R.; Teixeira, L.; Nunes da Silva, M.A.; Barreto, E.; Mattos, M.; et al. Lipid droplets fuel SARS-CoV-2 replication and production of inflammatory mediators. *PLoS Pathog.* **2020**, *16*, e1009127. [CrossRef]
147. Lee, W.; Ahn, J.H.; Park, H.H.; Kim, H.N.; Kim, H.; Yoo, Y.; Shin, H.; Hong, K.S.; Jang, J.G.; Park, C.G.; et al. COVID-19-activated SREBP2 disturbs cholesterol biosynthesis and leads to cytokine storm. *Signal Transduct. Target* **2020**, *5*, 186. [CrossRef] [PubMed]
148. Sohrabi, Y.; Reinecke, H.; Godfrey, R. Altered Cholesterol and Lipid Synthesis Mediates Hyperinflammation in COVID-19. *Trends Endocrinol. Metab.* **2021**, *32*, 132–134. [CrossRef] [PubMed]
149. Hyrina, A.; Meng, F.; McArthur, S.J.; Eivemark, S.; Nabi, I.R.; Jean, F. Human Subtilisin Kexin Isozyme-1 (SKI-1)/Site-1 Protease (S1P) regulates cytoplasmic lipid droplet abundance: A potential target for indirect-acting anti-dengue virus agents. *PLoS ONE* **2017**, *12*, e0174483. [CrossRef] [PubMed]
150. Blanchet, M.; Sureau, C.; Guevin, C.; Seidah, N.G.; Labonte, P. SKI-1/S1P inhibitor PF-429242 impairs the onset of HCV infection. *Antivir. Res.* **2015**, *115*, 94–104. [CrossRef]
151. Olmstead, A.D.; Knecht, W.; Lazarov, I.; Dixit, S.B.; Jean, F. Human subtilase SKI-1/S1P is a master regulator of the HCV Lifecycle and a potential host cell target for developing indirect-acting antiviral agents. *PLoS Pathog.* **2012**, *8*, e1002468. [CrossRef]
152. Merino-Ramos, T.; Jiménez de Oya, N.; Saiz, J.C.; Martín-Acebes, M.A. Antiviral Activity of Nordihydroguaiaretic Acid and Its Derivative Tetra-O-Methyl Nordihydroguaiaretic Acid against West Nile Virus and Zika Virus. *Antimicrob. Agents Chemother.* **2017**, *61*, e00376-17. [CrossRef]
153. Uchida, L.; Urata, S.; Ulanday, G.E.; Takamatsu, Y.; Yasuda, J.; Morita, K.; Hayasaka, D. Suppressive Effects of the Site 1 Protease (S1P) Inhibitor, PF-429242, on Dengue Virus Propagation. *Viruses* **2016**, *8*, 46. [CrossRef]
154. Kleinfelter, L.M.; Jangra, R.K.; Jae, L.T.; Herbert, A.S.; Mittler, E.; Stiles, K.M.; Wirchnianski, A.S.; Kielian, M.; Brummelkamp, T.R.; Dye, J.M.; et al. Haploid Genetic Screen Reveals a Profound and Direct Dependence on Cholesterol for Hantavirus Membrane Fusion. *mBio* **2015**, *6*, e00801. [CrossRef] [PubMed]
155. Pasquato, A.; Kunz, S. Novel drug discovery approaches for treating arenavirus infections. *Expert Opin. Drug Discov.* **2016**, *11*, 383–393. [CrossRef] [PubMed]
156. Elagoz, A.; Benjannet, S.; Mammarbassi, A.; Wickham, L.; Seidah, N.G. Biosynthesis and cellular trafficking of the convertase SKI-1/S1P: Ectodomain shedding requires SKI-1 activity. *J. Biol. Chem.* **2002**, *277*, 11265–11275. [CrossRef] [PubMed]
157. Pasquato, A.; Cendron, L.; Kunz, S. Cleavage of the Glycoprotein of Arenaviruses. In *Activation of Viruses by Host Proteases*; Springer: Cham, Switzerland, 2018; pp. 47–70.
158. Bestle, D.; Heindl, M.R.; Limburg, H.; Pilgram, O.; Moulton, H.; Stein, D.A.; Hardes, K.; Eickmann, M.; Dolnik, O.; Rohde, C.; et al. TMPRSS2 and furin are both essential for proteolytic activation of SARS-CoV-2 in human airway cells. *Life Sci. Alliance* **2020**, *3*. [CrossRef] [PubMed]
159. Anderson, E.D.; Molloy, S.S.; Jean, F.; Fei, H.; Shimamura, S.; Thomas, G. The ordered and compartment-specfific autoproteolytic removal of the furin intramolecular chaperone is required for enzyme activation. *J. Biol. Chem.* **2002**, *277*, 12879–12890. [CrossRef]
160. Ginefra, P.; Filippi, B.G.H.; Donovan, P.; Bessonnard, S.; Constam, D.B. Compartment-Specific Biosensors Reveal a Complementary Subcellular Distribution of Bioactive Furin and PC7. *Cell Rep.* **2018**, *22*, 2176–2189. [CrossRef]
161. Ye, J.; Rawson, R.B.; Komuro, R.; Chen, X.; Dave, U.P.; Prywes, R.; Brown, M.S.; Goldstein, J.L. ER stress induces cleavage of membrane-bound ATF6 by the same proteases that process SREBPs. *Mol. Cell* **2000**, *6*, 1355–1364. [CrossRef]
162. Yuan, J.; Cai, T.; Zheng, X.; Ren, Y.; Qi, J.; Lu, X.; Chen, H.; Lin, H.; Chen, Z.; Liu, M.; et al. Potentiating CD8(+) T cell antitumor activity by inhibiting PCSK9 to promote LDLR-mediated TCR recycling and signaling. *Protein Cell* **2021**, *12*, 240–260. [CrossRef]
163. He, Z.; Khatib, A.M.; Creemers, J.W.M. Loss of the proprotein convertase Furin in T cells represses mammary tumorigenesis in oncogene-driven triple negative breast cancer. *Cancer Lett.* **2020**, *484*, 40–49. [CrossRef] [PubMed]
164. Rufer, A.C. Drug discovery for enzymes. *Drug Discov. Today* **2021**, *26*, 875–886. [CrossRef] [PubMed]
165. Krammer, F.; Smith, G.J.D.; Fouchier, R.A.M.; Peiris, M.; Kedzierska, K.; Doherty, P.C.; Palese, P.; Shaw, M.L.; Treanor, J.; Webster, R.G.; et al. Influenza. *Nat. Rev. Dis. Prim.* **2018**, *4*, 3. [CrossRef] [PubMed]
166. Kido, H.; Okumura, Y.; Takahashi, E.; Pan, H.Y.; Wang, S.; Chida, J.; Le, T.Q.; Yano, M. Host envelope glycoprotein processing proteases are indispensable for entry into human cells by seasonal and highly pathogenic avian influenza viruses. *J. Mol. Genet. Med.* **2008**, *3*, 167–175. [CrossRef] [PubMed]

Article

The Protein Kinase Receptor Modulates the Innate Immune Response against Tacaribe Virus

Hector Moreno * and Stefan Kunz †

Institute of Microbiology, Lausanne University Hospital, CH-1011 Lausanne, Switzerland
* Correspondence: hector.moreno@chuv.ch; Tel.: +41-21-314-4100
† Deceased on 10 January 2020.

Abstract: The New World (NW) mammarenavirus group includes several zoonotic highly pathogenic viruses, such as Junin (JUNV) or Machupo (MACV). Contrary to the Old World mammarenavirus group, these viruses are not able to completely suppress the innate immune response and trigger a robust interferon (IFN)-I response via retinoic acid-inducible gene I (RIG-I). Nevertheless, pathogenic NW mammarenaviruses trigger a weaker IFN response than their nonpathogenic relatives do. RIG-I activation leads to upregulation of a plethora of IFN-stimulated genes (ISGs), which exert a characteristic antiviral effect either as lone effectors, or resulting from the combination with other ISGs or cellular factors. The dsRNA sensor protein kinase receptor (PKR) is an ISG that plays a pivotal role in the control of the mammarenavirus infection. In addition to its well-known protein synthesis inhibition, PKR further modulates the overall IFN-I response against different viruses, including mammarenaviruses. For this study, we employed Tacaribe virus (TCRV), the closest relative of the human pathogenic JUNV. Our findings indicate that PKR does not only increase IFN-I expression against TCRV infection, but also affects the kinetic expression and the extent of induction of Mx1 and ISG15 at both levels, mRNA and protein expression. Moreover, TCRV fails to suppress the effect of activated PKR, resulting in the inhibition of a viral titer. Here, we provide original evidence of the specific immunomodulatory role of PKR over selected ISGs, altering the dynamic of the innate immune response course against TCRV. The mechanisms for innate immune evasion are key for the emergence and adaptation of human pathogenic arenaviruses, and highly pathogenic mammarenaviruses, such as JUNV or MACV, trigger a weaker IFN response than nonpathogenic mammarenaviruses. Within the innate immune response context, PKR plays an important role in sensing and restricting the infection of TCRV virus. Although the mechanism of PKR for protein synthesis inhibition is well described, its immunomodulatory role is less understood. Our present findings further characterize the innate immune response in the absence of PKR, unveiling the role of PKR in defining the ISG profile after viral infection. Moreover, TCRV fails to suppress activated PKR, resulting in viral progeny production inhibition.

Keywords: protein kinase receptor (PKR); mammarenavirus; interferon; innate immune response; Mx1; ISG15; CCL5

Citation: Moreno, H.; Kunz, S. The Protein Kinase Receptor Modulates the Innate Immune Response against Tacaribe Virus. *Viruses* **2021**, *13*, 1313. https://doi.org/10.3390/v13071313

Academic Editors: Michael B. A. Oldstone and Juan C. De la Torre

Received: 25 May 2021
Accepted: 29 June 2021
Published: 7 July 2021

1. Introduction

Mammarenaviruses are a large genus of viruses divided into Old and New World arenavirus groups (OW and NW, respectively), according to antigenic properties, phylogeny, and geographic distribution [1]. Both groups include zoonotic viruses that are highly pathogenic to humans, such as Lassa (LASV), JUNV, MACV, and Guaranito virus [2]. The prototypic OW arenavirus Lymphocytic Choriominingitis virus (LCMV) is a neglected pathogen with world-wide distribution and clinical significance in immunocompromised individuals and pregnant women [3]. The highly diverse NW arenavirus group is further divided into four clades: A, B, C, and D. While several members of clade B are confirmed human pathogens, some clade D viruses also show potential for viral emergence [4,5].

Mammarenaviruses are enveloped, bi-segmented, negative-stranded viruses with a life cycle restricted to the cytosol [6]. The small genomic segment encodes the glycoprotein precursor and the viral nucleoprotein (NP), while the large segment codes for the matrix protein (Z) and the RNA-dependent RNA-polymerase (L). NP and L, together with the cis-acting sequences of the viral genome, are necessary for the virus ribonucleoprotein (RNP) complex formation, which is needed for the replication and transcription processes [7,8].

Mammarenavirus infection is typically detected by RIG-I-like receptors and Toll-like receptors. The activation of these pattern recognition receptors (PRRs) triggers an IFN-I immune response, upregulating a plethora of ISGs, which encode effector proteins that have an antiviral effect and induce a cellular antiviral status. Mammarenaviruses are capable of inhibiting the IFN-I response to different extents, via a 3′–5′ exoribonuclease NP domain [9–12], and highly pathogenic arenaviruses trigger a weaker induction of the IFN-I response than their nonpathogenic counterparts [12–16], suggesting that the capacity to overcome or suppress the IFN-I response is relevant for causing disease in humans. Indeed, the attenuated JUNV Candid#1 triggers a stronger IFN-β expression than the highly pathogenic Romero strain [17]. Moreover, the Z protein of pathogenic mammarenaviruses was shown to inhibit RIG-I, essential for the IFN-I response [16], summing to the immunosuppressive activity of NP. Despite the inhibitory effect of NP and Z over IFN-I response, and contrary to OW arenaviruses, NW arenaviruses fail to completely abolish IFN-I response in human cells [10,12,14,15,18].

In addition to the aforementioned host's PRRs, recent studies from others and us revealed that the dsRNA-PKR plays an important role during NW arenavirus infection [14,18,19]. PKR is an ISG that contributes to the enhancement of the IFN-I response against measles virus [20,21], West-Nile virus [22], or upon IFN-I treatment [23]. Upon detection and recognition of foreign dsRNA, PKR undergoes autophosphorylation and subsequently phosphorylates the α subunit of eIF2α, leading to the inhibition of protein cap-dependent translation [24]. Interestingly, a recent study showed that, in contrast to LASV, highly pathogenic NW arenaviruses accumulate dsRNA during infections [25], possibly leading to the observed colocalization of the viral RNP with RIG-I, the melanoma differentiation-associated protein 5 (MDA5), and phosphorylated PKR [19,26]. Moreover, it has been shown that the highly pathogenic arenaviruses, JUNV Romero Strain and MACV, but not LASV, induce higher IFN-β levels in PKR null cells than in non-transduced parental cells [18]. Nevertheless, previous results in our lab showed that PKR partially controls the infection by the nonhuman pathogenic TCRV, but had no impact on the infection by the attenuated JUNV Candid#1 strain [14].

Altogether, the literature suggests that the role of PKR in the IFN-I response might differ among pathogenic, attenuated, or nonpathogenic mammarenavirus infection. In the present study, we investigated and characterized the innate immune response triggered by nonpathogenic arenavirus TCRV in PKR KO cells, compared to parental cells subjected to an analogous mock clustered, regularly interspaced, short palindromic repeat–associated 9 (CRISPR/Cas9) engineering. Our results indicated that PKR changes the expression kinetic of Mx1 and ISG15, but not CCL5, whose expression is inhibited, maintaining a comparable pattern in presence and absence of PKR. Moreover, our findings indicated that activated PKR controls TCRV infection, inhibiting the viral progeny production.

2. Materials and Methods

2.1. Cells, Viruses and Infections

Scrambled A549 control (A549/Scr) and PKR KO A549 (A549/PKR KO) cells were obtained as described in [14]. Briefly, all cells were maintained in Dulbecco's modified Eagle medium with high glucose (4.5 mg/liter) and GlutaMAX (DMEM; Gibco BRL) with 10% (vol/vol) fetal calf serum (FCS) and held in an CO_2 incubator (37 °C and 5% (vol/vol) CO_2). TCRV (strain 11573) was plaque purified and propagated in VeroE6 and baby hamster kidney (BHK) cells, followed by PEG-precipitation and sucrose cushion purification, as described in [14].

For infections, cells were seeded 48 h in advance and counted before infection. The multiplicity of infection (MOI) was determined as described in each particular experiment. Before each experiment, cells were tested for mycoplasma contamination using a MycoAlert mycoplasma detection kit (Lonza, Basel, Switzerland). Inoculums were prepared by diluting the desired amount of virus in DMEM/10% FCS and incubated with cells for 90 min in a CO_2 incubator. Upon adsorption, the inoculums were removed and fresh DMEM/10%FCS was added to each well. For infections in rIFN-αA/D-stimulated cells, 24 h after seeding, cells were treated with 100 U/mL for 24 h. Cells were then infected as described above.

2.2. Antibodies and Reagents

Anti TCRV NP MA03-BE06 antibody [27] was obtained from BEI Resources (Manassas, VA, USA). Rabbit antibody against ISG15 was obtained from Cell Signaling Technology (Danvers, MA, USA). Rabbit antibody against Mx1 was purchased from Proteintech (Rosemont, IL, USA). Goat antibody against CCL5 was purchased from R&D Systems (Minneapolis, MN, USA). Antibody against Vinculin (EPR8185) was obtained from Abcam (Cambridge, UK). Recombinant human IFN (interferon-αA/D human; #I4401) was purchased from Sigma-Aldrich (St. Louis, MO, USA). Alexa Fluor-488 F(ab′)2 fragment of goat anti-mouse IgG was obtained from Life Technologies (Carlsbad, CA, USA). Polyclonal rabbit anti-mouse and donkey anti-goat antibody conjugated with horseradish peroxidase (HRP) were obtained from Dako (Santa Clara, CA, USA).

2.3. Immunofocus Assay (IFA)

For viral titer quantitation by IFA, supernatants were cleared from cellular debris by centrifugation at 1200 rpm for 3 min, and stored at −80 °C until analysis. Samples were 10-fold serially diluted in DMEM/10% FCS and used to infect previously prepared VeroE6 cells in 96-well plate format. After 16–20 h of infection, cells were washed with PBS, and fixed with 2% (wt/vol) formaldehyde for 30 min at room temperature. Then, cells were permeabilized with PBS/0.1% saponin/1% FCS (working solution) for 30 min at room temperature. MA03-BE06 antibody was used as the primary antibody to detect TCRV-NP, diluted 1:500 in working solution and applied to the cells for 1 h at room temperature in a rocking station. Alexa 488-congugated anti-mouse IgG1 was used as the secondary antibody, diluted 1:500 in working solution, and applied to cells for 45 min at room temperature in a rocking station. Before scoring the samples, cells were washed three times with PBS. Positive infectious foci were scored using an EVOS FLoid cell imaging station with a 20× Plan Fluorite Lens (Thermo Fisher Scientific).

2.4. RNA Extraction, RT, qPCR and RT2 Profiler

Samples collected for RNA extraction were kept in RNAlater (Sigma-Aldrich) at −20 °C until analysis. RNA for IFN-β quantitation was extracted with a NucleoSpin RNA kit (Macherey-Nagel, Düren, Germany) and eluted in 60 microliters of nuclease-free water, in accordance with the manufacturer's instructions. RNA for RT2 profiler assay was extracted with TRIzol reagent (Invitrogen), following the manufacturer's instructions and resuspending the RNA pellet in 20 microliters of nuclease-free water. Total RNA quantitation was performed with a Qubit 4.0 instrument (Thermo Fisher Scientific). A total of 0.5 micrograms of total RNA was used for reverse transcription reactions using a high-capacity cDNA reverse transcription kit (Applied Biosystems, Foster City, CA, USA), following the manufacturer's instructions. TaqMan probes specific targeting human IFN-β (Hs01077958_s1/FAM) and glyceraldehyde-3-phosphate dehydrogenase (GAPDH; Hs99999905_m1/VIC) were obtained from Applied Biosystems. Quantitative PCR (qPCR) was performed using a StepOne qPCR system (Applied Biosystems, Foster City, CA, USA), and relative gene expression to GAPDH was determined following the $2^{-\Delta\Delta}$Ct calculation (where Ct is threshold cycle). For transcriptome profiling, human antiviral response RT2 Profiler PCR array kits were used (PAHS-122ZG; Qiagen, Hilden, Germany). A total of

3.5 micrograms (quantified by Qubit 4.0) of total cellular RNA was used for the reverse transcription reaction (RT2 SYBR green qPCR mastermix; Qiagen, Hilden, Germany). RT2 Profiler PCR array 384-well plates were set up by a PIRO personal pipetting robot (Labgene, Châtel-Saint-Denis, Switzerland). All RNA samples were tested for quality and integrity in a fragment analyzer (Agilent Technologies, Santa Clara, CA, USA), selecting only those with an RNA quality score (RIN) of 10. All tested samples were tested negative for genomic DNA contamination. Samples with aberrant amplification curves or shifted or multiple melting peaks were discarded from the analysis. qPCR reaction was performed using a LightCycler 480 instrument II (Roche, Basel, Switzerland) in accordance with the manufacturer's instructions, and relative gene expression to GAPDH was determined following the $2^{-\Delta\Delta Ct}$ calculation (where Ct is threshold cycle).

2.5. Immunoblotting

Samples were collected and lysed in CelLytic by incubating on ice for 30 min. Then, samples were centrifuged for 15 min at 4 °C at 14.000 rpm, the supernatants were moved to a new clean tube, and stored at −80 °C until analysis. Samples were then loaded with Laemli buffer in Novex Value 4–20% Tris-Glycine precasted gels, and transferred to a nitrocellulose membrane. Membranes were stained with indicated antibodies diluted in 3% powder milk in PBS 0.1% Tween 20. Signals were acquired by an ImageQuant LAS 4000 Mini (GE Healthcare Lifesciences), and quantitation and analyses of Western blot results were performed with ImageJ software.

2.6. IFN Quantitation Bioassay

To measure the antiviral activity of IFN produced by A549/Scr and A549/PKR KO cells infected with TCRV, we performed an IFN-I bioassay, as described previously [12,14]. Briefly, VeroE6 were treated cells with UV-inactivated tissue culture supernatants from TCRV-infected A549/Scr and A549/PKR KO cells. As a control, we used titrated amounts of IFN-IIa (0, 10, 100, 1000, and 10,000 IU/mL) and subjected them to the same UV inactivation protocol as the tested samples. UV treatment was performed at 4 °C, in a rocking station at 10 cm from the UV irradiation source for 2 min. After 16 h of incubation, cells were infected with VSV, which is known to be highly susceptible to IFN [28]. After 8 h of infection, cells were fixed and assayed in IFA with specific antibodies against the VSV M protein, as reported previously [14]. The level of infection of VSV is therefore inversely proportional to the IFN produced.

3. Results

To investigate the role of PKR in the IFN-I response triggered by a NW arenavirus infection, we measured the levels of IFN-β mRNA in control and PKR knockout human lung epithelial A549 cells (A549/PKR KO) infected with TCRV and JUNV-Candid#1. A549 cells have been extensively used by others and us to recapitulate mammarenavirus infection and represent a reliable model of study (12, 14, 17, 19). To obtain the A549/PKR KO cells, we used CRISPR/Cas9 guide RNAs [29,30], and A549 cells subjected to analogous CRISPR/Cas9 editing with a scrambled guide RNA sequence (A549/Scr) as control cells. First, we infected A549/Scr and A549/PKR KO cells with TCRV at low MOI (0.01 PFU/cell) and collected total cellular RNA after 5 days of infection. As previously shown in [14], at this time after infection, the number of TCRV-infected cells in A549/Scr and A549/PKR KO cells are comparable (93.6 +/− 0.5% and 95 +/− 0.8%, respectively), but the viral titers were significantly higher in the absence of PKR. We observed lower levels of IFN-β transcripts in infected A549/PKR KO cells compared to infected A549/Scr control cells (Figure 1A). To faithfully address the biological consequences the IFN produced upon TCRV infection in A549/Scr and A549/PKR KO cells, we evaluated the antiviral effect of UV-inactivated supernatants from infected cells. To this aim, and as previously described [12,14], we used vesicular stomatitis virus (VSV) as a surrogate to quantify the amount of IFN produced (Figure 1B). These results were in agreement with previous studies on the role of PKR in the innate immune response against different viruses [20–22,31], as well as with the previously reported reduced IFN-I response against TCRV [10,12,14].

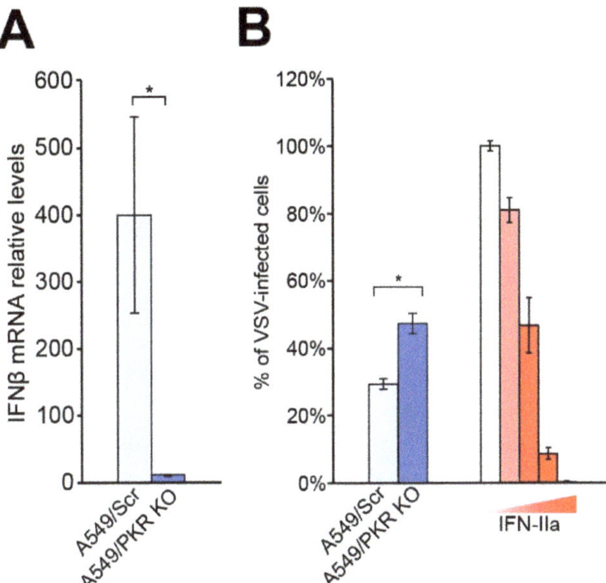

Figure 1. (**A**) IFN-β mRNA expression in A549/Scr and A549/PKR KO cells infected with TCRV. Cells were infected at MOI 0.01 PFU/Cell with TCRV, total RNA was collected 5 days after infection and subjected to RT-qPCR as described in Materials and Methods. Fold-induction was calculated by the $2^{-\Delta\Delta Ct}$ method. Error bars represent standard deviations (n = 6). (**B**) Detection of IFN activity by bioassay. Conditioned supernatants from TCRV-infected A549/Scr and A549/PKR KO cells were UV inactivated for 2 min and used to pretreat VeroE6 cells for 16 h. As a positive control, we used titrated amounts of IFN-IIa (10^4, 10^3, 10^2, 10, and 0 IU/mL) diluted in supernatant from uninfected cells. Pretreated cells were infected with VSV (300 PFU/well). Values were normalized to samples in absence of IFN. Error bars represent standard deviations (n = 3) of results from one representative experiment out of two independent experiments. * stands for p value < 0.001 in an ANOVA test.

Next, we further characterized the IFN-I response in PKR null cells by studying the expression profile of ISGs. We compared the expression of 66 ISGs in A549/Scr and in A549/PKR KO cells infected with TCRV. To this aim, we performed infections at low MOI (0.01 PFU/cell) and collected cellular total RNA after 5 days of infection. Coherently with previous results (Figure 1), we observed that depletion of PKR in A549 cells reduces IFN-β expression. The differences in the amplitude of IFN-β expression between single IFN-β RNA quantitation (Figure 1) and in the screening array (Table 1) are likely due to the use of different qPCR methods, as well as the employ of multiple housekeeping genes in the array, which renders a more accurate relative quantitation. Indeed, our results in the bioassay experiment (Figure 1B) and the accurate mRNA quantitation using multiple housekeeping genes are comparable. Interestingly, despite lower IFN-β mRNA levels, Mx1 and ISG15 increased their expression in A549/PKR KO cells (Table 1).

Table 1. ISG profile of A549/Scr and A549/PKR KO cells infected with TCRV. Cells were infected at MOI of 0.01 PFU/cell and collected 5 days after infection. Values correspond to fold-induction (calculated by the $2^{-\Delta\Delta Ct}$ method) and asterisks means statistical significance when compared to uninfected cells (* $p < 0.01$, ** $p < 0.001$) in two-way ANOVA test. For clarity, color intensity is proportional to up- or down-regulation (red and blue, respectively).

ISG Name (Gene ID)	A549/Scr	A549/PKR KO	ISG Name	A549/Scr	A549/PKR KO
CCL5 (6352)	12,673.81 **	5135.67 **	*RELA* (5970)	2.10	0.84
IFNB1 (3456)	5028.57 **	2020.63 **	*ATG5* (9474)	1.87	1.61
MX1 (4599)	1046.76 **	1753.85 **	*IRF3* (3661)	1.86	1.72
ISG15 (9636)	622.35 **	751.67 **	*APOBECC3G* (60489)	1.82	1.45
OAS2 (4939)	470.5 **	304.32 **	*TBK1* (29110)	1.76	2.85
IL6(3569)	142.55 **	32.25 **	*RIPK1* (8737)	1.74	0.81
IFIH1 (64135)	56.58 **	62.8 **	*CTSL* (1514)	1.72	2.27
IRF7 (3665)	35.41 **	14.54 *	*IL15* (3600)	1.71	1.26
DDX58 (23586)	30.11 **	38.58 **	*IL18* (3606)	1.70	1.31
CXCL8 (3576)	12.4 **	7.49 **	*CD80* (941)	1.56	2.82
TLR3 (7098)	10.65 **	12.05 **	*MAP2K1* (5604)	1.54	1.99
IL12A (3592)	9.34 **	5.27 **	*CASP10* (843)	1.52	0.89
STAT1 (6772)	9.25 **	8.39 **	*MAP3K7* (6885)	1.49	1.48
CYLD (1540)	5.67 **	4.23 **	*DDX3X* (1654)	1.41	1.06
TICAM1 (148022)	3.7 **	1.23	*IRF5* (3663)	1.41	1.10
NFKB1A (4792)	3.49 **	2.69	*IFNAR1* (15975)	1.37	1.05
IRAK1 (3654)	3.23	0.77	*MAPK1* (5594)	1.35	1.29
TRIM25 (7706)	3.22 **	1.64	*IKBKB* (3551)	1.30	0.64
NFKB1 (4790)	3.17 *	1.40	*MAPK8* (5599)	1.30	1.00
MYD88(17874)	3.11 **	2.86 **	*CASP8* (841)	1.26	1.08
TRADD (8717)	2.94	2.53	*SPP1* (6696)	1.21	2.07
CD40 (958)	2.8	1.45	*PIN1* (5300)	1.16	1.53
MAP2K3 (26397)	2.78 **	1.96 **	*SUGT1* (10910)	1.13	1.69
AZI2(64343)	2.73 *	2.31	*HSP90AA1* (3320)	1.05	1.27
TRAF6 (7189)	2.7 **	1.78 *	*CARD9* (64170)	1.04	0.78
CHUK(1147)	2.63 **	2.37	*MAPK14* (1432)	0.97	0.67
FADD (8772)	2.63 *	1.72	*MAPK3* (5595)	0.93	0.45
CTSS (1520)	2.41 **	2.61 **	*CTSB* (1508)	0.91	1.30
MAVS (57506)	2.4 **	2.26	*IL1B* (3553)	0.85	0.83
TRAF3 (7187)	2.36	1.34	*PYCARD* (29108)	0.73	1.13
CXCL9 (4283)	2.13	3.21	*TKFC* (26007)	0.72	0.34
MAP3K1 (4214)	2.12 **	0.91	*FOS* (14281)	0.61	0.28
JUN (3725)	2.11	0.59	*IFNA1* (3439)	0.61	0.37

Given the relevance of these effector ISGs in the IFN-I response, we monitored the protein levels of Mx1, ISG15, and CCL5 after TCRV infection. Our results show that the lack of PKR causes a delayed production of Mx1 and ISG15. However, concomitant with the higher gene expression observed (Table 1), Mx1 and ISG15 reach higher protein levels in A549/PKR KO cells than in A549/Scr control cells at late times after infection (Figure 2A). Moreover, the absence of PKR resulted in the overall inhibition of CCL5 expression, without causing any delay in the expression kinetic after TCRV infection (Figure 2B). In line with previous findings [14], we further confirmed that the absence of PKR led to a limited increase of TCRV NP expression during early times postinfection (Figure 2C). The effects of lacking PKR on the levels and dynamics expression of ISGs confirmed the pivotal role of PKR in the innate immune response triggered by TCRV infection.

TCRV NP expression is partially limited by PKR during early times after infection (Figure 2C and [14]), and the results above strongly suggest that PKR activation efficiently controls TCRV infection course via ISG activation. To better elucidate the consequences of activation of PKR during TCRV infection, we stimulated A549/Scr and A549/PKR KO cells with 100 IU/mL of rIFN-αA/D and then infected at MOI of 0.01 PFU/cell. Our results showed that TCRV reached significantly higher viral titers in the absence of activated PKR than in its presence during early times after infection (<3 days postinfection), albeit the production of viral progeny became comparable at later time points (Figure 3). These results indicated that, once activated, PKR partially restricted TCRV growth.

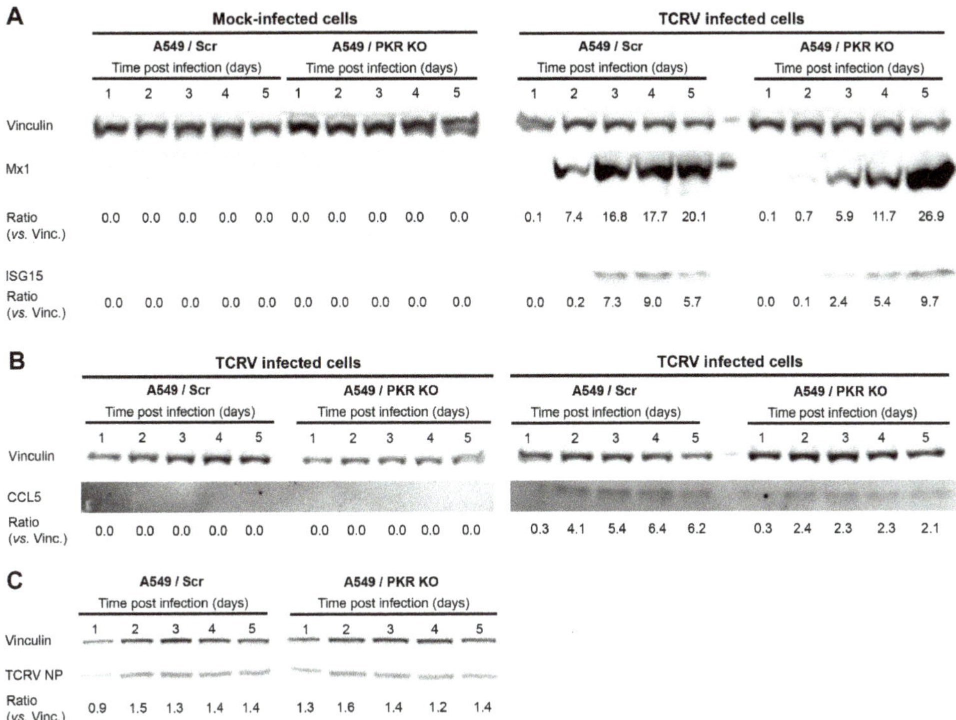

Figure 2. Protein expression of Mx1, ISG15, and CCL5 in TCRV-infected A549/Scr and A549/PKR cells. Cells were infected with TCRV at MOI of 0.01 PFU/cell and cell lysates were collected every 24 h. Total cellular proteins were probed for Mx1, ISG15 (**A**), and CCL5 (**B**), expression by Western blotting. Vinculin was included as a loading control. The ratios of Mx1, ISG15, and CCL5 versus Vinculin were calculated by densitometry at the corresponding day postinfection. One representative example out of three independent experiments is shown; (**C**) TCRV NP expression in A549/Scr and A549/PKR KO cells.

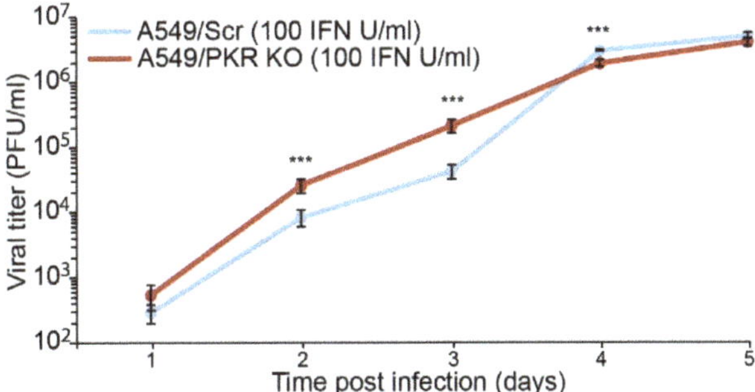

Figure 3. TCRV viral progeny production in IFN-stimulated A549/Scr and A549/PKR KO cells. Cells were stimulated with 100 IU/mL of rIFN-αA/D 24 h before infection. Cells were infected at MOI 0.01 PFU/cell and supernatants were assayed for viral titer by IFA. Error bars represent standard deviations ($n = 4$) and asterisks (***) means statistical significance ($p < 0.01$) in two-way ANOVA test.

4. Discussion

In the present study, we show that, despite causing a reduced expression of IFN-β (Figure 1), depletion of the dsRNA sensor PKR leads to changes in the levels and dynamics of production of ISGs, including Mx1 and ISG15 (Table 1, Figure 2). Previous reports already described PKR as an enhancer of the IFN-I response upon viral infection [20–22,31]. Interestingly, in the case of JUNV Candid#1, despite the differences in IFN-β mRNA levels in infected A549/Scr and A549/PKR KO cells, viral progeny production was not affected [14], suggesting that either the contribution of PKR against JUNV infection is not biologically relevant or that JUNV can deploy molecular mechanisms to overcome the host's innate immune response. Interestingly, a previous study showed that infections with the highly pathogenic JUNV Romero strain and MACV at high MOI (3 PFU/cell) in nontransduced A549 cells and A549/PKR KO cells resulted in an increase of the IFN-β expression [18], suggesting that attenuated and pathogenic JUNV strains may interact with PKR in different manners or efficiencies.

We previously described that TCRV infection is increased in A549/PKR KO cells at late time points, concomitantly with PKR activation in A549 cells [14]. Moreover, we found that similar to JUNV Candid#1, TCRV NP expression is subject to a very limited control of PKR during early stages of the infection (Figure 2) and [19], suggesting that PKR may also be activated during initial virus propagation. In this scenario, PKR could inhibit the early local viral propagation and viral protein synthesis, remaining undetectable by analysis of whole cell lysate, as only a small fraction of PKR is activated. In the absence of PKR, TCRV reaches higher viral titers only at late times postinfection [14], but this effect was observed at earlier times after infection when cells were prestimulated with IFN (Figure 3). These results suggest that a PKR antiviral effect may be effective only for a limited time after its activation. Furthermore, this observation suggests that, despite being able to affect initial viral protein production, the impact on viral progeny production can only occur with extensive PKR activation, either via induced IFN stimulation or via intrinsic viral detection.

Despite rendering lower IFN-β levels, TCRV infection in A549/PKR KO cells results in a delay of Mx1 and ISG15 expression, but also in increased mRNA and protein expression of both host factors at late time points (Table 1 and Figure 2). Importantly, reduced ISG15 and Mx1 expression occurs concomitantly with increased TCRV NP expression (Figure 2 and [14]), suggesting that the observed antiviral effect of PKR may be implemented with the participation of these ISGs. In contrast, CCL5 expression is reduced due to the absence of PKR at both gene and protein levels, and renders a comparable kinetic of expression. Therefore, PKR ablation seems to alter the ISG profile kinetic of Mx1 and ISG15, but not other ISGs, such as CCL5. Our results then suggest that the observed differences in viral progeny production in A549/PKR KO cells infected with TCRV but not with JUNV [14] might be the result of not only reduced IFN-β levels but also of changes in the expression pattern of specific ISGs and altered innate immune response.

CCL5 is a chemokine expressed in many cell types in response to viral infections and IFN-β, and plays a pivotal role in migration of effector and memory T cells [32,33]. CCL5 is a relevant player in the response against arenaviruses, and its absence in mice leads to the establishment of chronic infections of LCMV clone 13 [34]. Moreover, infection with the nonhuman pathogenic NW mammarenavirus Pichinde virus (PICV) p2 strain, which causes mild disease in guinea pigs, was followed by increased CCL5 expression at late time points, in contrast to the virulent PICV p18 strain [35]. The CCL5 reduction due to the lack of PKR may then contribute to worse disease outcome in mammarenavirus infections. In such scenarios, the weaker PKR activation observed in highly pathogenic NW arenaviruses would contribute to a lower CCL5 expression and increased virulence.

ISG15 is strongly induced by the IFN-I cascade and exerts its antiviral activity by being incorporated to nascent peptides in a process, similar to ubiquitination, termed ISGylation, which affects their stability [36]. The expression of ISG15 increases in response to many viral infections, including influenza A, Ebola, hepatitis B and C, human immunodeficiency virus 1, human papillomavirus, West Nile, and Zika [37]. Nevertheless, many other viruses, such as Middle East and severe acute respiratory syndromes, foot and mouth disease virus, or influenza B, deploy mechanisms to prevent the antiviral effect of ISG15 [38–42]. Interestingly, in addition to its role as an effector ISG, ISG15 also prevents the over amplification of the IFN-I cascade [41]. It is then plausible that the lower expression of ISG15 in the absence of PKR at early time points upon TCRV infection disrupts the regulatory feedback loop, failing to repress the IFN-I response at later times, which may result in the increased expression of Mx1 and ISG15. Although hemorrhagic syndrome caused by OW arenavirus is not associated to a cytokine storm [43,44], the symptomatology of highly pathogenic NW arenaviruses seems to correlate with activation of infected macrophages that leads to massive release of proinflammatory cytokines [45]. Furthermore, LASV, but not highly pathogenic NW arenaviruses, prevents the accumulation of the dsRNA danger signal which leads to PKR activation [25]. Therefore, it is conceivable that PKR activation and the subsequent dysregulation of ISG15 may alter the cytokine release pattern, with consequences in the symptomatology caused by mammarenaviruses. A detailed investigation is encouraged to determine the biological and clinical consequences of ISG15 altered expression upon mammarenavirus infection.

Mx1 is an effector ISG with GTPase activity and an antiviral effect against several viral infections, including influenza, bunyaviruses, and hantaviruses [46–49]. When activated, Mx1 oligomerizes and sequesters viral factors of LaCrosse virus, influenza A, or Thogoto virus, disturbing the viral life cycle [46,49,50]. Unlike its mouse orthologues, human Mx1 locates in the cytoplasm and is effective against a broad range of viruses, regardless their replication site [47,48]. To the current date, there is no reported inhibition of mammarenaviruses by Mx1. The different consequences of PKR on Mx1, ISG15, and CCL5 expression kinetics suggest that, although all these genes are overall upregulated upon viral infections, they are also differently tuned by additional host factors such as PKR.

A complete vision of the innate immune response and its modulation is crucial to understand the mechanisms underlying the pathogenicity of mammarenaviruses, and for the development of new antiviral strategies. The results presented here contribute to a better understanding of the dynamics in the onset of the innate immune response triggered by NW arenavirus infections and, in particular, of the key role of PKR on it. The change in the kinetic and temporal expression patterns of Mx1 and ISG15, but not of CCL5, demonstrates that PKR does not only modulate the amplitude of the IFN-I response, but also tunes the expression of selected ISGs over time throughout mammarenavirus infection. These findings highlight the importance of the ability to control PKR activation during an NW arenavirus infection, and the potential consequences for a better understanding of the hemorrhagic syndromes caused by highly pathogenic mammarenaviruses.

Author Contributions: Conceptualization, S.K. and H.M.; methodology, H.M.; validation, H.M.; formal analysis, H.M.; investigation, H.M.; resources, S.K. and H.M.; data curation, H.M.; writing—original draft preparation, H.M.; writing—review and editing, H.M.; visualization, H.M.; supervision, H.M.; project administration, H.M.; funding acquisition, S.K. Author Stefan Kunz was unable to confirm their authorship contributions. On their behalf, the corresponding author has reported their contributions to the best of their knowledge.

Funding: This research was supported by Swiss National Science Foundation grant SINERGIA Nr. CRSII3_160780/1 to S.K. and funds to S.K. from the University of Lausanne.

Institutional Review Board Statement: Not applicable.

Informed Consent Statement: Not applicable.

Data Availability Statement: All relevant data were within the manuscript.

Acknowledgments: Tragically, Stefan Kunz passed away on 10 January 2020. His memory will remain among all of us who shared part of our lives and enjoyed his company. Because of his enthusiasm, dedication, and motivation, he will be remembered as a wonderful person and scientist. We would like to thank Roberto Balbontín for his valuable and critical scientific feedback. We also would like to thank Gisa Gerold and Rebecca Møller for the A549/Scr and A549/PKR cells. The monoclonal anti-Junin virus NR-41860 (MA03-BE06) antibody was obtained from BEI Resources, NIAID, NIH. This research was supported by Swiss National Science Foundation grant SINERGIA Nr. CRSII3_160780/1 to S.K. and funds to S.K. from the University of Lausanne.

Conflicts of Interest: The authors declare no conflict of interest.

References

1. Radoshitzky, S.R.; Bao, Y.; Buchmeier, M.J.; Charrel, R.N.; Clawson, A.N.; Clegg, C.S.; DeRisi, J.L.; Emonet, S.; Gonzalez, J.-P.; Kuhn, J.H.; et al. Past, present, and future of arenavirus taxonomy. *Arch. Virol.* **2015**, *160*, 1851–1874. [CrossRef] [PubMed]
2. Geisbert, T.W.; Jahrling, P.B. Exotic emerging viral diseases: Progress and challenges. *Nat. Med.* **2004**, *10*, S110–S121. [CrossRef] [PubMed]
3. Buchmeier, M.J.; de la Torre, J.C.; Peters, C. Arenaviridae: The viruses and their replication. In *Fields Virology*, 4th ed.; Lippincott-Raven: Philadelphia, PA, USA, 2007; pp. 1791–1828.
4. Centers for Disease Control and Prevention. Fatal illnesses associated with a new world arenavirus—California, 1999–2000. *MMWR Morb. Mortal. Wkly. Rep.* **2000**, *49*, 709–711.
5. Moreno, H.; Rastrojo, A.; Pryce, R.; Fedeli, C.; Zimmer, G.; Bowden, T.A.; Gerold, G.; Kunz, S. A novel circulating tamiami mammarenavirus shows potential for zoonotic spillover. *PLoS Negl. Trop. Dis.* **2020**, *14*, e0009004. [CrossRef]
6. de la Torre, J.C. Molecular and cell biology of the prototypic arenavirus LCMV: Implications for understanding and combating hemorrhagic fever arenaviruses. *Ann. N. Y. Acad. Sci.* **2009**, *1171*, E57–E64. [CrossRef]
7. Iwasaki, M.; Ngo, N.; Cubitt, B.; de la Torre, J.C. Efficient Interaction between Arenavirus Nucleoprotein (NP) and RNA-Dependent RNA Polymerase (L) Is Mediated by the Virus Nucleocapsid (NP-RNA) Template. *J. Virol.* **2015**, *89*, 5734–5738. [CrossRef] [PubMed]
8. Knopp, K.A.; Ngo, T.; Gershon, P.D.; Buchmeier, M.J. Single nucleoprotein residue modulates arenavirus replication complex formation. *mBio* **2015**, *6*, e00524-15. [CrossRef]
9. Hastie, K.M.; Kimberlin, C.R.; Zandonatti, M.A.; MacRae, I.J.; Saphire, E.O. Structure of the Lassa virus nucleoprotein reveals a dsRNA-specific 3′ to 5′ exonuclease activity essential for immune suppression. *Proc. Natl. Acad. Sci. USA* **2011**, *108*, 2396–2401. [CrossRef]
10. Huang, C.; Kolokoltsova, O.A.; Yun, N.E.; Seregin, A.V.; Ronca, S.; Koma, T.; Paessler, S. Highly Pathogenic New World and Old World Human Arenaviruses Induce Distinct Interferon Responses in Human Cells. *J. Virol.* **2015**, *89*, 7079–7088. [CrossRef]
11. Jiang, X.; Huang, Q.; Wang, W.; Dong, H.; Ly, H.; Liang, Y.; Dong, C. Structures of arenaviral nucleoproteins with triphosphate dsRNA reveal a unique mechanism of immune suppression. *J. Biol. Chem.* **2013**, *288*, 16949–16959. [CrossRef]
12. Martinez-Sobrido, L.; Giannakas, P.; Cubitt, B.; Garcia-Sastre, A.; de la Torre, J.C. Differential inhibition of type I interferon induction by arenavirus nucleoproteins. *J. Virol.* **2007**, *81*, 12696–12703. [CrossRef]
13. Huang, Q.; Shao, J.; Lan, S.; Zhou, Y.; Xing, J.; Dong, C.; Liang, Y.; Ly, H. In vitro and in vivo characterizations of pichinde viral nucleoprotein exoribonuclease functions. *J. Virol.* **2015**, *89*, 6595–6607. [CrossRef]
14. Moreno, H.; Moller, R.; Fedeli, C.; Gerold, G.; Kunz, S. Comparison of the Innate Immune Responses to Pathogenic and Nonpathogenic Clade B New World Arenaviruses. *J. Virol.* **2019**, *93*. [CrossRef]
15. Rodrigo, W.W.; Ortiz-Riano, E.; Pythoud, C.; Kunz, S.; de la Torre, J.C.; Martinez-Sobrido, L. Arenavirus nucleoproteins prevent activation of nuclear factor kappa B. *J. Virol.* **2012**, *86*, 8185–8197. [CrossRef] [PubMed]
16. Xing, J.; Ly, H.; Liang, Y. The Z proteins of pathogenic but not nonpathogenic arenaviruses inhibit RIG-I-like receptor-dependent interferon production. *J. Virol.* **2015**, *89*, 2944–2955. [CrossRef] [PubMed]
17. Huang, C.; Kolokoltsova, O.A.; Yun, N.E.; Seregin, A.V.; Poussard, A.L.; Walker, A.G.; Brasier, A.R.; Zhao, Y.; Tian, B.; De La Torre, J.C.; et al. Junin virus infection activates the type I interferon pathway in a RIG-I-dependent manner. *PLoS Negl. Trop. Dis.* **2012**, *6*, e1659. [CrossRef] [PubMed]
18. Huang, C.; Kolokoltsova, O.A.; Mateer, E.J.; Koma, T.; Paessler, S. Highly Pathogenic New World Arenavirus Infection Activates the Pattern Recognition Receptor Protein Kinase R without Attenuating Virus Replication in Human Cells. *J. Virol.* **2017**, *91*, e01090-17. [CrossRef] [PubMed]
19. King, B.R.; Hershkowitz, D.; Eisenhauer, P.L.; Weir, M.E.; Ziegler, C.M.; Russo, J.; Bruce, E.A.; Ballif, B.A.; Botten, J. A Map of the Arenavirus Nucleoprotein-Host Protein Interactome Reveals that Junin Virus Selectively Impairs the Antiviral Activity of Double-Stranded RNA-Activated Protein Kinase (PKR). *J. Virol.* **2017**, *91*, e00763-17. [CrossRef]
20. McAllister, C.S.; Toth, A.M.; Zhang, P.; Devaux, P.; Cattaneo, R.; Samuel, C.E. Mechanisms of protein kinase PKR-mediated amplification of beta interferon induction by C protein-deficient measles virus. *J. Virol.* **2010**, *84*, 380–386. [CrossRef]
21. McAllister, C.S.; Taghavi, N.; Samuel, C.E. Protein Kinase PKR Amplification of Interferon beta Induction Occurs through Initiation Factor eIF-2 alpha-mediated Translational Control. *J. Biol. Chem.* **2012**, *287*, 36384–36392. [CrossRef]
22. Gilfoy, F.D.; Mason, P.W. West nile virus-induced interferon production is mediated by the double-stranded RNA-Dependent protein kinase PKR. *J. Virol.* **2007**, *81*, 11148–11158. [CrossRef]
23. Yang, Y.L.; Reis, L.F.; Pavlovic, J.; Aguzzi, A.; Schafer, R.; Kumar, A.; Williams, B.; Aguet, M.; Weissmann, C. Deficient signaling in mice devoid of double-stranded RNA-dependent protein kinase. *EMBO J.* **1995**, *14*, 6095–6106. [CrossRef]
24. Dey, M.; Cao, C.; Dar, A.C.; Tamura, T.; Ozato, K.; Sicheri, F.; Dever, T.E. Mechanistic link between PKR dimerization, autophosphorylation, and eIF2 alpha substrate recognition. *Cell* **2005**, *122*, 901–913. [CrossRef] [PubMed]

25. Mateer, E.J.; Maruyama, J.; Card, G.E.; Paessler, S.; Huang, C. Lassa Virus, but Not Highly Pathogenic New World Arenaviruses, Restricts Immunostimulatory Double-Stranded RNA Accumulation during Infection. *J. Virol.* **2020**, *94*. [CrossRef] [PubMed]
26. Mateer, E.J.; Paessler, S.; Huang, C. Visualization of Double-Stranded RNA Colocalizing with Pattern Recognition Receptors in Arenavirus Infected Cells. *Front. Cell Infect. Microbiol.* **2018**, *8*, 251. [CrossRef]
27. Sanchez, A.; Pifat, D.Y.; Kenyon, R.H.; Peters, C.J.; McCormick, J.B.; Kiley, M.P. Junin virus monoclonal antibodies: Characterization and cross-reactivity with other arenaviruses. *J. Gen. Virol.* **1989**, *70 Pt 5*, 1125–1132. [CrossRef]
28. Stojdl, D.F.; Lichty, B.D.; tenOever, B.R.; Paterson, J.M.; Power, A.T.; Knowles, S.; Marius, R.; Reynard, J.; Poliquin, L.; Atkins, H.; et al. VSV strains with defects in their ability to shutdown innate immunity are potent systemic anti-cancer agents. *Cancer Cell* **2003**, *4*, 263–275. [CrossRef]
29. Kim, S.; Koo, T.; Jee, H.G.; Cho, H.Y.; Lee, G.; Lim, D.G.; Shin, H.S.; Kim, J.-S. CRISPR RNAs trigger innate immune responses in human cells. *Genome Res.* **2018**, *28*, 367–373. [CrossRef] [PubMed]
30. Wienert, B.; Shin, J.; Zelin, E.; Pestal, K.; Corn, J.E. In vitro-transcribed guide RNAs trigger an innate immune response via the RIG-I pathway. *PLoS Biol.* **2018**, *16*, e2005840. [CrossRef] [PubMed]
31. Schulz, O.; Pichlmair, A.; Rehwinkel, J.; Rogers, N.C.; Scheuner, D.; Kato, H.; Takeuchi, O.; Akira, S.; Kaufman, R.J.; e Sousa, C.R. Protein kinase R contributes to immunity against specific viruses by regulating interferon mRNA integrity. *Cell Host Microb.* **2010**, *7*, 354–361. [CrossRef] [PubMed]
32. Schall, T.J.; Bacon, K.; Toy, K.J.; Goeddel, D.V. Selective attraction of monocytes and T lymphocytes of the memory phenotype by cytokine RANTES. *Nature* **1990**, *347*, 669–671. [CrossRef]
33. Kim, M.O.; Suh, H.S.; Brosnan, C.F.; Lee, S.C. Regulation of RANTES/CCL5 expression in human astrocytes by interleukin-1 and interferon-beta. *J. Neurochem.* **2004**, *90*, 297–308. [CrossRef]
34. Matloubian, M.; Concepcion, R.J.; Ahmed, R. CD4+ T cells are required to sustain CD8+ cytotoxic T-cell responses during chronic viral infection. *J. Virol.* **1994**, *68*, 8056–8063. [CrossRef]
35. Scott, E.P.; Aronson, J.F. Cytokine patterns in a comparative model of arenavirus haemorrhagic fever in guinea pigs. *J. Gen. Virol.* **2008**, *89*, 2569–2579. [CrossRef]
36. Zhang, D.; Zhang, D.E. Interferon-stimulated gene 15 and the protein ISGylation system. *J. Interferon Cytokine Res.* **2011**, *31*, 119–130. [CrossRef]
37. Morales, D.J.; Lenschow, D.J. The antiviral activities of ISG15. *J. Mol. Biol.* **2013**, *425*, 4995–5008. [CrossRef] [PubMed]
38. Lindner, H.A.; Lytvyn, V.; Qi, H.; Lachance, P.; Ziomek, E.; Menard, R. Selectivity in ISG15 and ubiquitin recognition by the SARS coronavirus papain-like protease. *Arch. Biochem. Biophys.* **2007**, *466*, 8–14. [CrossRef] [PubMed]
39. Mielech, A.M.; Kilianski, A.; Baez-Santos, Y.M.; Mesecar, A.D.; Baker, S.C. MERS-CoV papain-like protease has deISGylating and deubiquitinating activities. *Virology* **2014**, *450*, 64–70. [CrossRef] [PubMed]
40. Swatek, K.N.; Aumayr, M.; Pruneda, J.N.; Visser, L.J.; Berryman, S.; Kueck, A.F.; Geurink, P.P.; Ovaa, H.; van Kuppeveld, F.J.M.; Tuthill, T.J.; et al. Irreversible inactivation of ISG15 by a viral leader protease enables alternative infection detection strategies. *Proc. Natl. Acad. Sci. USA* **2018**, *115*, 2371–2376. [CrossRef]
41. Zhang, X.; Bogunovic, D.; Payelle-Brogard, B.; Francois-Newton, V.; Speer, S.D.; Yuan, C.; Volpi, S.; Li, Z.; Sanal, O.; Mansouri, D.; et al. Human intracellular ISG15 prevents interferon-alpha/beta over-amplification and auto-inflammation. *Nature* **2015**, *517*, 89–93. [CrossRef]
42. Zhao, C.; Sridharan, H.; Chen, R.; Baker, D.P.; Wang, S.; Krug, R.M. Influenza B virus non-structural protein 1 counteracts ISG15 antiviral activity by sequestering ISGylated viral proteins. *Nat. Commun.* **2016**, *7*, 12754. [CrossRef] [PubMed]
43. Baize, S.; Kaplon, J.; Faure, C.; Pannetier, D.; Georges-Courbot, M.C.; Deubel, V. Lassa virus infection of human dendritic cells and macrophages is productive but fails to activate cells. *J. Immunol.* **2004**, *172*, 2861–2869. [CrossRef] [PubMed]
44. Mahanty, S.; Bausch, D.G.; Thomas, R.L.; Goba, A.; Bah, A.; Peters, C.J.; Rollin, P. Low levels of interleukin-8 and interferon-inducible protein-10 in serum are associated with fatal infections in acute Lassa fever. *J. Infect. Dis.* **2001**, *183*, 1713–1721. [CrossRef]
45. Groseth, A.; Hoenen, T.; Weber, M.; Wolff, S.; Herwig, A.; Kaufmann, A.; Becker, S. Tacaribe Virus but Not Junin Virus Infection Induces Cytokine Release from Primary Human Monocytes and Macrophages. *PLoS Negl. Trop. D* **2011**, *5*, e1137. [CrossRef] [PubMed]
46. Dittmann, J.; Stertz, S.; Grimm, D.; Steel, J.; Garcia-Sastre, A.; Haller, O.; Kochs, G. Influenza A virus strains differ in sensitivity to the antiviral action of Mx-GTPase. *J. Virol.* **2008**, *82*, 3624–3631. [CrossRef] [PubMed]
47. Frese, M.; Kochs, G.; Feldmann, H.; Hertkorn, C.; Haller, O. Inhibition of bunyaviruses, phleboviruses, and hantaviruses by human MxA protein. *J. Virol.* **1996**, *70*, 915–923. [CrossRef]
48. Gordien, E.; Rosmorduc, O.; Peltekian, C.; Garreau, F.; Brechot, C.; Kremsdorf, D. Inhibition of hepatitis B virus replication by the interferon-inducible MxA protein. *J. Virol.* **2001**, *75*, 2684–2691. [CrossRef]
49. Kochs, G.; Haller, O. Interferon-induced human MxA GTPase blocks nuclear import of Thogoto virus nucleocapsids. *Proc. Natl. Acad. Sci. USA* **1999**, *96*, 2082–2086. [CrossRef]
50. Reichelt, M.; Stertz, S.; Krijnse-Locker, J.; Haller, O.; Kochs, G. Missorting of LaCrosse virus nucleocapsid protein by the interferon-induced MxA GTPase involves smooth ER membranes. *Traffic* **2004**, *5*, 772–784. [CrossRef]

Review

The Interplay between Bluetongue Virus Infections and Adaptive Immunity

Daniel Rodríguez-Martín, Andrés Louloudes-Lázaro , Miguel Avia, Verónica Martín , José M. Rojas and Noemí Sevilla *

Centro de Investigación en Sanidad Animal, Centro Nacional Instituto de Investigación y Tecnología Agraria y Alimentaria, Consejo Superior de Investigaciones Científicas (CISA-INIA-CSIC), Valdeolmos, 28130 Madrid, Spain; rodriguez.daniel@inia.es (D.R.-M.); andres.louloudes@inia.es (A.L.-L.); miguelavia87@gmail.com (M.A.); mgarcia.veronica@inia.es (V.M.); rojas.jose@inia.es (J.M.R.)
* Correspondence: sevilla@inia.es; Tel.: +34-916-20-2300; Fax: +34-916-202-247

Abstract: Viral infections have long provided a platform to understand the workings of immunity. For instance, great strides towards defining basic immunology concepts, such as MHC restriction of antigen presentation or T-cell memory development and maintenance, have been achieved thanks to the study of lymphocytic choriomeningitis virus (LCMV) infections. These studies have also shaped our understanding of antiviral immunity, and in particular T-cell responses. In the present review, we discuss how bluetongue virus (BTV), an economically important arbovirus from the *Reoviridae* family that affects ruminants, affects adaptive immunity in the natural hosts. During the initial stages of infection, BTV triggers leucopenia in the hosts. The host then mounts an adaptive immune response that controls the disease. In this work, we discuss how BTV triggers CD8$^+$ T-cell expansion and neutralizing antibody responses, yet in some individuals viremia remains detectable after these adaptive immune mechanisms are active. We present some unpublished data showing that BTV infection also affects other T cell populations such as CD4$^+$ T-cells or $\gamma\delta$ T-cells, as well as B-cell numbers in the periphery. This review also discusses how BTV evades these adaptive immune mechanisms so that it can be transmitted back to the arthropod host. Understanding the interaction of BTV with immunity could ultimately define the correlates of protection with immune mechanisms that would improve our knowledge of ruminant immunology.

Keywords: orbivirus; cytotoxic T-lymphocytes; T-helper cells; gamma-delta T-cells; ruminants; B-cells; T-cells

Citation: Rodríguez-Martín, D.; Louloudes-Lázaro, A.; Avia, M.; Martín, V.; Rojas, J.M.; Sevilla, N. The Interplay between Bluetongue Virus Infections and Adaptive Immunity. *Viruses* **2021**, *13*, 1511. https://doi.org/10.3390/v13081511

Academic Editors: Michael B. A. Oldstone and Juan C. De la Torre

Received: 5 June 2021
Accepted: 28 July 2021
Published: 31 July 2021

1. Introduction

Viral infections have long provided a platform to understand the workings of immunity. Early vaccination studies, dating back to the first immunizations in the 18th century by Edward Jenner that used cowpox lesion material to combat smallpox virus [1] and moving to the more systematic attenuation of the rabies virus performed by Louis Pasteur a century later [2], demonstrated that the immune system could be "trained" with less virulent pathogens to confer protection against these devastating diseases.

The study of pathogen infections has thus helped shape our understanding of multiple immunology notions. A prime example of this is the study of lymphocytic choriomeningitis virus (LCMV) infections that has revealed several immunology concepts (reviewed in [3–7]). For instance, studies of LCMV infections permitted the characterization of antigen recognition by T-cells [8–10] or understand memory development in T-cells [11,12]. Because the different strains of LCMV produce a wide spectrum of immune responses, ranging from acute viral clearance to chronic infections, studying this viral infection has also shed some light on viral induced immune dysfunctions. For example, LCMV "clone 13" persistent infection has been associated with the disruption of dendritic cell activity [13,14] and the induction of CD8$^+$ T-cell exhaustion [15]. Thus, studying the immunosuppressive effects of viral infection on immunity allows us to understand regulatory immune mechanisms as well as routes of viral escape that can be targeted for therapy.

These concepts are highly relevant for economically important ruminant viral diseases as they can help develop better vaccination protocols. Ruminant viral pathogens can also provide us with the opportunity to better understand ruminant immunology. In the present review, we will discuss how the arbovirus bluetongue virus produces disease in domestic ruminants and how the host immune system responds to the infection. We review and present data on bluetongue virus effects on circulating lymphoid cells population and discuss how the virus evades adaptive immunity.

2. Bluetongue Virus: The Prototypical Orbivirus

Bluetongue virus (BTV) belongs to the *Reoviridae* family and is the prototypical orbivirus. BTV is an arbovirus, i.e., a virus transmitted to the mammalian host by arthropods. Infections by arboviruses represent an increasing challenge in global health as climate change has altered the distribution of competent vectors throughout the world. As a consequence, diseases that once were restricted to subtropical regions can now be transmitted endemically by these invading vectors. BTV is transmitted by the bite of *Culicoides* spp midges and affects wild and domestic ruminants [16]. The redistribution of competent vectors from Northern Africa to the whole Mediterranean basin, as well as the competence of some endemic *Culicoides* spp in transmitting the virus means that BTV can now be considered endemic in Europe Southern latitudes [17]. BTV produces a disease of mandatory notification to the World Organization for Animal Health (OIE) that mainly affects sheep. BTV infections lead to loss of productivity through increased mortality and abortion rates, reduced fertility, decreases in milk and wool yield as well as restriction to animal trade which account for an estimated 3 billion $ annual losses globally [18].

BTV is a dsRNA virus that possesses 10 segments encoding for 7 structural proteins (VP) and at least 4 non-structural proteins (NS) (Figure 1) [19]. The genetic material is encapsidated in a two-layered protein capsid. The outer capsid constituted by the VP2 and VP5 proteins contains the main antigenic determinants of neutralizing antibodies. As a result of VP2 high variability, multiple serotypes of BTV have been defined (at least 27 to date) [20,21]. Little cross-protection in terms of sterilizing immunity exists between serotypes, thus limiting the efficacy of current vaccination strategies based on serotype-specific inactivated vaccines [22]. Because of its segmented genetic material, BTV can form reassortants [23], which further complicates its control in regions where multiple serotypes circulate. The segmented genetic material, the RNA polymerase VP1, the RNA capping enzyme and methyl transferase VP4, and the RNA helicase VP6 are enclosed within the inner capsid constituted by VP7 and VP3 [19]. VP7 is one of the most conserved antigenic determinants between serotypes and as a result seroconversion assays are often based around the detection of antibodies against this protein [24]. BTV non-structural proteins are involved in virus replication and in impairing the host response to infection. NS1 promotes viral protein expression [25]; NS2 is an RNA binding protein which is the main component of viral inclusion bodies [26]; while NS3 is involved in virion egress [27–29]. VP3, NS3, NS4, and the putative NS5 are involved in impairing the host cellular response to BTV by acting as interferon antagonist and/or by promoting cellular shut-off [30–35].

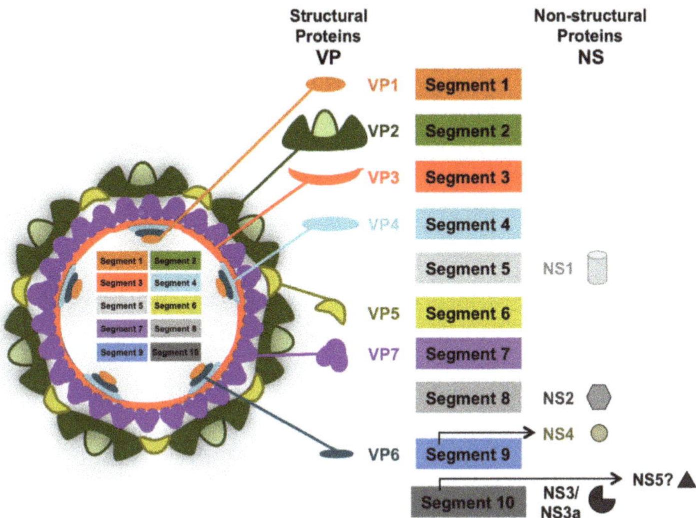

Figure 1. Schematic of bluetongue virus particle. BTV is a 10-segment dsRNA virus that belongs to the *Reoviridae* family. Its genetic material encodes for 7 structural proteins (VP) and 4–5 nonstructural proteins (NS). Segment 1 encodes for the RNA polymerase VP1. Segment 2 encodes for the highly variable outer capsid protein VP2. Segment 3 encodes for the inner core protein VP3. Segment 4 encodes for RNA capping enzyme and methyl transferase VP4. Segment 5 encodes for the nonstructural protein NS1 which forms cytoplasmic tubules during infection. Segment 6 encodes for the outer capsid protein VP5. Segment 7 encodes for the inner core protein VP7. Segment 8 encodes for the nonstructural protein NS2 which is part of the viral inclusion bodies. Segment 9 encodes for the RNA helicase VP6 and the non-structural protein NS4 which interferes with host immune response. Segment 10 encodes for the non-structural protein NS3 and its isoform NS3a which is involved in virion egress but also acts as an IFN antagonist. Segment 10 also putatively encodes for a fifth non-structural protein (NS5) that could be involved in host cellular shut-off.

3. The Control of BTV by Host Immunity

BTV infections in ruminants are often characterized by a biphasic viremia attributed to the IFN response [36,37]. The first viremia peak is likely controlled by the primary IFN response, but a second viremia peak is often detected once the IFN response subsides [36,37] and prior to the adaptive immune response taking place to clear the infection. Viremia in BTV infections can be long in some individuals and could be referred to as a state of semi-persistence [38,39]. This implies that the virus can effectively evade the effector arms of adaptive immunity, so that it can be transferred back to the arthropod host. Indeed, BTV has been shown to "hide" in erythrocytes, a mechanism that probably prolongs viremia [38].

Adoptive transfer experiments performed by Jeggo et al. helped establish that BTV protection is likely mediated by a combination of humoral and cellular immunity [40,41]. Transfer of neutralizing antibodies is effective at preventing re-infection from the same serotype, but confer limited protection towards different ones [42]. Jeggo et al. also showed that T cell adoptive transfer from BTV-recovered sheep protected from BTV challenge in monozygotic twins [41]. Cellular immunity in absence of neutralizing antibodies typically only provides partial protection [41,43], but it is cross-reactive between serotypes [44–49]. Indeed, vaccination based on BTV antigenic determinants that are conserved between serotypes (such as VP7 or NS1) have shown promising results in eliciting at least partial cross-serotype protection in animal models [44,46].

It is also important to mention that BTV infection severity varies greatly between susceptible species and even within the same host species [50]. Typically goat infection is asymptomatic, while cattle can develop moderate clinical signs in some cases. Sheep are more prone to develop clinically evident disease, but some animals can remain asymptomatic. This complicates BTV control as subclinical infected animals can act as reservoir for the disease. Control is even more troublesome because of the existence of wild ruminant hosts. Indeed, BTV is considered endemic in wild life in parts of North America and Africa [51,52], and studies in Europe have shown that wild ruminants populations are also infected during domestic outbreaks [53]. As such, deer and other wild ruminants such as mouflons or ibex could act as reservoir for the disease [54]. Differences in pathogenesis are also observed between BTV serotypes. For instance, the circulating BTV-8 strain responsible for the 2006 outbreak in Northern Europe produced frequent clinical infection in cattle [55].

4. Experimental BTV Infections of Natural Hosts and Prolonged Viremia

BTV infections are characterized by pyrexia, lack of appetite, and depression. In more severe cases, the disease manifests itself with conjunctivitis, congestion of the nasal and oral mucosa, and edema of the lips and face. These signs can develop to more prominent clinical features such as salivation mucopurulent nasal and oral discharge, ulcerations and hemorrhages in the nose lips and tongue, which sometimes progress to the cyanosis of the tongue that gave its name to the disease [24]. Scoring systems have been proposed to mark these signs [56–59], which can be useful to assess the degree of affectation in experimentally infected animals. Post-mortem histological evaluation can also be valuable to assess vascular lesions characteristic of the disease [60]. Most studies use temperature and clinical scoring of signs to macroscopically evaluate disease onset and progression [56–59]. In sheep experimentally infected with BTV-8, we were able to detect pyrexia and clinical signs such as nasal and oral congestion, depression, and loss of appetite which were scored based on the aforementioned systems [56–59] (Figure 2A,B).

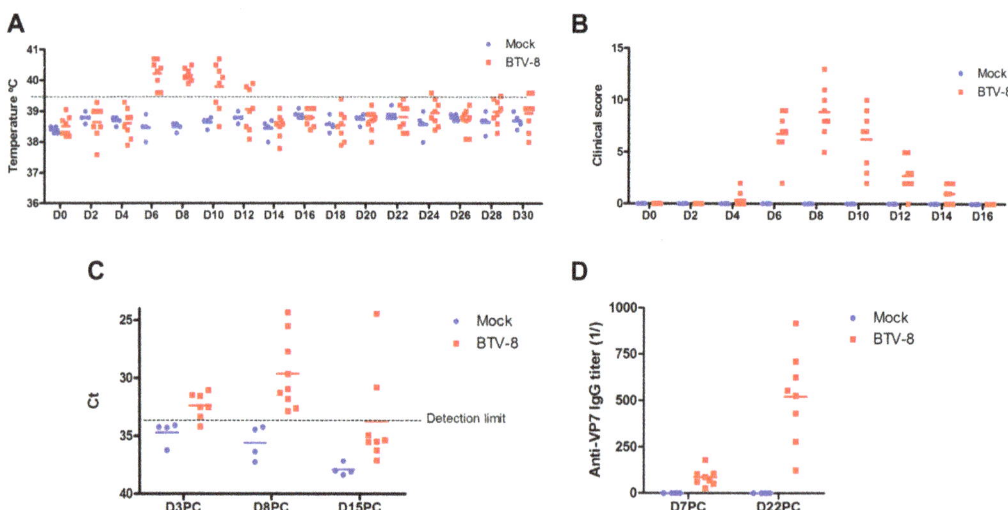

Figure 2. Assessment of BTV experimental infection in sheep: an example of BTV-8 infection in sheep. Sheep were experimentally infected intradermally with 2×10^6 pfu BTV-8 (n = 8) or mock infected as control (n = 4). (**A**,**B**) BTV-8 infected sheep developed (**A**) fever and (**B**) characteristic clinical signs (pyrexia, apathy, loss of appetite, depression, oro-mucal lesions . . .) which were scored. (**C**) qPCR showed the presence of viral RNA in blood in 7 out of 8 infected sheep from day 3 post-challenge (D3PC). BTV RNA was detected in blood in all sheep at day 8 post-challenge (D8PC), while viremia was maintained up to day 15 post-challenge (D15PC) in 2 animals. (**D**) Seroconversion of BTV-8 infected sheep assessed by anti-VP7 ELISA. Anti-VP7 IgG are detectable at day 7 post-challenge (D7PC). Antibody titers increased by day 22 post-challenge (D22PC) in all infected sheep but titers are widely spread ranging from 1:125 up to 1:918. Detailed methodology can be found in Supplemental File S1 and in the following works [43,45,61].

In addition to the macroscopic evaluation of the signs, it is now accepted that quantitative PCR (qPCR) of reverse transcribed RNA samples obtained from blood is a good approximation of viral replication [62]. Typically, BTV infections are characterized by an early peak in replication which is controlled by the IFN system, followed by a secondary peak once the IFN response subside and before adaptive immunity clear the infection [36,37]. We detected viral RNA in blood samples as early as day 3 post-challenge, which then peaked at day 8 post-challenge (Figure 2C). A feature of BTV infections is that viremia is prolonged in some individuals, which is thought to favor the transmission back to the vector [38,39]. The mechanisms behind this prolonged viremia remains unclear.

Based on nucleotide and amino-acid sequence identities, BTV serotypes have recently been classified as "classical" (serotypes 1–24) and "atypical" for recent isolates which are difficult to serotype by traditional serological methods [63–65]. While "classical" BTV outbreaks are of compulsory notification to the OIE, infections with atypical BTV are not [66]. Most of these atypical BTV serotypes have been identified from asymptomatic goats and produce only mild clinical infections in sheep [63]. Infections with these atypical serotypes also appear restricted mostly to small ruminants. Recently, atypical BTV-25 appears to produce "persistent" infections, as a goat flock monitored over 4 years showed on and off PCR positivity for this serotype [67]. Interestingly, BTV-25 is only partially neutralized by antibodies in these animals [67]. It is unknown whether this state of BTV "persistence" is due to low levels of replication in the host or to cyclic reinfections by competent *Culicoides* spp. because of impaired virus clearance, or to a combination of both mechanisms. This also raises an important point for infections with "classical" BTV serotypes. "Classical" BTV serotypes are unlikely to produce persistence [68,69], but prolonged viremia is observable in some animals. Although viral RNA can be detected up to day 222 post-BTV inoculation in some cases, it is not thought that "classical" BTV can infect the feeding vector after day 21 post-inoculation both in cattle and sheep [70]. It thus appears that both classical and atypical BTV possesses, yet to be clarified, escape mechanisms that allow for prolonged viremia to occur. Understanding these escape routes will probably improve the control of the disease.

Seroconversion as assessed by the presence of antibodies reactive to the VP7 protein can occur early in infection [60], although antibody titers start to become evident from day 7 post-challenge and increase at later stage of the infection (Figure 2D). Some serotypes such as BTV-15 do have however different antibody induction kinetics whereby antibody titers to VP7 become apparent only at later stages of infection [71,72]. In BTV-8 reinfection experiments in sheep, VP7-expressing peripheral blood mononuclear cells are periodically detected by flow cytometry, although qPCR studies did not indicate that productive replication cycles were taking place as Ct values are stable once past the primary peak of replication [73]. BTV RNA levels were detectable in this study in spite of the presence of a concomitant cellular response to the BTV-VP7 protein [73]. Another study also showed in cattle that BTV-8 RNA could be detected in blood 60 days after reinfection in spite of the presence of neutralizing antibodies [72]. These reports illustrate that studies are still required to elucidate the mechanisms that allow BTV to prolong its circulation in spite of the presence of neutralizing antibodies and/or the existence of cellular immunity. BTV affinity for erythrocyte membrane could contribute to the prolonged detection of RNA in blood [38], it does however not explain the differences in viremia duration that are detected between animals. To the best of our knowledge, no studies have attempted to pinpoints which factors contribute to this prolonged high viremia that can be detected in some individuals. It is likely that, as for the IFN response that BTV overcomes in ruminants, the virus has developed mechanisms that have yet to be elucidated to prolong its replication and circulation in blood so that it can be transmitted back to susceptible *Culicoides* spp. First, it is nonetheless important to describe the interactions between BTV and immunity in order to later discuss BTV immune evasion mechanisms. The present discussion will be mostly centered on BTV effects on lymphoid cells with a particular focus on cellular immunity.

5. Effects of BTV Infection on Peripheral Lymphoid Cells Populations

5.1. BTV Infections Produce Leucopenia

BTV infection is known to induce leucopenia [73–76] typically within the first 2–7 days after experimental infections. This effect can precede the onset of fever in animals [77]. Alterations in leucocyte populations not only happen with blood circulating cells [74], but also in the lymph nodes after BTV infection [78]. BTV is known to be transported from the biting site by conventional dendritic cells (DC) to the draining lymph node where further replication likely occurs (Figure 3A) [79]. From there, the virus is systemically disseminated and infects further organs such as lungs or spleen. BTV can infect endothelial cells [80], cells of the monocytic phagocytic lineage, and lymphocytes [81,82]. Endothelial cell infections contribute to the characteristic hemorrhagic lesions produced by the infection [83], whereas infection of lymphocytes could account for some of the leucopenic events observed during the early stages of infection. Indeed, BTV infection induces apoptosis in peripheral blood mononuclear cells (PBMC) during the viremia peak which could contribute to leucocyte depletion [76]. This produces a transient immunosuppressive state that can render the animal susceptible to opportunistic infections [76].

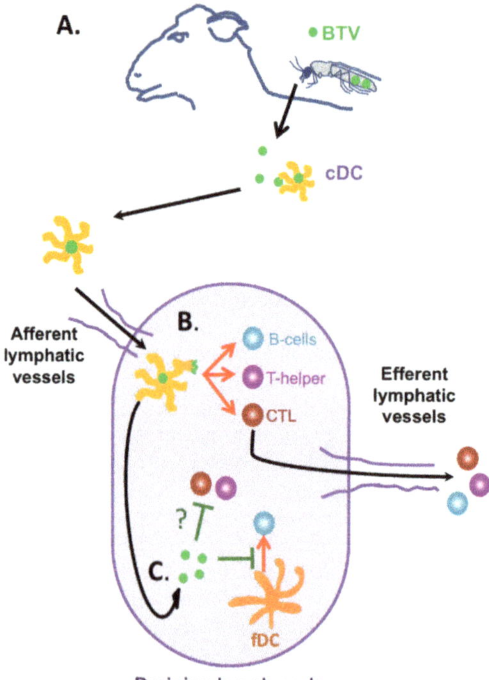

Figure 3. Overview of BTV effects on adaptive immunity. (**A**) BTV is transmitted by the bite of *Culicoides* spp midges to the ruminant host. The virus is captured by conventional dendritic cells (cDC), which transport it to the lymph node. (**B**) Once in the lymph node, cDC can deliver/present BTV antigens so that adaptive immunity is triggered. BTV is known to induce the activation of humoral immunity (B-cells) and cellular immunity (T-helper and cytotoxic T lymphocytes (CTL)). Both arms of adaptive immunity are required for protection from the virus. Once activated, B-cells, T-helper cells and CTL are released from the lymph node to help clear the viral infection. (**C**) BTV infection is known to impair adaptive immune responses. BTV can disrupt the function of follicular DC and consequently impairs antibody generation by B-cells. BTV also induces a transient hyporesponsiveness of T-cells to polyclonal stimuli. The mechanism of this effect is yet to be established (denoted as "**?**" in the figure).

Although BTV is considered to mostly produce acute infections, some animals can develop a chronic disease in which individuals suffer severe muscle loss and wool break. Some animals that appear to recover from the acute infection can also suddenly die, probably due to rapidly progressing pulmonary oedema [50]. Indeed, pneumonia and diarrhea signs in diseased animals are likely the result of opportunistic secondary infections [84] that take hold thanks to the transient immunosuppressive state produce by BTV infection. It is worth noting that the immunosuppressive period induced by BTV does not appear to be as prolonged or marked as in other ruminant viral diseases such as Peste des Petits Ruminants virus infections [85], in which the viral-induced acute immunosuppression leads to increased mortality in affected flocks due to secondary infections [86].

BTV has been described as pan-leucopenic, as primary replication has been reported in monocytes and neutrophils [77]. BTV has also been reported to affect lymphoid cell populations. Ellis et al. also showed decreases in $CD4^+$ T cells, $CD8^+$ T cells, and $CD5^+$ $CD4^-$ $CD8^-$ T-cells (which would include $\gamma\delta$ T-cells) following BTV-17 or BTV-10 infection in sheep [74]. They also showed that a similar reduction in these T-cell populations occurred in cattle during experimental BTV-17 infections [74]. Another report detected similar decreases in $CD4^+$ T-cell and $\gamma\delta$ T-cells population in BTV-1 and BTV-8 infections in sheep [87], which did not occur in vaccinated animals [88]. We studied by flow cytometry the variation in different peripheral blood mononuclear cell (PBMC) populations produced by BTV-8 infection (Figure 4A–F). We observed that by day 7 post-infection, the percentage of $CD4^+$ T-cells (Figure 4A), $CD8^+$ T-cells (Figure 4B), $\gamma\delta$ T-cells (Figure 4E) and B-cells (Figure 4C) in PBMC decreased in BTV-8 infected sheep when compared to mock infected counterparts. This effect was concomitant with the peak of viremia and clinical signs in infected animals (Figure 2). Taken together, these reports are therefore indicative that the pan-leucopenic effect of BTV infection also involves alterations in lymphoid cell populations.

BTV infections appear to mostly affect $CD4^+$ T cells and $\gamma\delta$ T-cells populations in the periphery. It has been shown that $\gamma\delta$ T-cells are recruited at the biting site of *Culicoides* spp midges [39]. Indeed, targeting of $\gamma\delta$ T-cells by BTV has been suggested as a mechanism of overwintering in hosts [39]. In ruminants, $\gamma\delta$ T-cell represent a significant proportion of T-cells that can reach 30% of PBMC in some adult animals [89]. This T-cell subset has been linked to responses to tuberculosis antigen in sheep and cattle [90,91], but may also play a role in antiviral responses in ruminants [92,93]. In spite of the relevance of this T-cell subset, little is known on the activity of $\gamma\delta$ T-cells during BTV infections. Several reports [39,74,87] as well as the data herein presented concur to show that $\gamma\delta$ T-cells may be targeted by BTV during the early stages of infection.

The effect of BTV infection on B-cells has been less studied. Sánchez-Cordón et al. reported no changes in B-cells population during the course of experimental infection with BTV-1 and BTV-8 [87]. In contrast, we are reporting here a prolonged decrease in the percentage of B-cells in infected sheep which lasted from day 5 to day 23 in a BTV-8 sheep infection (Figure 4C). BTV is known to disrupt B-cell responses by infecting follicular DC in the lymph node [94]. The decrease we observe in this experiment could indicate that BTV infects B-cells to limit their activity. Further work will be required to further elucidate this observation. NK cell percentage did not significantly change throughout the experiment (Figure 4F), indicating that lymphopenia was not sufficiently marked to trigger homeostatic NK cell expansion to compensate for cell loss [95]. This also indicates that BTV infection is not affecting this lymphoid cell population, although further analysis will be required to confirm this observation.

Overall, the pan-leukopenic effects of BTV infection appear to involve several lymphoid cell compartments. Indeed, the observation that apoptosis increased in PBMC as well as in spleen concurrent with the peak in viremia in BTV-23 infected sheep [76] suggests that BTV could be triggering lymphoid cell death and thus promoting an immunosuppressive state. Further work will be required to evaluate the exact contribution of this phenomenon to the pathogenesis of the disease.

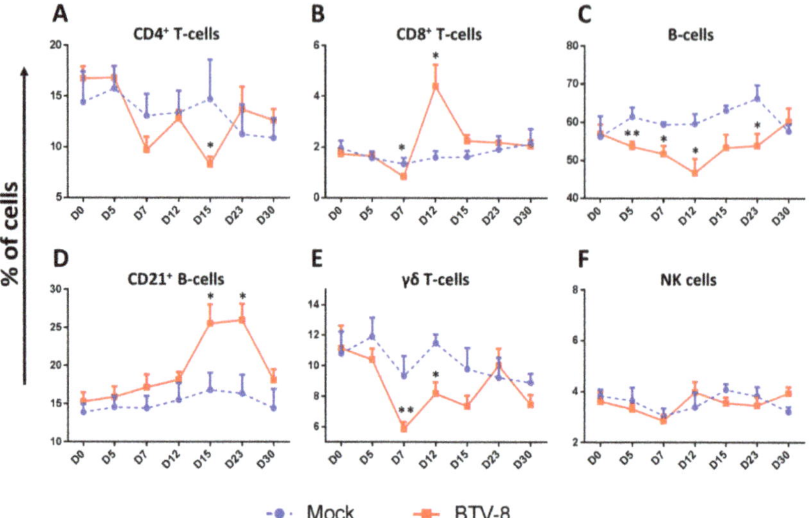

Figure 4. Flow cytometry analysis of lymphoid cell population in PBMC following BTV-8 infection. PBMC were obtained prior to infection (day 0), and at day 5, 7, 12, 15, 23, and 30 post-infection from BTV-8 or mock-infected sheep. (A–F) Flow cytometry analysis was performed to assess the percentage of CD4$^+$ T-cells, CD8$^+$ T-cells, B-cells, CD21$^+$ B-cells, $\gamma\delta$ T-cells, and NK cells in isolated PBMC. (**A**) CD4$^+$ T-cell percentage decreased at day 7 post-BTV infection and was significantly lower at day 12 post-BTV infection when compared to mock-infected sheep. (**B**) CD8$^+$ T-cell percentage decreased at day 7 post-infection and peaked at day 12 post-infection in BTV-infected sheep. (**C**) B-cell percentage decreased from day 5 up to day 23 post-infection in BTV infected sheep and only recovered to similar levels to mock-infected sheep by day 30. (**D**) Mature B-cells (CD21$^+$ B-cells) percentage peaked in BTV-infected sheep at day 15 to day 23 post-infection when compared to mock-infected counterparts. (**E**) $\gamma\delta$ T-cell percentage decreased by day 7 post-BTV infection and only recovered to levels similar to mock-infected animals by day 23 post-infection. (**F**) NK cell percentage did not appear to significantly change in BTV-infected sheep when compared to mock-infected animals. * $p < 0.05$; ** $p < 0.01$; Unpaired t-test (BTV-8 vs. Mock at indicated timepoint). Please refer to Supplemental File S1 for detailed methodology and Figure S1 for gating strategy.

5.2. BTV Infections Trigger Cellular and Humoral Immunity

Once the host overcomes the leucopenic phase of the infection, an adaptive immune response consisting in cellular and humoral components can be detected in infected animals (Figure 3B). As previously mentioned, Jeggo et al. showed in the 1980s using adoptive transfer experiments that both arms of adaptive immunity are required for BTV protection [40,41]. An increase in circulating CD8$^+$ T-cells has been reported for experimental BTV infections in several studies using different BTV serotypes [47,73,74,76,87]. We also found a peak in circulating CD8$^+$ T-cells at day 12 post-infection in the experiment described herein (Figure 4B). Barratt-Boyes et al. also showed in infected sheep that CD8$^+$ T-cell numbers in efferent lymph from the lymph node (LN) proximal to the infection site increased from day 10 to 12 when compared to the efferent lymph of the contralateral LN [78]. This likely indicates that anti-BTV CD8$^+$ T-cells are generated in the LN in response to the infection and that the expansion of these cells is also detectable in PBMC. Indeed, CD8$^+$ T-cell responses in vaccinated animals have been associated with protection in the natural host [43,47]. Thus, it appears that CD8$^+$ T-cell expansion forms an integral part of the adaptive immune response to BTV, which probably helps clear the infection.

Humoral immunity is also triggered by BTV infections (Figure 3B). Barratt-Boyes et al. showed that B-cells expand in the LN proximal to the BTV infection site in calves when compared to the contralateral LN of the same animal [78]. They also detected an increase in B-cells in efferent lymph from the proximal LN at day 10 to 12 [78]. We also detect an increase in CD21$^+$ B-cells at day 15 to 23 in PBMC (Figure 4D). The complement receptor 2 (CD21) is a marker of mature B cells that promotes retention of immune complexes [96]. The increase in circulating CD21$^+$ B-cells is concomitant to the increase in anti-VP7 IgG by day 22 (Figure 2D). It thus appears that in spite of the decreased percentage of circulating B-cells detected in the experiment herein described, the host can still mount an antibody response to BTV. Indeed, BTV is known to only produce a transient impairment of B cell responses, thus allowing the host to eventually mount a humoral response to the virus [94].

It should be noted that not all experimental infection studies give clear cut data when PBMC populations are analyzed [73,76,97] which is likely due to a combination of factors, such as isolate virulence, heterogeneity within breeds, susceptibility of breeds, and/or route and dose used for infection [98]. These alterations in PBMC populations after BTV infections are also likely to depend on the severity of the experimental infection. Overall, most studies concur to show that BTV affects multiple lymphoid populations in the early stages of infection which probably contribute to the pan-leucopenic effects of BTV infection. Ultimately, the ruminant host is capable of mounting a cellular and humoral immune response that helps clear the infection. Nonetheless, as previously discussed, BTV can prolong its circulation in the host so that it is transmitted back to the vector, which indicates that BTV possesses mechanisms that disrupt adaptive immunity. Some of the potential mechanisms are discussed in the final section of this review.

6. BTV Disruption of Adaptive Immunity

The mechanisms that allow BTV to prolong its viremia in spite of the presence of cellular and humoral immunity are poorly understood. Infective virus can circulate for at least 21 days in the host, while sensitive RT-qPCR techniques can pick up viral RNA presence for more than 200 days in some cases [70]. BTV is known to associate with erythrocytes [38] probably to facilitate its passage back to the *Culicoides* spp. Nonetheless, a mechanism of prolonged circulation based solely on the virus affinity for erythrocyte membrane is unlikely to fully explain the extended viremia particularly when cellular and humoral responses are elicited by the infection.

6.1. BTV Effects on Dendritic Cells

Broadly speaking, dendritic cells are professional antigen presenting cells that are essential for mounting adequate adaptive immune responses to pathogens (reviewed in [99]). Although these cells are part of the innate immune system, they represent a key link between innate and adaptive immunity. In their immature state, these sentinel cells can capture pathogen antigens in the periphery either through phagocytosis or even by becoming infected. They possess a myriad of pattern recognition receptors (PRR) that can recognize pathogen-associated molecular pattern (PAMP), such as viral genetic material. Once a PRR recognizes a PAMP, it triggers the activation of the DC which migrates to the proximal lymph node where it can present the pathogen antigens to naïve T-cells so that they can become activated. Similarly, specialized DC in secondary lymphoid organs (follicular DC) can capture and retain native antigen immune complexes so that antigen-specific B-cells can mature and differentiate to antibody-producing cells [100,101]. Unsurprisingly, viruses have evolved strategies that disrupt DC function and consequently the development of humoral and cellular immunity. For instance, disruption of T-cell antigen presentation is an integral part of LCMV immune evasion strategies [13,102], while measles virus targets multiple aspects of DC activation to promote T-cell hyporesponsiveness and immunosuppression [103].

Indeed, BTV is known to interact with DC during infection (Figure 3). It is now accepted that conventional DC (cDC) are responsible for transporting the virus from the biting site to the afferent LN where further viral replication can occur [79]. Once BTV reaches the LN, it can disrupt follicular DC activity [94]. BTV has the capacity to replicate in ruminants in presence of a concomitant antiviral IFN response (reviewed in [34]). We have shown that BTV presence in the LN triggered the expression of the IFN-stimulated gene Mx1 [94]. In spite of this, BTV could replicate and impair follicular DC activity resulting in delayed antibody response to a model antigen. Thus, BTV can target follicular DC to impair antibody responses.

The effect of BTV infection on cDC antigen presentation to T-cells is less well characterized. BTV is unable to productively infect sheep hematopoietic cells and thus could not impair in vitro differentiation of DC from hematopoietic precursors [104]. This resistance appears to be controlled by the host IFN response, since BTV can impair the differentiation of DC in type I IFN receptor-defective bone marrow precursors [104]. Upon infection, BTV increased the apoptosis and interfered with the upregulation of costimulatory molecules in these type I IFN receptor-defective DC [104], indicating that BTV-mediated impairment in IFN signaling could promote DC dysfunction. Hemati et al. showed that BTV is capable of productively infecting cDC in sheep [79]. Nonetheless, BTV-infected cDC are still capable of presenting BTV antigens to BTV-specific T cells obtained from the lymph of infected sheep [79]. Thus, BTV does not appear to affect cDC antigen presentation to antigen-experienced T-cells. Whether the virus affects cDC priming of T-cell responses remains to be elucidated.

6.2. *BTV Effects on Humoral Immunity*

There is some limited evidence that BTV can infect B-cells. Hemati et al. showed using NS2 immunofluorescence staining that some B-cells isolated from sheep lymph could be infected in vitro, although percentage of infected cells varied between BTV serotypes [79]. In contrast, another report using bovine PBMC indicates that B-cell infection by BTV is minimal [81]. The direct effect of BTV infection in B cells therefore remains an unanswered question.

As previously discussed, BTV can disrupt humoral responses by targeting follicular DC activity so that antibody responses are delayed (Figure 3C) [94]. This effect on the generation of antibody responses was still detected when the antigen was delivered 3 days post-infection showing that this mechanism probably provides BTV with a window of opportunity in which it can further its dissemination. In spite of the delayed antibody response produced by BTV infection, these antibody responses ultimately lead to the production of neutralizing antibodies [94]. Whether these response are long lived such as the ones elicited by sheep and cattle vaccination with inactivated virus that can be detected at least 7 years post-immunization remains to be determined [105,106]. It is also worth mentioning that IgG avidity to VP7 was impaired in sheep challenged with virulent BTV-8 as compared to animals that received an attenuated virus, indicating that BTV disruption of the antibody response could have long-term implication for BTV immunity [94].

6.3. *BTV Effects on Cellular Immunity*

The effects of BTV infection on cellular immunity are less well characterized. Infection of $\gamma\delta$ T-cells at the biting site has been proposed as a mechanism for prolonged viremia. $\gamma\delta$ T-cells can be persistently infected in vitro and are recruited at the *Culicoides* spp biting site in vivo [39]. Hemati et al. also detected infection in a small percentage of $\gamma\delta$ T-cells isolated from sheep lymph [79]. Flow cytometry analysis of bovine PBMC also showed that activated $\gamma\delta$ T-cells (IL-2 receptor$^+$ cells) could become more readily infected in vitro than resting counterparts [81]. Overall, $\gamma\delta$ T-cells could be a target for BTV during infection. The contribution of $\gamma\delta$ T-cells to protective cellular immunity or pathogenesis is nonetheless unclear in BTV infections.

Two immunofluorescence analysis of CD4$^+$ and CD8$^+$ T-cell in vitro infection concur to show that CD4$^+$ T-cells are more susceptible to infection than CD8$^+$ T-cells in sheep and cattle [79,81]. In cattle, Barratt-Boyes et al. also showed that activated T-cells (IL-2 receptor$^+$ cells) appear to be more susceptible to infection [81]. Further work will be required to better characterize the direct effects of BTV on T-cell activity, although infection of activated T-cells could represent an immunosuppressive mechanism of cellular immunity.

As previously discussed, CD8⁺ T-cell expansion is a common feature of BTV infections by days 10–15 (Figure 4B) [47,73,74,76,87] probably as a result of mounted anti-BTV T cell responses in the LN [78]. In spite of this expansion, primary anti-BTV responses, as assessed by IFN-γ production, are usually discreet at early timepoints. For instance, we only detected specific IFN-γ responses by ELISPOT in PBMC in 3 out of 8 infected sheep at day 9 post-challenge and in 2 out of 8 sheep at day 18 post-challenge (Figure 5). This is in line with previous work in which IFN-γ responses in PBMC became readily detectable upon reinfection [73]. It should be noted that these assays used inactivated virus as stimuli and thus response to non-structural proteins (such as NS1 which contain immunodominant epitopes [45,46]) are underestimated. Reinfection with BTV-8 produced a narrowing of the number of VP7 epitopes against which responses were detected [73]. This indicates that primary responses to BTV probably involve a broad repertoire of T cells, but that upon reinfection the cellular response tends to focus on a few dominant epitopes. Ellis et al. have also reported partial cellular response using proliferation assays with PBMC from BTV-17 infected sheep [74]. Two out of three sheep responded at day 7 and 14 post-infection by lymphoproliferation assays to purified BTV-17, although overall all three sheep responded at least at one time point. Calf PBMC responded similarly in these assays, indicating that ruminants can mount a cellular response to BTV. Interestingly, in all three sheep response to the T cell mitogen concanavalin-A was impaired by BTV infection when compared to day 0 samples [74]. This phenomenon was not observed with PBMC from infected calves, indicating that BTV may be more effective at impairing sheep cellular response.

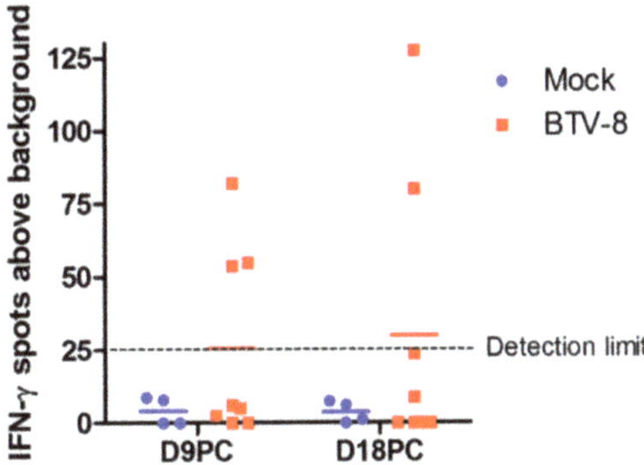

Figure 5. BTV-8 infection elicits anti-BTV IFN-γ producing cells. The presence of IFN-γ producing cells in sheep PBMC was assessed using ELISPOT assays. PBMC were stimulated with inactivated BTV-8 or mock cell lysate as negative control. All assays were performed at least in triplicates. Data are presented as average IFN-γ spots above background for 1×10^6 PBMC at the indicated timepoint (D9PC: day 9 post-challenge; D18PC: day 18 post-challenge). The dotted line indicates the detection limit of the assay. IFN-γ producing cells to BTV-8 were detected in 3 out of 8 infected sheep at D9PC and in 2 sheep out of 8 infected sheep at D18PC. No specific IFN-γ responses were detected in PBMC from mock-infected sheep. Detailed ovine IFN-γ ELISPOT methods can be found in Supplemental File S1 and in [45].

It thus appears that BTV can induce a transient immunosuppressive state in T-cells (Figure 3C). The mechanism through which BTV impairs T-cell reactivity has yet to be established. Several putative mechanisms could be at play. Activated bovine T-cells are reported to be more susceptible to BTV infection, thus BTV could potentially eliminate these cells preferentially. Other viruses, such as LCMV or measles virus, manipulate DC activity to induce an immunosuppressive T-cell state [13,102,103]. Whether BTV also disrupts the priming of naïve T cell responses by DC remains an open question. It should also be noted that BTV-specific T-cells isolated from lymph produce IFN-γ as well as the immunomodulatory cytokine IL-10 upon antigen recognition [79]. IL-10 is known to be a key regulator of inflammation and cellular immunity that can be highjacked in the course of some viral infections (reviewed in [107]). Whether BTV manipulate this pathway remains to be determined.

Additionally, BTV is known to impair IFN-γ signaling [31]. IFN-γ is critical to the development of Th1 responses that target intracellular pathogens such as viruses [108]. The BTV-NS3 and -NS4 proteins have been shown to impair IFN-γ signaling in reporter assays [31]. Whether other BTV proteins that act as IFN antagonist, such as VP3 [35], are also involved in impairing IFN-γ signaling remains to be determined. It is therefore plausible that immune cells susceptible to BTV infection may therefore be unable to respond adequately to IFN-γ stimulation, which in turn could impair the antiviral Th1 response dependent on this cytokine [109]. Overall, although BTV effects on cellular immunity are not fully characterized, several studies hint at BTV capacity to also interfere with this arm of adaptive immunity. Much work still remains to be done to better understand T-cell responses to BTV and how they shape the course of infection.

7. Conclusions

Just as deciphering LCMV effects on immunity has helped elucidate essential immunology concepts, studying BTV infections will shed some light on the workings of the ruminant immune system. The study of BTV infections has already helped characterize new mechanisms of viral escape at the cellular and molecular level [31,94]. Many pieces of the puzzles are still nonetheless missing to fully comprehend BTV infections. For instance, the factors that govern the differences in susceptibility between hosts are still unknown. Thus, evaluating BTV effects on immunity could not only improve our understanding of ruminant immunology, but could also help untangle more general immunological mechanisms.

Supplementary Materials: The following are available online at https://www.mdpi.com/article/10.3390/v13081511/s1, Supplemental File S1: Material and methods, Figure S1: Gating strategy for the flow cytometry analysis of lymphoid cell population in PBMC following BTV-8 Infection.

Author Contributions: Conceptualization, J.M.R. and N.S.; Sample collection and investigation, D.R.-M., A.L.-L., M.A., and J.M.R.; data analysis, D.R.-M., A.L.-L., and J.M.R.; writing—original draft preparation, J.M.R.; writing—review and editing, D.R.-M., A.L.-L., J.M.R., V.M., and N.S.; supervision, J.M.R., V.M., and N.S.; project administration, N.S.; funding acquisition, V.M. and N.S. All authors have read and agreed to the published version of the manuscript.

Funding: This work was funded by grants RTI2018-094616-B-100 from the Ministerio de Ciencia (Spain) and S2018/BAA-4370-PLATESA2 from Comunidad de Madrid (Fondo Europeo de Desarrollo Regional, FEDER). MA was funded by an FPI grant (BES-2013-066406). ALL was funded by an FPI grant (INIA).

Institutional Review Board Statement: All the animal experiments were carried out in a disease-secure isolation facility (BSL3) at the Centro de Investigación en Sanidad Animal (CISA), in strict accordance with the recommendations in the guidelines of the Code for Methods and Welfare Considerations in Behavioural Research with Animals (Directive 86/609EC; RD1201/2005) and all efforts were made to minimize suffering. Experiments were approved by the Committee on the Ethics of Animal Experiments (CEEA) (Permit number: 10/142792.9/12 approved on 15 July 2018) of the Spanish Instituto Nacional de Investigación y Tecnología Agraría y Alimentaria (INIA) and the "Comisión de ética estatal de bienestar animal" (Permit numbers: CBS2012/06 and PROEX 228/14).

Informed Consent Statement: Not applicable.

Data Availability Statement: Data are available upon request.

Acknowledgments: The authors wish to thank all the members of the New Strategies for the Control of Relevant Pathogens in Animal Health laboratory for technical assistance and helpful discussions.

Conflicts of Interest: The authors declare no conflict of interest.

References

1. Riedel, S. Edward Jenner and the history of smallpox and vaccination. *Bayl. Univ. Med. Cent. Proc.* **2005**, *18*, 21–25. [CrossRef] [PubMed]
2. Smith, K.A. Louis Pasteur, the father of immunology? *Front. Immunol.* **2012**, *3*, 68. [CrossRef] [PubMed]
3. Abdel-Hakeem, M.S. Viruses Teaching Immunology: Role of LCMV Model and Human Viral Infections in Immunological Discoveries. *Viruses* **2019**, *11*, 106. [CrossRef] [PubMed]
4. Oldstone, M.B. Biology and pathogenesis of lymphocytic choriomeningitis virus infection. *Curr. Top. Microbiol. Immunol.* **2002**, *263*, 83–117. [PubMed]
5. Zinkernagel, R.M. Lymphocytic choriomeningitis virus and immunology. *Curr. Top. Microbiol. Immunol.* **2002**, *263*, 1–5.
6. Zhou, X.; Ramachandran, S.; Mann, M.; Popkin, D.L. Role of lymphocytic choriomeningitis virus (LCMV) in understanding viral immunology: Past, present and future. *Viruses* **2012**, *4*, 2650–2669. [CrossRef] [PubMed]
7. Oldstone, M.B.A. Anatomy of Viral Persistence. *PLoS Pathog.* **2009**, *5*, e1000523. [CrossRef] [PubMed]
8. Zinkernagel, R.M.; Doherty, P.C. Restriction of in vitro T cell-mediated cytotoxicity in lymphocytic choriomeningitis within a syngeneic or semiallogeneic system. *Nature* **1974**, *248*, 701–702. [CrossRef]
9. Doherty, P.C.; Zinkernagel, R.M. H-2 compatibility is required for T-cell-mediated lysis of target cells infected with lymphocytic choriomeningitis virus. *J. Exp. Med.* **1975**, *141*, 502–507. [CrossRef] [PubMed]
10. Zinkernagel, R.M.; Doherty, P.C. H-2 compatability requirement for T-cell-mediated lysis of target cells infected with lymphocytic choriomeningitis virus. Different cytotoxic T-cell specificities are associated with structures coded for in H-2K or H-2D. *J. Exp. Med.* **1975**, *141*, 1427–1436. [CrossRef]
11. Lau, L.L.; Jamieson, B.D.; Somasundaram, T.; Ahmed, R. Cytotoxic T-cell memory without antigen. *Nature* **1994**, *369*, 648–652. [CrossRef] [PubMed]
12. Murali-Krishna, K.; Lau, L.L.; Sambhara, S.; Lemonnier, F.; Altman, J.; Ahmed, R. Persistence of memory CD8 T cells in MHC class I-deficient mice. *Science* **1999**, *286*, 1377–1381. [CrossRef]
13. Sevilla, N.; Kunz, S.; Holz, A.; Lewicki, H.; Homann, D.; Yamada, H.; Campbell, K.P.; de La Torre, J.C.; Oldstone, M.B. Immunosuppression and resultant viral persistence by specific viral targeting of dendritic cells. *J. Exp. Med.* **2000**, *192*, 1249–1260. [CrossRef]
14. Kunz, S.; Sevilla, N.; McGavern, D.B.; Campbell, K.P.; Oldstone, M.B. Molecular analysis of the interaction of LCMV with its cellular receptor [alpha]-dystroglycan. *J. Cell Biol.* **2001**, *155*, 301–310. [CrossRef] [PubMed]
15. Moskophidis, D.; Lechner, F.; Pircher, H.; Zinkernagel, R.M. Virus persistence in acutely infected immunocompetent mice by exhaustion of antiviral cytotoxic effector T cells. *Nature* **1993**, *362*, 758–761. [CrossRef] [PubMed]
16. Baylis, M.; O'Connell, L.; Mellor, P.S. Rates of bluetongue virus transmission between Culicoides sonorensis and sheep. *Med. Vet. Entomol.* **2008**, *22*, 228–237. [CrossRef] [PubMed]
17. Baylis, M.; Caminade, C.; Turner, J.; Jones, A.E. The role of climate change in a developing threat: The case of bluetongue in Europe. *Rev. Sci. Tech.* **2017**, *36*, 467–478. [CrossRef]
18. Rushton, J.; Lyons, N. Economic impact of Bluetongue: A review of the effects on production. *Vet. Ital* **2015**, *51*, 401–406.
19. Roy, P. Bluetongue virus structure and assembly. *Curr. Opin. Virol.* **2017**, *24*, 115–123. [CrossRef]
20. Maan, S.; Maan, N.S.; Belaganahalli, M.N.; Potgieter, A.C.; Kumar, V.; Batra, K.; Wright, I.M.; Kirkland, P.D.; Mertens, P.P.C. Development and Evaluation of Real Time RT-PCR Assays for Detection and Typing of Bluetongue Virus. *PLoS ONE* **2016**, *11*, e0163014. [CrossRef]
21. Zientara, S.; Sanchez-Vizcaino, J.M. Control of bluetongue in Europe. *Vet. Microbiol.* **2013**, *165*, 33–37. [CrossRef]
22. Breard, E.; Belbis, G.; Viarouge, C.; Nomikou, K.; Haegeman, A.; De Clercq, K.; Hudelet, P.; Hamers, C.; Moreau, F.; Lilin, T.; et al. Evaluation of adaptive immune responses and heterologous protection induced by inactivated bluetongue virus vaccines. *Vaccine* **2015**, *33*, 512–518. [CrossRef]
23. Shaw, A.E.; Ratinier, M.; Nunes, S.F.; Nomikou, K.; Caporale, M.; Golder, M.; Allan, K.; Hamers, C.; Hudelet, P.; Zientara, S.; et al. Reassortment between two serologically unrelated bluetongue virus strains is flexible and can involve any genome segment. *J. Virol.* **2013**, *87*, 543–557. [CrossRef]
24. Rojas, J.M.; Rodríguez-Martín, D.; Martín, V.; Sevilla, N. Diagnosing bluetongue virus in domestic ruminants: Current perspectives. *Vet. Med.* **2019**, *10*, 17–27. [CrossRef]
25. Boyce, M.; Celma, C.C.; Roy, P. Bluetongue virus non-structural protein 1 is a positive regulator of viral protein synthesis. *Virol. J.* **2012**, *9*, 178. [CrossRef] [PubMed]
26. Kar, A.K.; Bhattacharya, B.; Roy, P. Bluetongue virus RNA binding protein NS2 is a modulator of viral replication and assembly. *BMC Mol. Biol.* **2007**, *8*, 4. [CrossRef] [PubMed]
27. Wirblich, C.; Bhattacharya, B.; Roy, P. Nonstructural protein 3 of bluetongue virus assists virus release by recruiting ESCRT-I protein Tsg101. *J. Virol.* **2006**, *80*, 460–473. [CrossRef] [PubMed]
28. Celma, C.C.; Roy, P. A viral nonstructural protein regulates bluetongue virus trafficking and release. *J. Virol.* **2009**, *83*, 6806–6816. [CrossRef]
29. Labadie, T.; Jegouic, S.; Roy, P. Bluetongue Virus Nonstructural Protein 3 Orchestrates Virus Maturation and Drives Non-Lytic Egress via Two Polybasic Motifs. *Viruses* **2019**, *11*, 1107. [CrossRef] [PubMed]

30. Chauveau, E.; Doceul, V.; Lara, E.; Breard, E.; Sailleau, C.; Vidalain, P.O.; Meurs, E.F.; Dabo, S.; Schwartz-Cornil, I.; Zientara, S.; et al. NS3 of bluetongue virus interferes with the induction of type I interferon. *J. Virol.* **2013**, *87*, 8241–8246. [CrossRef]
31. Avia, M.; Rojas, J.M.; Miorin, L.; Pascual, E.; Van Rijn, P.A.; Martín, V.; García-Sastre, A.; Sevilla, N. Virus-induced autophagic degradation of STAT2 as a mechanism for interferon signaling blockade. *EMBO Rep.* **2019**, *20*, e48766. [CrossRef] [PubMed]
32. Ratinier, M.; Shaw, A.E.; Barry, G.; Gu, Q.; Di Gialleonardo, L.; Janowicz, A.; Varela, M.; Randall, R.E.; Caporale, M.; Palmarini, M. Bluetongue Virus NS4 Protein Is an Interferon Antagonist and a Determinant of Virus Virulence. *J. Virol.* **2016**, *90*, 5427–5439. [CrossRef]
33. Stewart, M.; Hardy, A.; Barry, G.; Pinto, R.M.; Caporale, M.; Melzi, E.; Hughes, J.; Taggart, A.; Janowicz, A.; Varela, M.; et al. Characterization of a second open reading frame in genome segment 10 of bluetongue virus. *J. Gen. Virol.* **2015**, *96*, 3280–3293. [CrossRef] [PubMed]
34. Rojas, J.M.; Avia, M.; Martín, V.; Sevilla, N. Inhibition of the IFN Response by Bluetongue Virus: The Story So Far. *Front. Microbiol.* **2021**, *12*, 692069. [CrossRef]
35. Pourcelot, M.; Amaral Moraes, R.; Fablet, A.; Bréard, E.; Sailleau, C.; Viarouge, C.; Postic, L.; Zientara, S.; Caignard, G.; Vitour, D. The VP3 Protein of Bluetongue Virus Associates with the MAVS Complex and Interferes with the RIG-I-Signaling Pathway. *Viruses* **2021**, *13*, 230. [CrossRef] [PubMed]
36. MacLachlan, N.J.; Thompson, J. Bluetongue virus-induced interferon in cattle. *Am. J. Vet. Res.* **1985**, *46*, 1238–1241. [PubMed]
37. Foster, N.M.; Luedke, A.J.; Parsonson, I.M.; Walton, T.E. Temporal relationships of viremia, interferon activity, and antibody responses of sheep infected with several bluetongue virus strains. *Am. J. Vet. Res.* **1991**, *52*, 192–196.
38. Barratt-Boyes, S.M.; MacLachlan, N.J. Dynamics of viral spread in bluetongue virus infected calves. *Vet. Microbiol.* **1994**, *40*, 361–371. [CrossRef]
39. Takamatsu, H.; Mellor, P.S.; Mertens, P.P.C.; Kirkham, P.A.; Burroughs, J.N.; Parkhouse, R.M.E. A possible overwintering mechanism for bluetongue virus in the absence of the insect vector. *J. Gen. Virol.* **2003**, *84 Pt 1*, 227–235. [CrossRef]
40. Jeggo, M.H.; Wardley, R.C.; Taylor, W.P. Role of neutralising antibody in passive immunity to bluetongue infection. *Res. Vet. Sci.* **1984**, *36*, 81–86. [CrossRef]
41. Jeggo, M.H.; Wardley, R.C.; Brownlie, J. A study of the role of cell-mediated immunity in bluetongue virus infection in sheep, using cellular adoptive transfer techniques. *Immunology* **1984**, *52*, 403–410.
42. Zulu, G.B.; Venter, E.H. Evaluation of cross-protection of bluetongue virus serotype 4 with other serotypes in sheep. *J. S. Afr. Vet. Assoc.* **2014**, *85*, 1041. [CrossRef] [PubMed]
43. Martin, V.; Pascual, E.; Avia, M.; Pena, L.; Valcarcel, F.; Sevilla, N. Protective Efficacy in Sheep of Adenovirus-Vectored Vaccines against Bluetongue Virus Is Associated with Specific T Cell Responses. *PLoS ONE* **2015**, *10*, e0143273. [CrossRef]
44. Rojas, J.M.; Barba-Moreno, D.; Avia, M.; Sevilla, N.; Martín, V. Vaccination With Recombinant Adenoviruses Expressing the Bluetongue Virus Subunits VP7 and VP2 Provides Protection Against Heterologous Virus Challenge. *Front. Vet. Sci.* **2021**, *8*, 645561. [CrossRef] [PubMed]
45. Rojas, J.M.; Pena, L.; Martin, V.; Sevilla, N. Ovine and murine T cell epitopes from the non-structural protein 1 (NS1) of bluetongue virus serotype 8 (BTV-8) are shared among viral serotypes. *Vet. Res.* **2014**, *45*, 30. [CrossRef]
46. Marín-López, A.; Calvo-Pinilla, E.; Barriales, D.; Lorenzo, G.; Brun, A.; Anguita, J.; Ortego, J. CD8 T Cell Responses to an Immunodominant Epitope within the Nonstructural Protein NS1 Provide Wide Immunoprotection against Bluetongue Virus in IFNAR(-/-) Mice. *J. Virol.* **2018**, *92*, e00938-18. [CrossRef]
47. Umeshappa, C.S.; Singh, K.P.; Pandey, A.B.; Singh, R.P.; Nanjundappa, R.H. Cell-mediated immune response and cross-protective efficacy of binary ethylenimine-inactivated bluetongue virus serotype-1 vaccine in sheep. *Vaccine* **2010**, *28*, 2522–2531. [CrossRef]
48. Anderson, J.; Hägglund, S.; Bréard, E.; Riou, M.; Zohari, S.; Comtet, L.; Olofson, A.S.; Gélineau, R.; Martin, G.; Elvander, M.; et al. Strong protection induced by an experimental DIVA subunit vaccine against bluetongue virus serotype 8 in cattle. *Vaccine* **2014**, *32*, 6614–6621. [CrossRef] [PubMed]
49. Jeggo, M.H.; Wardley, R.C.; Brownlie, J.; Corteyn, A.H. Serial inoculation of sheep with two bluetongue virus types. *Res. Vet. Sci.* **1986**, *40*, 386–392. [CrossRef]
50. Maclachlan, N.J.; Drew, C.P.; Darpel, K.E.; Worwa, G. The pathology and pathogenesis of bluetongue. *J. Comp. Pathol.* **2009**, *141*, 1–16. [CrossRef] [PubMed]
51. Gerdes, G.H. A South African overview of the virus, vectors, surveillance and unique features of bluetongue. *Vet. Ital.* **2004**, *40*, 39–42.
52. Stallknecht, D.E.; Howerth, E.W. Epidemiology of bluetongue and epizootic haemorrhagic disease in wildlife: Surveillance methods. *Vet. Ital.* **2004**, *40*, 203–207. [PubMed]
53. García-Bocanegra, I.; Arenas-Montes, A.; Lorca-Oró, C.; Pujols, J.; González, M.Á.; Napp, S.; Gómez-Guillamón, F.; Zorrilla, I.; Miguel, E.S.; Arenas, A. Role of wild ruminants in the epidemiology of bluetongue virus serotypes 1, 4 and 8 in Spain. *Vet. Res.* **2011**, *42*, 88. [CrossRef] [PubMed]
54. Lorca-Oró, C.; López-Olvera, J.R.; Ruiz-Fons, F.; Acevedo, P.; García-Bocanegra, I.; Oleaga, Á.; Gortázar, C.; Pujols, J. Long-Term Dynamics of Bluetongue Virus in Wild Ruminants: Relationship with Outbreaks in Livestock in Spain, 2006–2011. *PLoS ONE* **2014**, *9*, e100027. [CrossRef] [PubMed]

55. Elbers, A.R.; Backx, A.; Meroc, E.; Gerbier, G.; Staubach, C.; Hendrickx, G.; van der Spek, A.; Mintiens, K. Field observations during the bluetongue serotype 8 epidemic in 2006. I. Detection of first outbreaks and clinical signs in sheep and cattle in Belgium, France and the Netherlands. *Prev. Vet. Med.* **2008**, *87*, 21–30. [CrossRef]
56. Perrin, A.; Albina, E.; Bréard, E.; Sailleau, C.; Promé, S.; Grillet, C.; Kwiatek, O.; Russo, P.; Thiéry, R.; Zientara, S.; et al. Recombinant capripoxviruses expressing proteins of bluetongue virus: Evaluation of immune responses and protection in small ruminants. *Vaccine* **2007**, *25*, 6774–6783. [CrossRef]
57. Caporale, M.; Di Gialleonorado, L.; Janowicz, A.; Wilkie, G.; Shaw, A.; Savini, G.; Van Rijn, P.A.; Mertens, P.; Di Ventura, M.; Palmarini, M. Virus and host factors affecting the clinical outcome of bluetongue virus infection. *J. Virol.* **2014**, *88*, 10399–10411. [CrossRef] [PubMed]
58. Van Gennip, R.G.P.; van de Water, S.G.P.; Maris-Veldhuis, M.; van Rijn, P.A. Bluetongue viruses based on modified-live vaccine serotype 6 with exchanged outer shell proteins confer full protection in sheep against virulent BTV8. *PLoS ONE* **2012**, *7*, e44619.
59. Huismans, H.; van der Walt, N.T.; Cloete, M.; Erasmus, B.J. Isolation of a capsid protein of bluetongue virus that induces a protective immune response in sheep. *Virology* **1987**, *157*, 172–179. [CrossRef]
60. Sánchez-Cordón, P.J.; Pleguezuelos, F.J.; Pérez de Diego, A.C.; Gómez-Villamandos, J.C.; Sánchez-Vizcaíno, J.M.; Cerón, J.J.; Tecles, F.; Garfia, B.; Pedrera, M. Comparative study of clinical courses, gross lesions, acute phase response and coagulation disorders in sheep inoculated with bluetongue virus serotype 1 and 8. *Vet. Microbiol.* **2013**, *166*, 184–194. [CrossRef]
61. Rojas, J.M.; Rodriguez-Calvo, T.; Pena, L.; Sevilla, N. T cell responses to bluetongue virus are directed against multiple and identical CD4+ and CD8+ T cell epitopes from the VP7 core protein in mouse and sheep. *Vaccine* **2011**, *29*, 6848–6857. [CrossRef]
62. Toussaint, J.F.; Sailleau, C.; Breard, E.; Zientara, S.; De Clercq, K. Bluetongue virus detection by two real-time RT-qPCRs targeting two different genomic segments. *J. Virol. Methods* **2007**, *140*, 115–123. [CrossRef]
63. Chaignat, V.; Worwa, G.; Scherrer, N.; Hilbe, M.; Ehrensperger, F.; Batten, C.; Cortyen, M.; Hofmann, M.; Thuer, B. Toggenburg Orbivirus, a new bluetongue virus: Initial detection, first observations in field and experimental infection of goats and sheep. *Vet. Microbiol.* **2009**, *138*, 11–19. [CrossRef]
64. Maan, S.; Maan, N.S.; Nomikou, K.; Batten, C.; Antony, F.; Belaganahalli, M.N.; Samy, A.M.; Reda, A.A.; Al-Rashid, S.A.; El Batel, M.; et al. Novel bluetongue virus serotype from Kuwait. *Emerg. Infect. Dis.* **2011**, *17*, 886–889. [CrossRef] [PubMed]
65. Zientara, S.; Sailleau, C.; Viarouge, C.; Höper, D.; Beer, M.; Jenckel, M.; Hoffmann, B.; Romey, A.; Bakkali-Kassimi, L.; Fablet, A.; et al. Novel bluetongue virus in goats, Corsica, France, 2014. *Emerg. Infect. Dis.* **2014**, *20*, 2123–2125. [CrossRef]
66. World Organization for Animal Health (OIE). Terrestrial Manual. Chapter 3.1.3. Bluetongue (Infection with Bluetongue Virus). In *Manual of Diagnostic Tests and Vaccines for Terrestrial Animals*; International Office of Epizootics: Paris, France, 2018.
67. Ries, C.; Domes, U.; Janowetz, B.; Böttcher, J.; Burkhardt, K.; Miller, T.; Beer, M.; Hoffmann, B. Isolation and Cultivation of a New Isolate of BTV-25 and Presumptive Evidence for a Potential Persistent Infection in Healthy Goats. *Viruses* **2020**, *12*, 983. [CrossRef] [PubMed]
68. Lunt, R.A.; Melville, L.; Hunt, N.; Davis, S.; Rootes, C.L.; Newberry, K.M.; Pritchard, L.I.; Middleton, D.; Bingham, J.; Daniels, P.W.; et al. Cultured skin fibroblast cells derived from bluetongue virus-inoculated sheep and field-infected cattle are not a source of late and protracted recoverable virus. *J. Gen. Virol.* **2006**, *87 Pt 12*, 3661–3666. [CrossRef]
69. Maclachlan, N.J.; Henderson, C.; Schwartz-Cornil, I.; Zientara, S. The immune response of ruminant livestock to bluetongue virus: From type I interferon to antibody. *Virus Res.* **2014**, *182*, 71–77. [CrossRef]
70. Bonneau, K.R.; DeMaula, C.D.; Mullens, B.A.; MacLachlan, N.J. Duration of viraemia infectious to Culicoides sonorensis in bluetongue virus-infected cattle and sheep. *Vet. Microbiol.* **2002**, *88*, 115–125. [CrossRef]
71. Wang, L.F.; Kattenbelt, J.A.; Gould, A.R.; Pritchard, L.I.; Crameri, G.S.; Eaton, B.T. Major core protein VP7 of Australian bluetongue virus serotype 15: Sequence and antigenicity divergence from other BTV serotypes. *J. Gen. Virol.* **1994**, *75 Pt 9*, 2421–2425. [CrossRef]
72. Martinelle, L.; Dal Pozzo, F.; Sarradin, P.; Van Campe, W.; De Leeuw, I.; De Clercq, K.; Thys, C.; Thiry, E.; Saegerman, C. Experimental bluetongue virus superinfection in calves previously immunized with bluetongue virus serotype 8. *Vet. Res.* **2016**, *47*, 73. [CrossRef] [PubMed]
73. Rojas, J.M.; Rodriguez-Calvo, T.; Sevilla, N. Recall T cell responses to bluetongue virus produce a narrowing of the T cell repertoire. *Vet. Res.* **2017**, *48*, 38. [CrossRef] [PubMed]
74. Ellis, J.A.; Luedke, A.J.; Davis, W.C.; Wechsler, S.J.; Mecham, J.O.; Pratt, D.L.; Elliott, J.D. T Lymphocyte Subset Alterations Following Bluetongue Virus Infection in Sheep and Cattle. *Vet. Immunol. Immunopathol.* **1990**, *24*, 49–67. [CrossRef]
75. Mahrt, C.R.; Osburn, B.I. Experimental bluetongue virus infection of sheep; effect of previous vaccination: Clinical and immunologic studies. *Am. J. Vet. Res.* **1986**, *47*, 1191–1197.
76. Umeshappa, C.S.; Singh, K.P.; Nanjundappa, R.H.; Pandey, A.B. Apoptosis and immuno-suppression in sheep infected with bluetongue virus serotype-23. *Vet. Microbiol.* **2010**, *144*, 310–318. [CrossRef]
77. Parsonson, I.M. Pathology and pathogenesis of bluetongue infections. *Curr. Top. Microbiol. Immunol.* **1990**, *162*, 119–141.
78. Barratt-Boyes, S.M.; Rossitto, P.V.; Taylor, B.C.; Ellis, J.A.; MacLachlan, N.J. Response of the regional lymph node to bluetongue virus infection in calves. *Vet. Immunol. Immunopathol.* **1995**, *45*, 73–84. [CrossRef]
79. Hemati, B.; Contreras, V.; Urien, C.; Bonneau, M.; Takamatsu, H.H.; Mertens, P.P.; Bréard, E.; Sailleau, C.; Zientara, S.; Schwartz-Cornil, I. Bluetongue virus targets conventional dendritic cells in skin lymph. *J. Virol.* **2009**, *83*, 8789–8799. [CrossRef]

80. Russell, H.; O'Toole, D.T.; Bardsley, K.; Davis, W.C.; Ellis, J.A. Comparative effects of bluetongue virus infection of ovine and bovine endothelial cells. *Vet. Pathol.* **1996**, *33*, 319–331. [CrossRef]
81. Barratt-Boyes, S.M.; Rossitto, P.V.; Stott, J.L.; MacLachlan, N.J. Flow cytometric analysis of in vitro bluetongue virus infection of bovine blood mononuclear cells. *J. Gen. Virol.* **1992**, *73 Pt 8*, 1953–1960. [CrossRef]
82. Ellis, J.A.; Coen, M.L.; MacLachlan, N.J.; Wilson, W.C.; Williams, E.S.; Leudke, A.J. Prevalence of bluetongue virus expression in leukocytes from experimentally infected ruminants. *Am. J. Vet. Res.* **1993**, *54*, 1452–1456.
83. Howerth, E.W. Cytokine release and endothelial dysfunction: A perfect storm in orbivirus pathogenesis. *Vet. Ital.* **2015**, *51*, 275–281.
84. Diniz Matos, A.C.; Alvarez Balaro, M.F.; Maldonado Coelho Guedes, M.I.; Costa, E.A.; Ca Mara Rosa, J.L.; Costa, A.T.; Branda, O.F.; Portela Lobato, Z.L. Epidemiology of a Bluetongue outbreak in a sheep flock in Brazil. *Vet. Ital.* **2016**, *52*, 325–331.
85. Rojas, J.M.; Moreno, H.; Valcarcel, F.; Pena, L.; Sevilla, N.; Martin, V. Vaccination with recombinant adenoviruses expressing the peste des petits ruminants virus F or H proteins overcomes viral immunosuppression and induces protective immunity against PPRV challenge in sheep. *PLoS ONE* **2014**, *9*, e101226. [CrossRef]
86. Kumar, N.; Maherchandani, S.; Kashyap, S.K.; Singh, S.V.; Sharma, S.; Chaubey, K.K.; Ly, H. Peste des petits ruminants virus infection of small ruminants: A comprehensive review. *Viruses* **2014**, *6*, 2287–2327. [CrossRef]
87. Sanchez-Cordon, P.J.; Perez de Diego, A.C.; Gomez-Villamandos, J.C.; Sanchez-Vizcaino, J.M.; Pleguezuelos, F.J.; Garfia, B.; del Carmen, P.; Pedrera, M. Comparative analysis of cellular immune responses and cytokine levels in sheep experimentally infected with bluetongue virus serotype 1 and 8. *Vet. Microbiol.* **2015**, *177*, 95–105. [CrossRef] [PubMed]
88. Pérez de Diego, A.C.; Sánchez-Cordón, P.J.; de las Heras, A.I.; Sánchez-Vizcaíno, J.M. Characterization of the immune response induced by a commercially available inactivated bluetongue virus serotype 1 vaccine in sheep. *Sci. World J.* **2012**, *2012*, 147158. [CrossRef] [PubMed]
89. Baldwin, C.L.; Yirsaw, A.; Gillespie, A.; Le Page, L.; Zhang, F.; Damani-Yokota, P.; Telfer, J.C. $\gamma\delta$ T cells in livestock: Responses to pathogens and vaccine potential. *Transbound. Emerg. Dis.* **2020**, *67* (Suppl. 2), 119–128. [CrossRef] [PubMed]
90. Evans, C.W.; Lund, B.T.; McConnell, I.; Bujdoso, R. Antigen recognition and activation of ovine gamma delta T cells. *Immunology* **1994**, *82*, 229–237.
91. Vesosky, B.; Turner, O.C.; Turner, J.; Orme, I.M. Gamma interferon production by bovine gamma delta T cells following stimulation with mycobacterial mycolylarabinogalactan peptidoglycan. *Infect. Immun.* **2004**, *72*, 4612–4618. [CrossRef]
92. Endsley, J.J.; Quade, M.J.; Terhaar, B.; Roth, J.A. Bovine viral diarrhea virus type 1- and type 2-specific bovine T lymphocyte-subset responses following modified-live virus vaccination. *Vet. Ther.* **2002**, *3*, 364–372.
93. Luján, L.; Begara, I.; Collie, D.D.; Watt, N.J. CD8+ lymphocytes in bronchoalveolar lavage and blood: In vivo indicators of lung pathology caused by maedi-visna virus. *Vet. Immunol. Immunopathol.* **1995**, *49*, 89–100. [CrossRef]
94. Melzi, E.; Caporale, M.; Rocchi, M.; Martin, V.; Gamino, V.; di Provvido, A.; Marruchella, G.; Entrican, G.; Sevilla, N.; Palmarini, M. Follicular dendritic cell disruption as a novel mechanism of virus-induced immunosuppression. *Proc. Natl. Acad. Sci. USA* **2016**, *113*, E6238–E6247. [CrossRef]
95. Sun, J.C.; Lanier, L.L. NK cell development, homeostasis and function: Parallels with $CD8^+$ T cells. *Nat. Rev. Immunol.* **2011**, *11*, 645–657. [CrossRef] [PubMed]
96. Roozendaal, R.; Carroll, M.C. Complement receptors CD21 and CD35 in humoral immunity. *Immunol. Rev.* **2007**, *219*, 157–166. [CrossRef] [PubMed]
97. Bitew, M.; Ravishankar, C.; Chakravarti, S.; Kumar Sharma, G.; Nandi, S. Comparative Evaluation of T-Cell Immune Response to BTV Infection in Sheep Vaccinated with Pentavalent BTV Vaccine When Compared to Un-Vaccinated Animals. *Vet. Med. Int.* **2019**, *2019*, 8762780. [CrossRef] [PubMed]
98. Coetzee, P.; van Vuuren, M.; Venter, E.H.; Stokstad, M. A review of experimental infections with bluetongue virus in the mammalian host. *Virus Res.* **2014**, *182*, 21–34. [CrossRef]
99. Eisenbarth, S.C. Dendritic cell subsets in T cell programming: Location dictates function. *Nat. Rev. Immunol.* **2019**, *19*, 89–103. [CrossRef] [PubMed]
100. Kranich, J.; Krautler, N.J. How Follicular Dendritic Cells Shape the B-Cell Antigenome. *Front. Immunol.* **2016**, *7*, 225. [CrossRef]
101. Heesters, B.A.; Myers, R.C.; Carroll, M.C. Follicular dendritic cells: Dynamic antigen libraries. *Nat. Rev. Immunol.* **2014**, *14*, 495–504. [CrossRef] [PubMed]
102. Sevilla, N.; McGavern, D.B.; Teng, C.; Kunz, S.; Oldstone, M.B. Viral targeting of hematopoietic progenitors and inhibition of DC maturation as a dual strategy for immune subversion. *J. Clin. Investig.* **2004**, *113*, 737–745. [CrossRef] [PubMed]
103. Schneider-Schaulies, S.; Klagge, I.M.; ter Meulen, V. Dendritic cells and measles virus infection. *Curr. Top. Microbiol. Immunol.* **2003**, *276*, 77–101.
104. Rodriguez-Calvo, T.; Rojas, J.M.; Martin, V.; Sevilla, N. Type I interferon limits the capacity of bluetongue virus to infect hematopoietic precursors and dendritic cells in vitro and in vivo. *J. Virol.* **2014**, *88*, 859–867. [CrossRef] [PubMed]
105. Hilke, J.; Strobel, H.; Woelke, S.; Stoeter, M.; Voigt, K.; Moeller, B.; Bastian, M.; Ganter, M. Presence of Antibodies against Bluetongue Virus (BTV) in Sheep 5 to 7.5 Years after Vaccination with Inactivated BTV-8 Vaccines. *Viruses* **2019**, *11*, 533. [CrossRef]
106. Ries, C.; Beer, M.; Hoffmann, B. BTV antibody longevity in cattle five to eight years post BTV-8 vaccination. *Vaccine* **2019**, *37*, 2656–2660. [CrossRef] [PubMed]

107. Rojas, J.M.; Avia, M.; Martín, V.; Sevilla, N. IL-10: A Multifunctional Cytokine in Viral Infections. *J. Immunol. Res.* **2017**, *2017*, 6104054. [CrossRef]
108. Schoenborn, J.R.; Wilson, C.B. Regulation of interferon-gamma during innate and adaptive immune responses. *Adv. Immunol.* **2007**, *96*, 41–101.
109. Hu, X.; Ivashkiv, L.B. Cross-regulation of Signaling Pathways by Interferon-γ: Implications for Immune Responses and Autoimmune Diseases. *Immunity* **2009**, *31*, 539–550. [CrossRef]

 *viruses* 

Review

Type I Interferon Induction and Exhaustion during Viral Infection: Plasmacytoid Dendritic Cells and Emerging COVID-19 Findings

Trever T. Greene and Elina I. Zuniga *

Division of Biological Sciences, University of California, San Diego, CA 92093, USA; ttgreene@ucsd.edu
* Correspondence: eizuniga@ucsd.edu

Abstract: Type I Interferons (IFN-I) are a family of potent antiviral cytokines that act through the direct restriction of viral replication and by enhancing antiviral immunity. However, these powerful cytokines are a caged lion, as excessive and sustained IFN-I production can drive immunopathology during infection, and aberrant IFN-I production is a feature of several types of autoimmunity. As specialized producers of IFN-I plasmacytoid (p), dendritic cells (DCs) can secrete superb quantities and a wide breadth of IFN-I isoforms immediately after infection or stimulation, and are the focus of this review. Notably, a few days after viral infection pDCs tune down their capacity for IFN-I production, producing less cytokines in response to both the ongoing infection and unrelated secondary stimulations. This process, hereby referred to as "pDC exhaustion", favors viral persistence and associates with reduced innate responses and increased susceptibility to secondary opportunistic infections. On the other hand, pDC exhaustion may be a compromise to avoid IFN-I driven immunopathology. In this review we reflect on the mechanisms that initially induce IFN-I and subsequently silence their production by pDCs during a viral infection. While these processes have been long studied across numerous viral infection models, the 2019 coronavirus disease (COVID-19) pandemic has brought their discussion back to the fore, and so we also discuss emerging results related to pDC-IFN-I production in the context of COVID-19.

Keywords: IFN-I; plasmacytoid dendritic cells; viral infection; LCMV; influenza; HIV-1; HCV; HBV; COVID-19; SARS-CoV-2

Citation: Greene, T.T.; Zuniga, E.I. Type I Interferon Induction and Exhaustion during Viral Infection: Plasmacytoid Dendritic Cells and Emerging COVID-19 Findings. *Viruses* **2021**, *13*, 1839. https://doi.org/10.3390/v13091839

Academic Editors: Michael B. A. Oldstone and Juan C. De la Torre

Received: 27 June 2021
Accepted: 1 September 2021
Published: 15 September 2021

Publisher's Note: MDPI stays neutral with regard to jurisdictional claims in published maps and institutional affiliations.

1. Introduction

The Type I interferon (IFN-I) family are antiviral and antineoplastic cytokines critical for the control of most types of viral infection [1,2]. This family includes 13 subtypes of IFN-α in humans (14 in mice) one IFN-β as well as a handful of other gene products (reviewed in [2]). While all cell types can produce IFN-I, plasmacytoid (p) dendritic cells (DCs) produce IFN-I and other interferons at exceptional levels, including all 13 subtypes of IFN-α, IFN-β, and 3 subtypes of IFN-λ, and IFN-τ (reviewed in [3–5] and described in [6]). As such, pDCs promote control of multiple types of viral infection (reviewed in [4]). However, despite the critical importance of IFN-I to control viral infections the consequences of IFN-I are not exclusively beneficial. Excessive and/or late IFN-I is associated with increased mortality in several animal models of viral infection (e.g., Influenza [7], arenavirus mediated hemorrhagic fever [8], MERS and SARS-CoV-1 infections in mice [9,10]). Furthermore, persistent IFN-I signaling in a mouse model of chronic viral infection promotes T cell exhaustion and compromises viral control [11,12]. Although it is notable that the addition of recombinant IFN-I early in the same infection promotes viral clearance [13]. Finally, aberrant IFN-I signaling is a signature of many autoimmune disorders several of which have been associated with activation of pDCs (reviewed in [14]).

Despite the importance of pDC derived IFN-I for the control of many viruses, after their initial activation pDCs rapidly lose their capacity to produce these antiviral mediators, a state that we refer to as "pDC exhaustion" (Reviewed in [15]). While this pDC exhaustion can favor prolonged viral replication and opportunistic secondary infections [16–18], it may have evolved as a "default" behavior to avoid the potentially harmful effects of sustained IFN-I production. Additionally, this adaptation may also help protect against interferonopathies in the absence of infection.

Here we provide a review of the available literature on pDC IFN-I induction and the mechanisms that support and oppose this important function of pDCs. First, we describe the pathways that promote or suppress pDC IFN-I production on a per-cell basis including modulators of signaling, inhibitory receptors, cytokines, and metabolism. Afterwards, we describe mechanisms that reduce or enhance pDC numbers. Together both types of regulation help to determine the availability of pDC derived IFN-I upon a viral infection.

The COVID-19 pandemic caused by SARS Coronavirus-2 (SARS-CoV-2) has led to rapidly evolving investigation of this disease. These studies have revealed that many of the features of other acute and potentially deadly infections are conserved in COVID-19, including critical roles for IFN-I in both protection and disease. Furthermore, COVID-19 has been associated directly with suppression of pDC IFN-I on a per cell basis [19], as well as with reduced pDC numbers [19–22]. At the end of this review, we describe emerging literature on pDCs, IFN-I, and COVID-19 and discuss how this relates to the established framework in the field. We will avoid a full review of pDC development, the roles of pDCs in tolerogenic responses, and the role of pDC IFN-I production in cancer. We suggest the reader consults these recent reviews as resources for those topics [5,15,23,24].

2. pDC IFN-I Production after Viral Infections

2.1. Uptake of Nucleic Acids

The first step in recognition of a viral infection by a pDC is the uptake of nucleic acids, and their delivery to toll-like receptor (TLR) containing endosomes. While the uptake of stimulatory oligonucleotides used to study pDC function in vitro primarily occurs through clathrin mediated endocytosis [25], viral material uptake can reach the TLR-containing endosomes through endocytosis or autophagy [26,27]. Furthermore, the viral material sensed can be functional or defective viral particles [28,29], exosomes [30], and even productive infection and replication within the pDC themselves [26]. Additionally, Fc receptors can facilitate uptake of immunocomplexed nucleic acids into pDCs [31] and this can trigger their IFN-I production [32]. Indeed, this process has been implicated in abnormal IFN-I elevation in systemic lupus erythematosus (SLE) [32,33]. In contrast, FcγIIB-based internalization of immune complexes has been shown to oppose IFN-I production in pDCs, and this is associated with reduced IFN-I production in response to Sendai virus (SV) infection when virus-specific-antibodies are present, suggesting this route could exist as a way to quell IFN-I responses [31]. Recently, it has also become well-established that, in some cases, pDCs require cell-cell contact with infected cells to robustly produce IFN-I [34–41], and recent observations in vitro support the idea that an "interferogenic synapse" can develop between pDCs and infected cells [42]. How stimulatory material is exchanged in this synapse, and its potential function in vivo still needs further study. Ultimately, the precise route for nucleic acid uptake depends on the type of viral infection and context, and this route can influence the magnitude of pDC IFN-I responses [31,34].

2.2. Signaling by TLR7/TLR9

The primary mechanism by which pDCs are known to recognize exogenous and endogenous RNA and DNA is through TLRs 7 and 9, respectively. These TLRs are highly expressed in pDCs and stimulation of either can lead to high levels of IFN-I production (Reviewed in ref [43]). TLR7 and TLR9 signal through a combination of MyD88-NF-kB and MyD88-IRF7 pathways (Reviewed in refs [44,45]). The relative use of these signaling pathways depends on the subcellular compartments in which the specific TLR is located and varies with the stimulus [46–48]. For example, the multimeric oligonucleotide family of CpG-A localizes to early endosomes where they preferentially activate the MYD88-IRF7 pathway inducing high levels of IFN-I [46,47]. In contrast the monomeric oligonucleotide family of CpG-B localizes to an endolysosomal compartment [46,47]. In the last-mentioned case the MYD88-NF-kB pathway is preferentially activated and leads to production of pro-inflammatory cytokines such as tumor necrosis factor alpha (TNFα) [46]. In pDCs, but not in conventional (c) DCs, the trafficking of CpG-A to early endosomes and their cytokine production relies on signaling through Src kinases Lyn and Fyn [49]. This connection between differential trafficking and signaling has been used to explain the long-standing observation that CpG-A induces higher levels of IFN-I, while CpG-B induces more pro-inflammatory cytokines [50,51]. It is important to note that neither of these cases is absolute as CpG-B induces some production of IFN-I while CpG-A also induces pro-inflammatory cytokines.

2.3. *TLR2/12*

While most of the work in pDCs focuses on TLR7 and TLR9, pDCs can also express TLR2 and TLR12 [52,53]. As with TLR7 and TLR9, recognition of *Toxoplasma gondii* profilin by TLR12 is sufficient to stimulate pDCs to produce IL-12. These observations, along with the observation that TLR12 deficient mice are more susceptible to *T. gondii* infection than wild type animals, suggest a potential role for pDCs in host resistance to this pathogen [52]. However, not all TLRs expressed in pDC stimulate the production of IFN-I or pro-inflammatory cytokines. Indeed, the commensal associated molecule polysaccharide A (PSA), which is sensed via TLR2 [54], has been shown to drive an immunoregulatory phenotype in pDCs that protects mice against 2,4,6-trinitrobenzenesulfonic acid (TNBS) induced colitis [53]. Further research into the roles for these and other TLRs expressed by pDCs are an exciting avenue of future work and will provide deeper insight into how pDC responses are regulated in a variety of contexts.

2.4. *Cytosolic Nucleic Acid Sensors*

In addition to TLRs pDCs express several cytosolic sensors of nucleic acids, some of which have been established to induce pDC activation and IFN-I production. The cytosolic nucleic acid receptors DHX36 and DHX9 have been shown to bind CpG-A and CpG-B and induce IFN-I and pro-inflammatory cytokine production, respectively [55]. On the other hand, it has been proposed that the cytosolic sensor retinoic acid-inducible gene I (RIG-I) does not induce IFN-I production in pDCs. This assertion is based on IFN-I levels in response to Newcastle disease virus (NDV), which did not change in RIG-I deficient pDCs [56]. However, it has since been shown that in some cases RIG-I can contribute to IFN-I production in pDCs. Specifically, adaptors of RIG-I help drive IFN-I production in pDCs in response to NDV in the absence of IFN-I restriction of viral growth [57]. Additionally, RIG-I may drive pDC responses to the cell-free yellow fever live vaccine (YF-17D) [34]. Though notably, the amount of IFN-I produced as the result of RIG-I signaling is less than that produced in response to TLR7 signaling [34].

Recently the cyclic GMP-AMP synthase (cGAS)-stimulator of IFN genes (STING) pathway has also emerged as a method by which pDCs can recognize and respond to insult. In this pathway the enzyme cGAS recognizes cytosolic dsDNA and makes cyclic GMP-AMP (cGAMP). Cyclic dinucleotides can then be recognized by STING which induces IFN-I expression (Reviewed in ref [58]). Human pDCs express both cGAS and STING and can produce IFN-I in response to both cytosolic dsDNA as well as the STING ligands cGAMP, dAMP, and diGMP [59]. STING stimulation of IFN-I in human pDCs associates primarily with nuclear translocation of IRF3 (in contrast to IRF7 for TLR7 and TLR9) [59,60]. Importantly, potentially as a result of these differences, stimulation with cGAMP induces significantly lower levels of IFN-α, IFN-β, and IFN-λ, that is also produced with slower kinetics, as compared to CpG-A stimulation of TLR9 [60]. Furthermore, STING pre-activation reduces pDC IFN-I production in response to TLR9 ligation [60].

3. Suppression of pDC IFN-I Production after Viral Infections

3.1. pDC Functional Exhaustion

Within a few days after an in vivo viral infection the capacity of pDCs to produce IFN-I is severely reduced [16,61], a phenotype that we now refer to as pDC exhaustion. This was first described after acute and chronic LCMV infection in mice, and coincided with a dramatic attenuation of systemic IFN-I early after its peak at day one post infection [16]. pDC exhaustion was also shown to associate with compromised Natural Killer cell activation, Interferon-γ production and cytotoxicity in response to an unrelated secondary virus [16]. Lee et al. confirmed these findings, and additionally showed pDC exhaustion associated with deficient cross-priming of T cells to an unrelated antigen [61]. Importantly, the pDC exhaustion phenotype described in vivo is consistent with other studies showing that pDCs from patients infected with HIV [62–65], HCV [66–70] and HBV [71–73] as well as macaques infected with SIV [74,75] exhibited reduced per-cell production of IFN-I when re-stimulated ex vivo. Additionally, in HIV-infected humans systemic IFN-I is attenuated early after infection [76], and reduced pDC numbers and/or loss of their ex vivo IFN-I production are correlated with opportunistic infections [18,77–79]. Remarkably, this is true even when considered independently of CD4 T cell numbers [77]. It is additionally important to note that IFN-I exhaustion in pDCs is sometimes associated with reduced capacity to produce other cytokines such as TNFα [17]. pDC exhaustion, however, is primarily studied in the context of IFN-I and it is unclear whether other interferons that can also be produced by pDCs (i.e., IFN-λ, and IFN-τ) are likewise suppressed in this condition.

More recently, it has been established that, analogous to T cell receptor (TCR) induction of CD8 T cell exhaustion, in vivo pDC exhaustion in LCMV-infected mice is dependent on TLR7, which reduces the expression of E2-2, a transcription factor essential for pDC development and function [80], via cell-intrinsic signaling in pDCs [17]. Evidence from HIV infection also suggests that persistent stimulation may promote pDC exhaustion as ex vivo IFN-I production by pDCs is reduced in patients with high viral load [62,63], recovers during the administration of successful antiretroviral therapy [65], but becomes suppressed again upon interruption of antiretroviral treatment [63]. Additionally, E2-2 expression is reduced in pDCs from HIV viremic patients [17], and in a human pDC cell line when stimulated with TLR ligands for 2 days [81]. Taken together, these data suggest that persistent TLR engagement may promote pDC IFN-I exhaustion in mice and human pDCs, analogous to what has been described for CD8 T cell exhaustion after chronic stimulation through the TCR [82].

3.2. Inhibitory Receptors

IFN-I production in pDCs can be modified by receptors expressed on their surface. Diverse families of receptors such as BDCA2 [83], Siglec-H [84], ILT7 [85], FcεRI [86], DCIR [87], Tim-3 [64], CD28 [88], E-cadherin [89], and Nkp44 [90] all have established rolls in modulating pDC cytokine production downstream of ligation. Much of this evidence exists ex vivo and the role of most of these receptors in pDC function during in vivo viral infection is not yet established. Even for the few that have been investigated in vivo little is known. Mice deficient in CD28 have enhanced serum IFN-I and increased IFN-I transcript levels in pDCs after either LCMV or MCMV infection, and this associates with improved control of MCMV [88]. Similarly, Siglec-H deficient mice also have increased systemic IFN-I after MCMV infection [91]. In addition, in HIV patients a functionally distinct subset of pDC expressing Tim-3 has been identified, correlating expression of this marker to distinctly reduced IFN-I production capacity in human infection [64].

As blockade of inhibitory receptors has been an effective treatment strategy for the modulation of immune cell functions, it is tempting to speculate that surface inhibitory receptors could be promising targets for enhancing pDC function. For example, the asthma drug Omiluzimab, which acts by sequestering free IgE, drives downregulation of FcεRI (a receptor for IgE) on mast cells and basophils [92]. However, as a side-effect Omiluzimab treatment also enhances IFN-α production capacity in pDCs [93]. This enhancement associated with reduced frequency of rhinovirus detection in a cohort of asthma patients treated with Omiluzimab [94], suggesting that targeting pDC inhibitory receptors has the potential to improve human anti-viral responses in vivo. However, pDC inhibitory receptors may also provide pathogens routes to modulate pDC function. For example, HIV protein Vpu has been shown to subvert anti-viral pDC functions by redistributing BST2 on infected cells. This permits viral exit while still engaging the inhibitory receptor ILT7 on pDCs to suppress IFN-I production [95]. Thus, the multiple inhibitory receptors expressed by pDCs may offer mechanisms by which both hosts, and pathogens can modulate pDC-derived IFN-I in the context of infections.

3.3. Cytokines and Hormones

Many soluble factors can modulate pDC function. Negative regulators of pDC IFN-I production include Transforming Growth Factor Beta (TGFβ), Interleukin (IL)-10, prostaglandin E2 (PGE2), and TNFα [96–100]. While positive regulators include IL-4 [97], IL-7 [97], IL-15 [97], Estrogen [101], and IFN-I itself [97,102–104]. As with inhibitory receptors the functions of most of these soluble molecules are primarily established in vitro or ex vivo and the roles these factors play and the mechanisms by which they modulate pDC function during viral infection in vivo are not fully elucidated.

The fact that IFN-I itself enhances pDC IFN-I production is worth noting as availability of autocrine or paracrine IFN-I signaling can enhance the percentage of IFN-I producing pDCs [97,102–104], although IFN-I also induces pDC death in HSV-1 infected mice [105]. The role of TNFα is also notable as pDCs also produce TNFα upon TLR [106] or FCεR1 engagement [86,107] and thus this may represent another autocrine and/or paracrine negative feedback loop. Integrating the potential autocrine and/or paracrine effects of IFN-I and TNFα may offer a partial explanation of differences between CpG-B and CpG-A stimulation. For example, it would be predicted that a pDC which starts producing IFN-I would receive autocrine signals that drive it toward further production of IFN-I. In contrast, a pDC producing primarily TNFα would receive feedback that opposes IFN-I production. This has been partially proposed before although these studies considered only the impacts of IFN-I positive feedback [104] and TNFα negative feedback [108,109] in isolation. Further study, and perhaps mathematical modeling may shed light on how to disentangle these feedback loops, as well as the established intrinsic differences in cell signaling, which support this phenomenon.

It is also worth noting the positive impact of Estrogen on pDC function [101]. This is one part of a larger convergence of factors that drive increased pDC responses in people with more than one X chromosome [101,110–113]. A phenomenon also partially driven by TLR7, which is expressed on the X chromosome but resistant to X chromosome inactivation [114]. Indeed, both X chromosome content and sex hormone levels contribute to increased IFN-I production in mice [111]. Furthermore, a study inclusive of both cis and transgender male and female humans recently demonstrated that both chromosome number and sex hormone levels associate independently with pDC functional capacity in humans as well [115].

3.4. Metabolic Requirements in pDCs

IFN-I responses in pDC require large-scale and rapid synthesis of IFN-I mRNA/protein. The metabolic requirements for this process are just beginning to be uncovered. It has been established that pDCs require both Oxidative metabolism (OxPhos) as well as glycolysis to maintain their function [116–118]. Furthermore, mammalian target of rapamycin (mTOR), a central regulator of metabolism, is essential for pDC IFN-I production [119]. While multiple studies agree that pDCs adjust their metabolism after stimulation the details of these changes are still being established, Bajwa et al. show that at 24 h post stimulation with influenza, lactate terminal glycolysis is enhanced in pDCs while OxPhos is slightly suppressed [116]. In contrast Hurley et al. show that 6 hr of influenza or HSV exposure drives a slight increase in oxygen consumption rate (OCR) indicating increased OxPhos [118].

It is likely that the choice of distinct time-points in these studies is responsible for the observed differences. Metabolic changes after stimulation may also provide feedback loops that help control and tune down IFN-I production. Indeed, as high levels of lactate can inhibit pDC IFN-I production [120], then it might be expected that over time enhanced levels of lactate terminal glycolysis after stimulation as described in Bajwa et al. [116] could inhibit pDC IFN-I production. Furthermore, Wu et al. show that IFN-I signaling can enhance basal OCR in both pDCs and epithelial cells, and show oxidative metabolism is essential to pDC function [117]. Thus, this could be part of the mechanism by which IFN-I enhances its own production. Further studies will be necessary to work out the precise details explaining how changes in pDC metabolism in response to environmental cues may tune their function. Furthermore, it will be especially important to assess how pDCs adapt their metabolism after in vivo infection.

4. Reduction of pDC Numbers after Viral Infections

In addition to the direct suppression of pDC IFN-I production by the above mechanisms pDC-derived-IFN-I is also compromised in many infections by a reduction in pDC numbers. This has been well established for HIV-1 [18,62–65,121,122], HCV [66–70,123–125] and HBV [71–73] infections, in humans. Reduced numbers of pDCs have also been reported during SIV infection in macaques [74,75,126,127], and LCMV, CMV, HSV, and VSV in mice [13,16,17,61]. The mechanisms by which pDC numbers may be altered include reductions in development from bone marrow (BM) progenitors, increased apoptosis, changes in pDC proliferative capacity, and conversion of pDCs into conventional (c)DC-like cells, as discussed below.

4.1. Compromised pDC Development from Bone Marrow Progenitors

As pDCs are typically short-lived (persisting for only ~3 days in the periphery of C57Bl/6 mice) [128,129] it is expected the entirety of the pDC population will turn over in the course of any infection lasting longer than a few days. However, in non-resolving infections pDC exhaustion and reduced numbers of pDCs are maintained long-term [16]. One possibility is that pDC progenitors are modulated by the infection in a way that they generate fewer and functionally deficient pDCs. Indeed, pDC generation from BM is reduced for long term in mice infected with a persistent LCMV variant, and transiently decreased after infection with an acute LCMV strain [17]. This associates with a numerical reduction in pDC progenitors in the BM, but also extends to a reduced capacity of these progenitors to produce pDCs and reduced expression of E2-2 [17]. Furthermore, the few pDCs generated from BM progenitors from infected mice develop to produce less IFN-I when stimulated ex vivo [17]. Reduced E2-2 in both the progenitors and matured DCs from these cultures raises the possibility that a transcriptional state inherited from progenitors may contribute to the exhaustion phenotype in their pDC progeny. It should, however, be considered as a caveat that virus does persist in these BM cultures and thus may provide persistent TLR7 stimulation to newly generated pDCs driving the observed exhaustion phenotype. Additionally, as advances are made in understanding the development of pDCs, it has become apparent that the BM pDC progenitors analyzed in this study were heterogeneous [130]. Thus, it will now be necessary to distinguish whether a specific depletion of E2-2 high pDC committed progenitors or a general suppression of E2-2 expression across multiple progenitor populations is responsible for these observations.

4.2. Apoptosis

Increased levels of apoptosis in pDCs have been observed in HIV and SIV infections in humans and macaques respectively and HSV infection in mice [75,105,126,131]. For HSV-1 infection in mice IFN-I signaling is at least partially responsible for this effect [105], suggesting that IFN-I driven apoptosis is part of a negative feedback loop limiting pDC IFN-I production after infection. It is also worth considering a role for glucocorticoids (GCs), which can be enhanced after viral infection [132,133] and can also promote pDC apoptosis [134–136]. However, as TLR stimulation opposes GC induced apoptosis, it is unclear whether increased GC levels during infection would ultimately result in increased pDC apoptosis in vivo and so further study will be needed to evaluate this.

4.3. Proliferation

Although during persistent LCMV infection pDC development is compromised, and apoptosis is increased, pDC numbers in peripheral lymphoid tissues are not universally decreased after infection. Indeed, at day 10 after LCMV infection there is an increase in pDC numbers in the spleens of infected mice [17]. This corresponded to an expansion of $CD4^-CCR9^-$ and $CD4^-CCR9^+$ pDC subsets with increased proliferative capacity [17]. While $CD4^-CCR9^-$ and $CD4^-CCR9^+$ pDCs have been described in the BM [128,137,138], where they do exhibit proliferative potential as precursors of $CD4^+CCR9^+$ pDCs, splenic pDCs are predominantly $CCR9^+$ at steady state and do not proliferate [128,137,138]. In LCMV infection this gain in proliferative subsets is dependent on both IFN-I and TLR7 signaling, in contrast to functional exhaustion which requires only the latter [17]. Notably, the cell-cycle marker Ki67 has also been reported to be enhanced in pDCs from SIV infected macaques [75,126,127]. As development from the BM is compromised at this time, this pool of proliferating pDC in the spleen (and potentially other lymphoid tissues) may perpetuate, at least partially, the pool of functionally exhausted pDCs during infection.

4.4. pDC Conversion during Viral Infection

Both acute and chronic LCMV infection as well as the direct administration of PolyI:C or IFN-I drives pDCs to differentiate into cells resembling $CD11b^+$ cDCs (cDC2s). These pDC-derived-cDCs exhibit phenotypic and functional features of cDC2s both in vitro [139] and in vivo [140]. While it was first observed in viral infection, pDC conversion has also been reported after stimulation with GM-CSF or intestinal epithelium supernatant [137], in $Ly49Q^-$ pDC from PolyI:C injected mice [141], and even after pDC transfer in steady-state conditions [138], suggesting this phenomenon may also regulate pDC numbers at steady state.

Recent subdivision of DC subsets has identified a potential intermediate stage in the transition between pDC and cDC2. High dimensional analysis of the DC compartments of humans and mice identified a "bridge" population termed transitional (t)DCs that occupy a spectrum between pDC and cDC2 [142,143]. This population is formed as an intermediate between pDC and cDC2 like fates in in vitro culture [144], shows intermediate E2-2 expression and IFN-I production when compared to pure pDCs and cDC2, and maps inside the recently described populations of "non-canonical DCs" [5] (i.e., Axl+ DC in humans [142,145] and $CX3CR1^+CD8^+$ cDC in mice [146,147]) that exhibit features of both pDCs and cDCs. Importantly, tDCs accumulate in the lungs of influenza virus infected mice [143]. This is in line with a model where there is increased differentiation from pDC, through tDC, to cDC2 after infection. While these results are in line with a role for pDC differentiation into tDC/cDC2 as a mechanism for reduced pDC numbers during infection, further work will be necessary to definitively show whether the transition between pDC and cDC2 like cells, that was first reported during in vivo viral infection [139,140], uses tDC as an intermediate, and whether disrupting this phenomenon can restore pDC numbers.

5. pDCs, IFN-I, SARS-CoV-2, and COVID-19

The COVID-19 pandemic caused by SARS-CoV-2 has led to rapidly evolving investigation of this disease and the virus that causes it. Although pDC have been shown to be protective against beta coronavirus infection in mice [148], the direct protective impact of pDC in SARS-CoV-2 infection has not yet been assessed. Still, much work has been done in a short amount of time to discover how IFN-I regulates SARS-CoV-2 infection and associated COVID-19, how this novel virus is sensed by pDCs, and how infection with SARS-CoV-2 modulates pDC population dynamics and function. Here we summarize the most recent work on these subjects as of the writing of this review.

5.1. *The Complex Role of IFN-I in COVID-19*

The precise role of IFN-I in COVID-19 patients is complex and still emerging. COVID-19 patients that have deficiencies in IFN-I responses as a result of germline mutations or anti-IFN-I autoantibodies are more susceptible to severe disease [149–153]. This is in line with the beneficial role of IFN-I in the control of a variety of viral infections. Furthermore, male identified patients (which typically have low levels of estrogen and only one X chromosome encoding TLR7), are significantly more likely to be hospitalized and develop severe COVID-19 [154], and this could be partly a result of reduced IFN-I production by their pDCs. Notably, use of recombinant IFN-β as a therapeutic for COVID-19 has shown promise in some trials, but not others [155–157], and a retrospective study described that early treatment of COVID-19 patients with IFN-α decreased mortality while late treatment resulted in the opposite outcome [158]. In addition, several reports have associated low levels of systemic IFN and reduced IFN-I signatures with severe disease [22,159] while other studies have found positive correlations among IFN-I and severe disease [160,161], although the later may be a consequence of more sustained viral replication that induces more IFN-I in severe cases.

5.2. *Sensing of SARS-CoV-2 by pDC*

It has been established that pDCs can produce IFN-I in response to stimulation with SARS-CoV-2 in vitro [162]. Much work has been done triangulating the PRR used by pDCs to sense SARS-CoV-2. For example, it has been shown that SARS-CoV-2 possesses more ssRNA fragments capable of being recognized by TLR7 and TLR8 when compared to SARS-CoV [163]. Furthermore, a study looking at humans carrying loss of function mutations in UNC93b or IRAK4, which are respectively essential for TLR endosomal localization and signaling [164,165], has shown that their pDCs fail to produce cytokines in response to SARS-CoV-2 exposure [162]. In the same line, SARS-CoV-2, like other coronaviruses, is an RNA virus and TLR7 is essential for pDC sensing of beta-coronavirus in mice [148]. A definitive role for TLR7 in SARS-CoV-2 was recently demonstrated in patients with familial deficiency for TLR7 function. These patients show little to no production of IFN-α2 in their pDCs after stimulation with SARS-CoV-2, though interestingly production of several other inflammatory cytokines was not significantly impacted in this case [153]. Together these studies provide compelling evidence that pDCs sense SARS-CoV2 via TLR7. Consistent with this, familial deficiency in TLR7 leads to increased susceptibility to severe COVID-19 [151,166]. It is important to note that pDC sensing of SARS-CoV-2 is likely to be independent of SARS-CoV-2 infection of pDCs, since human pDCs do not express the SARS-CoV-2 entry receptor ACE2 or support productive replication [162].

5.3. *Reduced pDC Numbers in SARS-CoV-2 Infection*

It is also now well established that, as in other human infections, pDCs are reduced in the peripheral blood of COVID-19 patients [19–22]. The mechanism underlying this phenomenon has not been fully established, but one study has shown that circulating pDCs from COVID-19 patients have an increased apoptosis signature [22]. As expected, increased pDC apoptosis signatures are negatively correlated with pDC numbers, suggesting increased apoptosis may be responsible for decreased pDC numbers in COVID-19 patients [22]. Intriguingly, increased apoptosis in pDCs also positively correlated with severe disease. Given that pDC numbers also correlate with IFN-I signatures in the same patients, this data is suggestive of a model in which pDC are protective against severe COVID-19, but increased apoptosis limits pDC numbers and IFN-I availability in severe disease [22].

5.4. *Decreased pDC Function in SARS-CoV-2 Infection*

It has been recently shown that pDC from the peripheral blood of COVID-19 patients show reduced IFN-I production in a per-cell basis [19]. Many plausible models exist that might explain this. Several cytokines, including IL-10 and TNFα (discussed above for their potential to suppress pDC function) are elevated in COVID-19 patients [160,167]. Additionally, in COVID-19 patients pDCs show reduced phosphorylation of S6 protein [19], suggesting that mTOR signaling, which supports IFN-I production in pDCs [119], may be compromised in this context. Furthermore, reduction of E2-2 expression upon TLR7 stimulation, as mentioned above for pDC exhaustion in other viral infections [17], is a likely mechanism driving pDC hypo-functionality in COVID-19 patients, although E2-2 levels have not been studied in pDC from SARS-CoV-2 infected cohorts.

5.5. Proposed Model for the Biology and Role of pDCs in SARS-CoV-2 Infection

Overall, the aforementioned studies are compatible with a model in which, at early times post-infection, pDCs are sensing SARS-CoV-2 through TLR7, which drives the production of IFN-I, and contains SARS-CoV-2 replication (Figure 1A, left). At later time points post-infection, reduction in pDC numbers (exacerbated in severe cases due to increased apoptosis) and decreased pDC function (exhaustion), may lead to reduced pDC-derived IFN-I (Figure 1A, right). Given that early pDC-derived IFN-I is expected to be beneficial for the host to control SARS-CoV-2, this immune function is likely intact in patients with mild disease (Figure 1B, left). On the other hand, COVID-19 severe cases may show diverse profiles of pDC-derived-IFN-I (Figure 1B, right). This may range from defective pDC IFN-I production (e.g., patients with TLR7 loss of function mutations) to intact or even enhanced pDC responses in patients infected with high viral inoculum or fast-spreading variants, or in individuals with deficiencies in other innate defense mechanisms. In the latter patients, sustained viral replication would perpetuate IFN-I levels from a variety of infected cells thereby leading to increased IFN-I signatures. While Arunachalam et al. did not observe any correlation between severity of disease and pDC intrinsic loss of function in COVID-19 patients (albeit with a limited cohort) [19], it is still likely that enhancing or sustaining pDC-derived-IFN-I during the early phase of SARS-CoV-2 infection would help contain the viral spread and favor a rapid resolution of the disease. Future work aiming at better defining the mechanisms by which pDC numbers are reduced and how pDC function is suppressed in COVID-19 patients, may allow us to design therapies to restore/enhance pDC-derived IFN-I early enough to quickly contain viral spread.

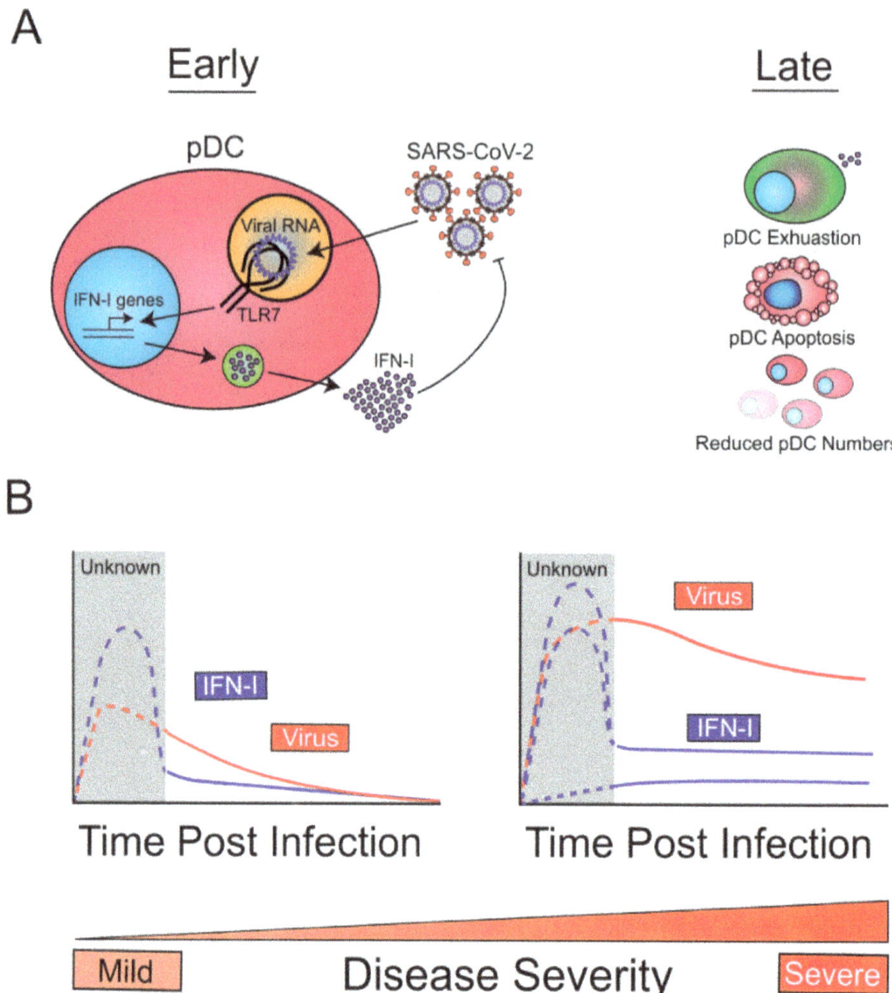

Figure 1. Biology of pDCs during SARS-CoV-2 Infection. (**A**, left) pDC are likely to uptake viral material independent of cell-intrinsic infection, and to sense SARS-CoV-2 RNA via TLR7. This first step stimulates the production of IFN-I which promotes early viral control. (**A**, right) Late after infection blood pDCs become hypofunctional (exhausted), undergo apoptosis and show reduced numbers, and all this may limit pDC derived IFN-I. (**B**, left) pDC from patients with mild disease are expected to produce high IFN-I levels promoting the control of SARS-CoV-2 early after infection. (**B**, right) On the other hand, individuals with severe disease course may exhibit varying degrees of pDC-IFN-I production. The latter may range from low pDC-derived IFN-I in patients with defective TLR7 signaling to normal or even enhanced pDC-derived IFN-I in patients with defects in other immune pathways, or infected with a high viral inoculum or a fast-replicating variant. In such individuals the initial IFN-I wave is expected to be present but insufficient to contain the virus, which may result in sustained stimulation of IFN-I production by a variety of cell types. Importantly, the viral loads or IFN-I levels have not been measured in patients at early times post infection, prior to symptom development, and so the levels of this initial IFN-I response are hypothetical (dotted lines).

6. Discussion

IFN-I are powerful cytokines for the control of viruses, but their power also has the potential to harm their host. The diverse mechanisms of IFN-I control in pDCs likely represent an evolved abundance of caution surrounding the power of pDCs to produce

IFN-I. Therefore, these suppressive systems are analogous to PD1 expression in CD8 T cells, which prevents autoimmunity and immunopathology, but also favors sustained infections or tumors [168]. They likely represent a natural mechanism of immune regulation more than direct viral strategies to suppress immune responses. This is emphasized most recently in COVID-19 infection where IFN-I can protect the host when administered early [158], and IFN-I deficiencies are associated with severe disease [149–151], but, in contrast, late IFN-I treatment may be detrimental [158].

The commonalities that COVID-19 shows with previously-studied infections reemphasizes the importance of the knife's-edge control that evolution has maintained with interferon. These commonalities also emphasize that the work that has been done understanding IFN-I regulation in other systems offers opportunities to quickly develop strategies to combat novel viral diseases. Beyond viral infection, our understanding of these mechanisms also has implications for the treatment of cancer as pDCs also become exhausted within the tumor microenvironment (Reviewed in [15]). Similarly, for autoimmunity and chronic inflammatory diseases where harnessing of the IFN-I natural braking systems may provide novel therapies. Altogether, the power of pDCs to produce IFN-I is paired by evolution with multiple pathways by which to contain it. By working toward a deeper understanding of those pathways we can harness pDCs for the treatment of a wide spectrum of human illnesses.

Author Contributions: Conceptualization T.T.G. and E.I.Z.; writing—original draft preparation, T.T.G. and E.I.Z.; writing—review and editing, T.T.G. and E.I.Z.; funding acquisition, E.I.Z. Both authors have read and agreed to the published version of the manuscript.

Funding: Research in the Zuniga lab is supported by NIH grants AI145314, AI081923, AI113923 and AI132122.

Institutional Review Board Statement: Not applicable.

Informed Consent Statement: Not applicable.

Data Availability Statement: Not applicable.

Acknowledgments: We thank members of the Zuniga lab for fruitful discussions. As space was limited, we were unable to cite all available studies, and so apologize to any authors who feel their studies were not adequately represented in our accounting.

Conflicts of Interest: The authors declare no conflict of interest.

References

1. Zitvogel, L.; Galluzzi, L.; Kepp, O.; Smyth, M.J.; Kroemer, G. Type I interferons in anticancer immunity. *Nat. Rev. Immunol.* **2015**, *15*, 405–414. [CrossRef]
2. McNab, F.; Mayer-Barber, K.; Sher, A.; Wack, A.; O'Garra, A. Type I interferons in infectious disease. *Nat. Rev. Immunol.* **2015**, *15*, 87–103. [CrossRef]
3. Fitzgerald-Bocarsly, P.; Feng, D. The role of type I interferon production by dendritic cells in host defense. *Biochimie* **2007**, *89*, 843–855. [CrossRef] [PubMed]
4. Swiecki, M.; Colonna, M. The multifaceted biology of plasmacytoid dendritic cells. *Nat. Rev. Immunol.* **2015**, *15*, 471–485. [CrossRef] [PubMed]
5. Reizis, B. Plasmacytoid Dendritic Cells: Development, Regulation, and Function. *Immunity* **2019**, *50*, 37–50. [CrossRef] [PubMed]
6. Ito, T.; Kanzler, H.; Duramad, O.; Cao, W.; Liu, Y.-J. Specialization, kinetics, and repertoire of type 1 interferon responses by human plasmacytoid predendritic cells. *Blood* **2006**, *107*, 2423–2431. [CrossRef] [PubMed]
7. Davidson, S.; Crotta, S.; McCabe, T.M.; Wack, A. Pathogenic potential of interferon $\alpha\beta$ in acute influenza infection. *Nat. Commun.* **2014**, *5*, 3864. [CrossRef] [PubMed]
8. Oldstone, M.B.A.; Ware, B.C.; Horton, L.E.; Welch, M.J.; Aiolfi, R.; Zarpellon, A.; Ruggeri, Z.M.; Sullivan, B.M. Lymphocytic choriomeningitis virus Clone 13 infection causes either persistence or acute death dependent on IFN-1, cytotoxic T lymphocytes (CTLs), and host genetics. *Proc. Natl. Acad. Sci. USA* **2018**, *115*, E7814–E7823. [CrossRef]
9. Channappanavar, R.; Fehr, A.R.; Vijay, R.; Mack, M.; Zhao, J.; Meyerholz, D.K.; Perlman, S. Dysregulated Type I Interferon and Inflammatory Monocyte-Macrophage Responses Cause Lethal Pneumonia in SARS-CoV-Infected Mice. *Cell Host Microbe* **2016**, *19*, 181–193. [CrossRef]

10. Channappanavar, R.; Fehr, A.R.; Zheng, J.; Wohlford-Lenane, C.; Abrahante, J.E.; Mack, M.; Sompallae, R.; McCray, P.B.; Meyerholz, D.K.; Perlman, S. IFN-I response timing relative to virus replication determines MERS coronavirus infection outcomes. *J. Clin. Investig.* **2019**, *129*, 3625–3639. [CrossRef]
11. Teijaro, J.R.; Ng, C.; Lee, A.M.; Sullivan, B.M.; Sheehan, K.C.F.; Welch, M.; Schreiber, R.D.; de la Torre, J.C.; Oldstone, M.B.A. Persistent LCMV Infection Is Controlled by Blockade of Type I Interferon Signaling. *Science* **2013**, *340*, 207–211. [CrossRef]
12. Wilson, E.B.; Yamada, D.H.; Elsaesser, H.; Herskovitz, J.; Deng, J.; Cheng, G.; Aronow, B.J.; Karp, C.L.; Brooks, D.G. Blockade of chronic type I interferon signaling to control persistent LCMV infection. *Science* **2013**, *340*, 202–207. [CrossRef] [PubMed]
13. Wang, Y.; Swiecki, M.; Cella, M.; Alber, G.; Schreiber, R.D.; Gilfillan, S.; Colonna, M. Timing and Magnitude of Type I Interferon Responses by Distinct Sensors Impact CD8 T Cell Exhaustion and Chronic Viral Infection. *Cell Host Microbe* **2012**, *11*, 631–642. [CrossRef]
14. Lee-Kirsch, M.A. The Type I Interferonopathies. *Annu. Rev. Med.* **2017**, *68*, 297–315. [CrossRef]
15. Greene, T.T.; Jo, Y.R.; Zuniga, E.I. Infection and cancer suppress pDC derived IFN-I. *Curr. Opin. Immunol.* **2020**, *66*, 114–122. [CrossRef] [PubMed]
16. Zuniga, E.I.; Liou, L.-Y.; Mack, L.; Mendoza, M.; Oldstone, M.B.A. Persistent virus infection inhibits type I interferon production by plasmacytoid dendritic cells to facilitate opportunistic infections. *Cell Host Microbe* **2008**, *4*, 374–386. [CrossRef] [PubMed]
17. Macal, M.; Jo, Y.; Dallari, S.; Chang, A.Y.; Dai, J.; Swaminathan, S.; Wehrens, E.J.; Fitzgerald-Bocarsly, P.; Zúñiga, E.I. Self-Renewal and Toll-like Receptor Signaling Sustain Exhausted Plasmacytoid Dendritic Cells during Chronic Viral Infection. *Immunity* **2018**, *48*, 730–744.e5. [CrossRef] [PubMed]
18. Soumelis, V.; Scott, I.; Gheyas, F.; Bouhour, D.; Cozon, G.; Cotte, L.; Huang, L.; Levy, J.A.; Liu, Y.-J. Depletion of circulating natural type 1 interferon-producing cells in HIV-infected AIDS patients. *Blood* **2001**, *98*, 906–912. [CrossRef]
19. Arunachalam, P.S.; Wimmers, F.; Mok, C.K.P.; Perera, R.A.P.M.; Scott, M.; Hagan, T.; Sigal, N.; Feng, Y.; Bristow, L.; Tsang, O.T.Y.; et al. Systems biological assessment of immunity to mild versus severe COVID-19 infection in humans. *Science* **2020**, *369*, 1210–1220. [CrossRef]
20. Peruzzi, B.; Bencini, S.; Capone, M.; Mazzoni, A.; Maggi, L.; Salvati, L.; Vanni, A.; Orazzini, C.; Nozzoli, C.; Morettini, A.; et al. Quantitative and qualitative alterations of circulating myeloid cells and plasmacytoid DC in SARS-CoV-2 infection. *Immunology* **2020**, *161*, 345–353. [CrossRef]
21. Laing, A.G.; Lorenc, A.; del Barrio, I.D.M.; Das, A.; Fish, M.; Monin, L.; Muñoz-Ruiz, M.; McKenzie, D.R.; Hayday, T.S.; Francos-Quijorna, I.; et al. A dynamic COVID-19 immune signature includes associations with poor prognosis. *Nat. Med.* **2020**, *26*, 1623–1635. [CrossRef] [PubMed]
22. Liu, C.; Martins, A.J.; Lau, W.W.; Rachmaninoff, N.; Chen, J.; Imberti, L.; Mostaghimi, D.; Fink, D.L.; Burbelo, P.D.; Dobbs, K.; et al. Time-resolved systems immunology reveals a late juncture linked to fatal COVID-19. *Cell* **2021**, *184*, 1836–1857.e22. [CrossRef] [PubMed]
23. Barrat, F.J.; Su, L. A pathogenic role of plasmacytoid dendritic cells in autoimmunity and chronic viral infection. *J. Exp. Med.* **2019**, *216*, 1974–1985. [CrossRef]
24. Demoulin, S.; Herfs, M.; Delvenne, P.; Hubert, P. Tumor microenvironment converts plasmacytoid dendritic cells into immunosuppressive/tolerogenic cells: Insight into the molecular mechanisms. *J. Leukoc. Biol.* **2013**, *93*, 343–352. [CrossRef]
25. Latz, E.; Schoenemeyer, A.; Visintin, A.; Fitzgerald, K.A.; Monks, B.G.; Knetter, C.F.; Lien, E.; Nilsen, N.J.; Espevik, T.; Golenbock, D.T. TLR9 signals after translocating from the ER to CpG DNA in the lysosome. *Nat. Immunol.* **2004**, *5*, 190–198. [CrossRef]
26. Lee, H.K.; Lund, J.M.; Ramanathan, B.; Mizushima, N.; Iwasaki, A. Autophagy-dependent viral recognition by plasmacytoid dendritic cells. *Science* **2007**, *315*, 1398–1401. [CrossRef] [PubMed]
27. Beignon, A.S.; McKenna, K.; Skoberne, M.; Manches, O.; DaSilva, I.; Kavanagh, D.G.; Larsson, M.; Gorelick, R.J.; Lifson, J.D.; Bhardwaj, N. Endocytosis of HIV-1 activates plasmacytoid dendritic cells via Toll-like receptor-viral RNA interactions. *J. Clin. Investig.* **2005**, *115*, 3265–3275. [CrossRef]
28. Lund, J.; Sato, A.; Akira, S.; Medzhitov, R.; Iwasaki, A. Toll-like Receptor 9–mediated Recognition of Herpes Simplex Virus-2 by Plasmacytoid Dendritic Cells. *J. Exp. Med.* **2003**, *198*, 513–520. [CrossRef]
29. Feldman, S.B.; Ferraro, M.; Zheng, H.M.; Patel, N.; Gould-Fogerite, S.; Fitzgerald-Bocarsly, P. Viral Induction of Low Frequency Interferon-α Producing Cells. *Virology* **1994**, *204*, 1–7. [CrossRef]
30. Dreux, M.; Garaigorta, U.; Boyd, B.; Décembre, E.; Chung, J.; Whitten-Bauer, C.; Wieland, S.; Chisari, F.V. Short-range exosomal transfer of viral RNA from infected cells to plasmacytoid dendritic cells triggers innate immunity. *Cell Host Microbe* **2012**, *12*, 558–570. [CrossRef]
31. Flores, M.; Chew, C.; Tyan, K.; Huang, W.Q.; Salem, A.; Clynes, R. FcγRIIB Prevents Inflammatory Type I IFN Production from Plasmacytoid Dendritic Cells during a Viral Memory Response. *J. Immunol.* **2015**, *194*, 4240–4250. [CrossRef]
32. Båve, U.; Magnusson, M.; Eloranta, M.-L.; Perers, A.; Alm, G.V.; Rönnblom, L. FcγRIIa Is Expressed on Natural IFN-α-Producing Cells (Plasmacytoid Dendritic Cells) and Is Required for the IFN-α Production Induced by Apoptotic Cells Combined with Lupus IgG. *J. Immunol.* **2003**, *171*, 3296–3302. [CrossRef] [PubMed]
33. Barrat, F.J.; Meeker, T.; Gregorio, J.; Chan, J.H.; Uematsu, S.; Akira, S.; Chang, B.; Duramad, O.; Coffman, R.L. Nucleic acids of mammalian origin can act as endogenous ligands for Toll-like receptors and may promote systemic lupus erythematosus. *J. Exp. Med.* **2005**, *202*, 1131–1139. [CrossRef]

34. Bruni, D.; Chazal, M.; Sinigaglia, L.; Chauveau, L.; Schwartz, O.; Desprès, P.; Jouvenet, N. Viral entry route determines how human plasmacytoid dendritic cells produce type i interferons. *Sci. Signal.* **2015**, *8*, ra25. [CrossRef]
35. Décembre, E.; Assil, S.; Hillaire, M.L.B.; Dejnirattisai, W.; Mongkolsapaya, J.; Screaton, G.R.; Davidson, A.D.; Dreux, M. Sensing of Immature Particles Produced by Dengue Virus Infected Cells Induces an Antiviral Response by Plasmacytoid Dendritic Cells. *PLoS Pathog.* **2014**, *10*, e1004434. [CrossRef] [PubMed]
36. Feng, Z.; Li, Y.; McKnight, K.L.; Hensley, L.; Lanford, R.E.; Walker, C.M.; Lemon, S.M. Human pDCs preferentially sense enveloped hepatitis A virions. *J. Clin. Investig.* **2015**, *125*, 169–176. [CrossRef] [PubMed]
37. García-Nicolás, O.; Auray, G.; Sautter, C.A.; Rappe, J.C.F.; McCullough, K.C.; Ruggli, N.; Summerfield, A. Sensing of Porcine Reproductive and Respiratory Syndrome Virus-Infected Macrophages by Plasmacytoid Dendritic Cells. *Front. Microbiol.* **2016**, *7*, 771. [CrossRef] [PubMed]
38. Lepelley, A.; Louis, S.; Sourisseau, M.; Law, H.K.W.; Pothlichet, J.; Schilte, C.; Chaperot, L.; Plumas, J.; Randall, R.E.; Si-Tahar, M.; et al. Innate Sensing of HIV-Infected Cells. *PLoS Pathog.* **2011**, *7*, e1001284. [CrossRef]
39. Wieland, S.F.; Takahashi, K.; Boyd, B.; Whitten-Bauer, C.; Ngo, N.; de la Torre, J.-C.; Chisari, F.V. Human Plasmacytoid Dendritic Cells Sense Lymphocytic Choriomeningitis Virus-Infected Cells In Vitro. *J. Virol.* **2014**, *88*, 752–757. [CrossRef]
40. Yun, T.J.; Igarashi, S.; Zhao, H.; Perez, O.A.; Pereira, M.R.; Zorn, E.; Shen, Y.; Goodrum, F.; Rahman, A.; Sims, P.A.; et al. Human plasmacytoid dendritic cells mount a distinct antiviral response to virus-infected cells. *Sci. Immunol.* **2021**, *6*, eabc7302. [CrossRef]
41. Takahashi, K.; Asabe, S.; Wieland, S.; Garaigorta, U.; Gastaminza, P.; Isogawa, M.; Chisari, F.V. Plasmacytoid dendritic cells sense hepatitis C virus-infected cells, produce interferon, and inhibit infection. *Proc. Natl. Acad. Sci. USA* **2010**, *107*, 7431–7436. [CrossRef] [PubMed]
42. Assil, S.; Coléon, S.; Dong, C.; Décembre, E.; Sherry, L.; Allatif, O.; Webster, B.; Dreux, M. Plasmacytoid Dendritic Cells and Infected Cells Form an Interferogenic Synapse Required for Antiviral Responses. *Cell Host Microbe* **2019**, *25*, 730–745.e6. [CrossRef] [PubMed]
43. Gilliet, M.; Cao, W.; Liu, Y.-J. Plasmacytoid dendritic cells: Sensing nucleic acids in viral infection and autoimmune diseases. *Nat. Rev. Immunol.* **2008**, *8*, 594–606. [CrossRef]
44. Blasius, A.L.; Beutler, B. Intracellular Toll-like Receptors. *Immunity* **2010**, *32*, 305–315. [CrossRef]
45. Kawai, T.; Akira, S. Toll-like Receptors and Their Crosstalk with Other Innate Receptors in Infection and Immunity. *Immunity* **2011**, *34*, 637–650. [CrossRef] [PubMed]
46. Honda, K.; Ohba, Y.; Yanai, H.; Hegishi, H.; Mizutani, T.; Takaoka, A.; Taya, C.; Taniguchi, T. Spatiotemporal regulation of MyD88-IRF-7 signalling for robust type-I interferon induction. *Nature* **2005**, *434*, 1035–1040. [CrossRef]
47. Guiducci, C.; Ott, G.; Chan, J.H.; Damon, E.; Calacsan, C.; Matray, T.; Lee, K.D.; Coffman, R.L.; Barrat, F.J. Properties regulating the nature of the plasmacytoid dendritic cell response to Toll-like receptor 9 activation. *J. Exp. Med.* **2006**, *203*, 1999–2008. [CrossRef] [PubMed]
48. Sasai, M.; Linehan, M.M.; Iwasaki, A. Bifurcation of toll-like receptor 9 signaling by adaptor protein 3. *Science* **2010**, *329*, 1530–1534. [CrossRef]
49. Dallari, S.; Macal, M.; Loureiro, M.E.; Jo, Y.; Swanson, L.; Hesser, C.; Ghosh, P.; Zuniga, E.I. Src family kinases Fyn and Lyn are constitutively activated and mediate plasmacytoid dendritic cell responses. *Nat. Commun.* **2017**, *8*, 14830. [CrossRef]
50. Kadowaki, N.; Antonenko, S.; Liu, Y.-J. Distinct CpG DNA and Polyinosinic-Polycytidylic Acid Double-Stranded RNA, Respectively, Stimulate CD11c − Type 2 Dendritic Cell Precursors and CD11c + Dendritic Cells to Produce Type I IFN. *J. Immunol.* **2001**, *166*, 2291–2295. [CrossRef]
51. Kerkmann, M.; Rothenfusser, S.; Hornung, V.; Towarowski, A.; Wagner, M.; Sarris, A.; Giese, T.; Endres, S.; Hartmann, G. Activation with CpG-A and CpG-B Oligonucleotides Reveals Two Distinct Regulatory Pathways of Type I IFN Synthesis in Human Plasmacytoid Dendritic Cells. *J. Immunol.* **2003**, *170*, 4465–4474. [CrossRef]
52. Koblansky, A.A.; Jankovic, D.; Oh, H.; Hieny, S.; Sungnak, W.; Mathur, R.; Hayden, M.S.; Akira, S.; Sher, A.; Ghosh, S. Recognition of Profilin by Toll-like Receptor 12 Is Critical for Host Resistance to Toxoplasma gondii. *Immunity* **2013**, *38*, 119–130. [CrossRef] [PubMed]
53. Dasgupta, S.; Erturk-Hasdemir, D.; Ochoa-Reparaz, J.; Reinecker, H.C.; Kasper, D.L. Plasmacytoid dendritic cells mediate anti-inflammatory responses to a gut commensal molecule via both innate and adaptive mechanisms. *Cell Host Microbe* **2014**, *15*, 413–423. [CrossRef] [PubMed]
54. Round, J.L.; Lee, S.M.; Li, J.; Tran, G.; Jabri, B.; Chatila, T.A.; Mazmanian, S.K. The Toll-like receptor pathway establishes commensal gut colonization. *Science* **2011**, *332*, 974. [CrossRef] [PubMed]
55. Kim, T.; Pazhoor, S.; Bao, M.; Zhang, Z.; Hanabuchi, S.; Facchinetti, V.; Bover, L.; Plumas, J.; Chaperot, L.; Qin, J.; et al. Aspartate-glutamate-alanine-histidine box motif (DEAH)/RNA helicase a helicases sense microbial DNA in human plasmacytoid dendritic cells. *Proc. Natl. Acad. Sci. USA* **2010**, *107*, 15181–15186. [CrossRef]
56. Kato, H.; Sato, S.; Yoneyama, M.; Yamamoto, M.; Uematsu, S.; Matsui, K.; Tsujimura, T.; Takeda, K.; Fujita, T.; Takeuchi, O.; et al. Cell type-specific involvement of RIG-I in antiviral response. *Immunity* **2005**, *23*, 19–28. [CrossRef]
57. Kumagai, Y.; Kumar, H.; Koyama, S.; Kawai, T.; Takeuchi, O.; Akira, S. Cutting Edge: TLR-Dependent Viral Recognition Along with Type I IFN Positive Feedback Signaling Masks the Requirement of Viral Replication for IFN-α Production in Plasmacytoid Dendritic Cells. *J. Immunol.* **2009**, *182*, 3960–3964. [CrossRef] [PubMed]

58. Chen, Q.; Sun, L.; Chen, Z.J. Regulation and function of the cGAS-STING pathway of cytosolic DNA sensing. *Nat. Immunol.* **2016**, *17*, 1142–1149. [CrossRef]
59. Bode, C.; Fox, M.; Tewary, P.; Steinhagen, A.; Ellerkmann, R.K.; Klinman, D.; Baumgarten, G.; Hornung, V.; Steinhagen, F. Human plasmacytoid dendritic cells elicit a Type I Interferon response by sensing DNA via the cGAS-STING signaling pathway. *Eur. J. Immunol.* **2016**, *46*, 1615–1621. [CrossRef]
60. Deb, P.; Dai, J.; Singh, S.; Kalyoussef, E.; Fitzgerald-Bocarsly, P. Triggering of the cGAS–STING Pathway in Human Plasmacytoid Dendritic Cells Inhibits TLR9-Mediated IFN Production. *J. Immunol.* **2020**, *205*, 223–236. [CrossRef]
61. Lee, L.N.; Burke, S.; Montoya, M.; Borrow, P. Multiple Mechanisms Contribute to Impairment of Type 1 Interferon Production during Chronic Lymphocytic Choriomeningitis Virus Infection of Mice. *J. Immunol.* **2009**, *182*, 7178–7189. [CrossRef]
62. Feldman, S.; Stein, D.; Amrute, S.; Denny, T.; Garcia, Z.; Kloser, P.; Sun, Y.; Megjugorac, N.; Fitzgerald-Bocarsly, P. Decreased Interferon-α Production in HIV-Infected Patients Correlates with Numerical and Functional Deficiencies in Circulating Type 2 Dendritic Cell Precursors. *Clin. Immunol.* **2001**, *101*, 201–210. [CrossRef]
63. Tilton, J.C.; Manion, M.M.; Luskin, M.R.; Johnson, A.J.; Patamawenu, A.A.; Hallahan, C.W.; Cogliano-Shutta, N.A.; Mican, J.M.; Davey, R.T.; Kottilil, S.; et al. Human Immunodeficiency Virus Viremia Induces Plasmacytoid Dendritic Cell Activation In Vivo and Diminished Alpha Interferon Production In Vitro. *J. Virol.* **2008**, *82*, 3997–4006. [CrossRef] [PubMed]
64. Schwartz, J.A.; Clayton, K.L.; Mujib, S.; Zhang, H.; Rahman, A.K.M.N.; Liu, J.; Yue, F.Y.; Benko, E.; Kovacs, C.; Ostrowski, M.A. Tim-3 is a Marker of Plasmacytoid Dendritic Cell Dysfunction during HIV Infection and Is Associated with the Recruitment of IRF7 and p85 into Lysosomes and with the Submembrane Displacement of TLR9. *J. Immunol.* **2017**, *198*, 3181–3194. [CrossRef] [PubMed]
65. Zhang, Z.; Fu, J.; Zhao, Q.; He, Y.; Jin, L.; Zhang, H.; Yao, J.; Zhang, L.; Wang, F.-S. Differential Restoration of Myeloid and Plasmacytoid Dendritic Cells in HIV-1-Infected Children after Treatment with Highly Active Antiretroviral Therapy. *J. Immunol.* **2006**, *176*, 5644–5651. [CrossRef]
66. Kanto, T.; Inoue, M.; Miyatake, H.; Sato, A.; Sakakibara, M.; Yakushijin, T.; Oki, C.; Itose, I.; Hiramatsu, N.; Takehara, T.; et al. Reduced Numbers and Impaired Ability of Myeloid and Plasmacytoid Dendritic Cells to Polarize T Helper Cells in Chronic Hepatitis C Virus Infection. *J. Infect. Dis.* **2004**, *190*, 1919–1926. [CrossRef]
67. Ulsenheimer, A.; Gerlach, J.T.; Jung, M.-C.; Gruener, N.; Wächtler, M.; Backmund, M.; Santantonio, T.; Schraut, W.; Heeg, M.H.J.; Schirren, C.A.; et al. Plasmacytoid dendritic cells in acute and chronic hepatitis C virus infection. *Hepatology* **2005**, *41*, 643–651. [CrossRef]
68. Goutagny, N.; Vieux, C.; Decullier, E.; Ligeoix, B.; Epstein, A.; Trépo, C.; Couzigou, P.; Inchauspé, G.; Bain, C. Quantification and Functional Analysis of Plasmacytoid Dendritic Cells in Patients with Chronic Hepatitis C Virus Infection. *J. Infect. Dis.* **2004**, *189*, 1646–1655. [CrossRef] [PubMed]
69. Lai, W.K.; Curbishley, S.M.; Goddard, S.; Alabraba, E.; Shaw, J.; Youster, J.; McKeating, J.; Adams, D.H. Hepatitis C is associated with perturbation of intrahepatic myeloid and plasmacytoid dendritic cell function. *J. Hepatol.* **2007**, *47*, 338–347. [CrossRef] [PubMed]
70. Dolganiuc, A.; Chang, S.; Kodys, K.; Mandrekar, P.; Bakis, G.; Cormier, M.; Szabo, G. Hepatitis C Virus (HCV) Core Protein-Induced, Monocyte-Mediated Mechanisms of Reduced IFN-α and Plasmacytoid Dendritic Cell Loss in Chronic HCV Infection. *J. Immunol.* **2006**, *177*, 6758–6768. [CrossRef]
71. van der Molen, R.G.; Sprengers, D.; Binda, R.S.; de Jong, E.C.; Niesters, H.G.M.; Kusters, J.G.; Kwekkeboom, J.; Janssen, H.L.A. Functional impairment of myeloid and plasmacytoid dendritic cells of patients with chronic hepatitis B. *Hepatology* **2004**, *40*, 738–746. [CrossRef]
72. Xie, Q.; Shen, H.-C.; Jia, N.-N.; Wang, H.; Lin, L.-Y.; An, B.-Y.; Gui, H.-L.; Guo, S.-M.; Cai, W.; Yu, H.; et al. Patients with chronic hepatitis B infection display deficiency of plasmacytoid dendritic cells with reduced expression of TLR9. *Microbes Infect.* **2009**, *11*, 515–523. [CrossRef] [PubMed]
73. Duan, X.-Z.; Wang, M.; Li, H.-W.; Zhuang, H.; Xu, D.; Wang, F.-S. Decreased Frequency and Function of Circulating Plasmocytoid Dendritic Cells (pDC) in Hepatitis B Virus Infected Humans. *J. Clin. Immunol.* **2004**, *24*, 637–646. [CrossRef]
74. Malleret, B.; Manéglier, B.; Karlsson, I.; Lebon, P.; Nascimbeni, M.; Perié, L.; Brochard, P.; Delache, B.; Calvo, J.; Andrieu, T.; et al. Primary infection with simian immunodeficiency virus: Plasmacytoid dendritic cell homing to lymph nodes, type I interferon, and immune suppression. *Blood* **2008**, *112*, 4598–4608. [CrossRef]
75. Bruel, T.; Dupuy, S.; Démoulins, T.; Rogez-Kreuz, C.; Dutrieux, J.; Corneau, A.; Cosma, A.; Cheynier, R.; Dereuddre-Bosquet, N.; Le Grand, R.; et al. Plasmacytoid Dendritic Cell Dynamics Tune Interferon-Alfa Production in SIV-Infected Cynomolgus Macaques. *PLoS Pathog.* **2014**, *10*, e1003915. [CrossRef]
76. Stacey, A.R.; Norris, P.J.; Qin, L.; Haygreen, E.A.; Taylor, E.; Heitman, J.; Lebedeva, M.; DeCamp, A.; Li, D.; Grove, D.; et al. Induction of a Striking Systemic Cytokine Cascade prior to Peak Viremia in Acute Human Immunodeficiency Virus Type 1 Infection, in Contrast to More Modest and Delayed Responses in Acute Hepatitis B and C Virus Infections. *J. Virol.* **2009**, *83*, 3719–3733. [CrossRef] [PubMed]
77. Siegal, F.P.; Fitzgerald-Bocarsly, P.; Holland, B.K.; Shodell, M. Interferon-α generation and immune reconstitution during antiretroviral therapy for human immunodeficiency virus infection. *AIDS* **2001**, *15*, 1603–1612. [CrossRef] [PubMed]
78. Lopez, C.; Fitzgerald, P.A.; Siegal, F.P. Severe Acquired Immune Deficiency Syndrome in Male Homosexuals: Diminished Capacity to Make Interferon-a in Vitro Associated with Severe Opportunistic Infections. *J. Infect. Dis.* **1983**, *148*, 962–966. [CrossRef]

79. Siegal, F.P.; Lopez, C.; Fitzgerald, P.A.; Shah, K.; Baron, P.; Leiderman, I.Z.; Imperato, D.; Landesman, S. Opportunistic infections in acquired immune deficiency syndrome result from synergistic defects of both the natural and adaptive components of cellular immunity. *J. Clin. Investig.* **1986**, *78*, 115–123. [CrossRef] [PubMed]
80. Cisse, B.; Caton, M.L.; Lehner, M.; Maeda, T.; Scheu, S.; Locksley, R.; Holmberg, D.; Zweier, C.; den Hollander, N.S.; Kant, S.G.; et al. Transcription factor E2-2 is an essential and specific regulator of plasmacytoid dendritic cell development. *Cell* **2008**, *135*, 37–48. [CrossRef]
81. Ghosh, H.S.; Cisse, B.; Bunin, A.; Lewis, K.L.; Reizis, B. Continuous Expression of the Transcription Factor E2-2 Maintains the Cell Fate of Mature Plasmacytoid Dendritic Cells. *Immunity* **2010**, *33*, 905–916. [CrossRef] [PubMed]
82. Mueller, S.N.; Ahmed, R. High antigen levels are the cause of T cell exhaustion during chronic viral infection. *Proc. Natl. Acad. Sci. USA* **2009**, *106*, 8623–8628. [CrossRef] [PubMed]
83. Dzionek, A.; Sohma, Y.; Nagafune, J.; Cella, M.; Colonna, M.; Facchetti, F.; Günther, G.; Johnston, I.; Lanzavecchia, A.; Nagasaka, T.; et al. BDCA-2, a Novel Plasmacytoid Dendritic Cell–specific Type II C-type Lectin, Mediates Antigen Capture and Is a Potent Inhibitor of Interferon α/β Induction. *J. Exp. Med.* **2001**, *194*, 1823–1834. [CrossRef] [PubMed]
84. Blasius, A.L.; Cella, M.; Maldonado, J.; Takai, T.; Colonna, M. Siglec-H is an IPC-specific receptor that modulates type I IFN secretion through DAP12. *Blood* **2006**, *107*, 2474–2476. [CrossRef]
85. Cao, W.; Rosen, D.B.; Ito, T.; Bover, L.; Bao, M.; Watanabe, G.; Yao, Z.; Zhang, L.; Lanier, L.L.; Liu, Y.-J. Plasmacytoid dendritic cell–specific receptor ILT7–FcεRIγ inhibits Toll-like receptor–induced interferon production. *J. Exp. Med.* **2006**, *203*, 1399–1405. [CrossRef]
86. Novak, N.; Allam, J.-P.; Hagemann, T.; Jenneck, C.; Laffer, S.; Valenta, R.; Kochan, J.; Bieber, T. Characterization of FcεRI-bearing CD123+ blood dendritic cell antigen-2+ plasmacytoid dendritic cells in atopic dermatitis. *J. Allergy Clin. Immunol.* **2004**, *114*, 364–370. [CrossRef]
87. Meyer-Wentrup, F.; Benitez-Ribas, D.; Tacken, P.J.; Punt, C.J.A.; Figdor, C.G.; de Vries, I.J.M.; Adema, G.J. Targeting DCIR on human plasmacytoid dendritic cells results in antigen presentation and inhibits IFN-α production. *Blood* **2008**, *111*, 4245–4253. [CrossRef]
88. Macal, M.; Tam, M.A.; Hesser, C.; Di Domizio, J.; Leger, P.; Gilliet, M.; Zuniga, E.I. CD28 Deficiency Enhances Type I IFN Production by Murine Plasmacytoid Dendritic Cells. *J. Immunol.* **2016**, *196*, 1900–1909. [CrossRef]
89. Bi, E.; Li, R.; Bover, L.C.; Li, H.; Su, P.; Ma, X.; Huang, C.; Wang, Q.; Liu, L.; Yang, M.; et al. E-cadherin expression on multiple myeloma cells activates tumor-promoting properties in plasmacytoid DCs. *J. Clin. Investig.* **2018**, *128*, 4821–4831. [CrossRef]
90. Fuchs, A.; Cella, M.; Kondo, T.; Colonna, M. Paradoxic inhibition of human natural interferon-producing cells by the activating receptor NKp44. *Blood* **2005**, *106*, 2076–2082. [CrossRef]
91. Puttur, F.; Arnold-Schrauf, C.; Lahl, K.; Solmaz, G.; Lindenberg, M.; Mayer, C.T.; Gohmert, M.; Swallow, M.; van Helt, C.; Schmitt, H.; et al. Absence of Siglec-H in MCMV Infection Elevates Interferon Alpha Production but Does Not Enhance Viral Clearance. *PLoS Pathog.* **2013**, *9*, e1003648. [CrossRef]
92. Thomson, N.C.; Chaudhuri, R. Omalizumab: Clinical use for the management of asthma. *Clin. Med. Insights Circ. Respir. Pulm. Med.* **2011**, *6*, 27–40. [CrossRef]
93. Gill, M.A.; Liu, A.H.; Calatroni, A.; Krouse, R.Z.; Shao, B.; Schiltz, A.; Gern, J.E.; Togias, A.; Busse, W.W. Enhanced plasmacytoid dendritic cell antiviral responses after omalizumab. *J. Allergy Clin. Immunol.* **2018**, *141*, 1735–1743.e9. [CrossRef] [PubMed]
94. Esquivel, A.; Busse, W.W.; Calatroni, A.; Togias, A.G.; Grindle, K.G.; Bochkov, Y.A.; Gruchalla, R.S.; Kattan, M.; Kercsmar, C.M.; Hershey, G.K.; et al. Effects of Omalizumab on Rhinovirus Infections, Illnesses, and exacerbations of asthma. *Am. J. Respir. Crit. Care Med.* **2017**, *196*, 985–992. [CrossRef] [PubMed]
95. Bego, M.G.; Côté, É.; Aschman, N.; Mercier, J.; Weissenhorn, W.; Cohen, É.A. Vpu Exploits the Cross-Talk between BST2 and the ILT7 Receptor to Suppress Anti-HIV-1 Responses by Plasmacytoid Dendritic Cells. *PLoS Pathog.* **2015**, *11*, e1005024. [CrossRef]
96. Labidi-Galy, S.I.; Sisirak, V.; Meeus, P.; Gobert, M.; Treilleux, I.; Bajard, A.; Combes, J.-D.; Faget, J.; Mithieux, F.; Cassignol, A.; et al. Quantitative and functional alterations of plasmacytoid dendritic cells contribute to immune tolerance in ovarian cancer. *Cancer Res.* **2011**, *71*, 5423–5434. [CrossRef]
97. Gary-Gouy, H.; Lebon, P.; Dalloul, A.H. Type I Interferon Production by Plasmacytoid Dendritic Cells and Monocytes Is Triggered by Viruses, but the Level of Production Is Controlled by Distinct Cytokines. *J. Interf. Cytokine Res.* **2002**, *22*, 653–659. [CrossRef] [PubMed]
98. Bekeredjian-Ding, I.; Schäfer, M.; Hartmann, E.; Pries, R.; Parcina, M.; Schneider, P.; Giese, T.; Endres, S.; Wollenberg, B.; Hartmann, G. Tumour-derived prostaglandin E 2 and transforming growth factor-β synergize to inhibit plasmacytoid dendritic cell-derived interferon-α. *Immunology* **2009**, *128*, 439–450. [CrossRef]
99. Li, L.; Liu, S.; Zhang, T.; Pan, W.; Yang, X.; Cao, X. Splenic Stromal Microenvironment Negatively Regulates Virus-Activated Plasmacytoid Dendritic Cells through TGF-β. *J. Immunol.* **2008**, *180*, 2951–2956. [CrossRef]
100. Son, Y.; Ito, T.; Ozaki, Y.; Tanijiri, T.; Yokoi, T.; Nakamura, K.; Takebayashi, M.; Amakawa, R.; Fukuhara, S. Prostaglandin E2 is a negative regulator on human plasmacytoid dendritic cells. *Immunology* **2006**, *119*, 36–42. [CrossRef]
101. Seillet, C.; Laffont, S.; Trémollières, F.; Rouquié, N.; Ribot, C.; Arnal, J.F.; Douin-Echinard, V.; Gourdy, P.; Guéry, J.C. The TLR-mediated response of plasmacytoid dendritic cells is positively regulated by estradiol in vivo through cell-intrinsic estrogen receptor α signaling. *Blood* **2012**, *119*, 454–464. [CrossRef] [PubMed]

102. Asselin-Paturel, C.; Brizard, G.; Chemin, K.; Boonstra, A.; O'Garra, A.; Vicari, A.; Trinchieri, G. Type I interferon dependence of plasmacytoid dendritic cell activation and migration. *J. Exp. Med.* **2005**, *201*, 1157–1167. [CrossRef] [PubMed]
103. Wimmers, F.; Subedi, N.; van Buuringen, N.; Heister, D.; Vivié, J.; Beeren-Reinieren, I.; Woestenenk, R.; Dolstra, H.; Piruska, A.; Jacobs, J.F.M.; et al. Single-cell analysis reveals that stochasticity and paracrine signaling control interferon-alpha production by plasmacytoid dendritic cells. *Nat. Commun.* **2018**, *9*, 3317. [CrossRef] [PubMed]
104. Kim, S.; Kaiser, V.; Beier, E.; Bechheim, M.; Guenthner-Biller, M.; Ablasser, A.; Berger, M.; Endres, S.; Hartmann, G.; Hornung, V. Self-priming determines high type I IFN production by plasmacytoid dendritic cells. *Eur. J. Immunol.* **2014**, *44*, 807–818. [CrossRef] [PubMed]
105. Swiecki, M.; Wang, Y.; Vermi, W.; Gilfillan, S.; Schreiber, R.D.; Colonna, M. Type I interferon negatively controls plasmacytoid dendritic cell numbers in vivo. *J. Exp. Med.* **2011**, *208*, 2367–2374. [CrossRef]
106. Kadowaki, N.; Antonenko, S.; Lau, J.Y.-N.; Liu, Y.-J. Natural Interferon α/β–Producing Cells Link Innate and Adaptive Immunity. *J. Exp. Med.* **2000**, *192*, 219–226. [CrossRef]
107. Schroeder, J.T.; Bieneman, A.P.; Xiao, H.; Chichester, K.L.; Vasagar, K.; Saini, S.; Liu, M.C. TLR9- and FcεRI-Mediated Responses Oppose One Another in Plasmacytoid Dendritic Cells by Down-Regulating Receptor Expression. *J. Immunol.* **2005**, *175*, 5724–5731. [CrossRef]
108. Palucka, A.K.; Blanck, J.P.; Bennett, L.; Pascual, V.; Banchereau, J. Cross-regulation of TNF and IFN-α in autoimmune diseases. *Proc. Natl. Acad. Sci. USA* **2005**, *102*, 3372–3377. [CrossRef] [PubMed]
109. Psarras, A.; Antanaviciute, A.; Alase, A.; Carr, I.; Wittmann, M.; Emery, P.; Tsokos, G.C.; Vital, E.M. TNF-α Regulates Human Plasmacytoid Dendritic Cells by Suppressing IFN-α Production and Enhancing T Cell Activation. *J. Immunol.* **2021**, *206*, 785–796. [CrossRef] [PubMed]
110. Meier, A.; Chang, J.J.; Chan, E.S.; Pollard, R.B.; Sidhu, H.K.; Kulkarni, S.; Wen, T.F.; Lindsay, R.J.; Orellana, L.; Mildvan, D.; et al. Sex differences in the Toll-like receptor-mediated response of plasmacytoid dendritic cells to HIV-1. *Nat. Med.* **2009**, *15*, 955–959. [CrossRef] [PubMed]
111. Laffont, S.; Rouquié, N.; Azar, P.; Seillet, C.; Plumas, J.; Aspord, C.; Guéry, J.-C. X-Chromosome Complement and Estrogen Receptor Signaling Independently Contribute to the Enhanced TLR7-Mediated IFN-α Production of Plasmacytoid Dendritic Cells from Women. *J. Immunol.* **2014**, *193*, 5444–5452. [CrossRef]
112. Berghöfer, B.; Frommer, T.; Haley, G.; Fink, L.; Bein, G.; Hackstein, H. TLR7 Ligands Induce Higher IFN-α Production in Females. *J. Immunol.* **2006**, *177*, 2088–2096. [CrossRef]
113. Griesbeck, M.; Ziegler, S.; Laffont, S.; Smith, N.; Chauveau, L.; Tomezsko, P.; Sharei, A.; Kourjian, G.; Porichis, F.; Hart, M.; et al. Sex Differences in Plasmacytoid Dendritic Cell Levels of IRF5 Drive Higher IFN- Production in Women. *J. Immunol.* **2015**, *195*, 5327–5336. [CrossRef]
114. Souyris, M.; Cenac, C.; Azar, P.; Daviaud, D.; Canivet, A.; Grunenwald, S.; Pienkowski, C.; Chaumeil, J.; Mejía, J.E.; Guéry, J.C. TLR7 escapes X chromosome inactivation in immune cells. *Sci. Immunol.* **2018**, *3*, eaap8855. [CrossRef] [PubMed]
115. Webb, K.; Peckham, H.; Radziszewska, A.; Menon, M.; Oliveri, P.; Simpson, F.; Deakin, C.T.; Lee, S.; Ciurtin, C.; Butler, G.; et al. Sex and pubertal differences in the type 1 interferon pathway associate with both X chromosome number and serum sex hormone concentration. *Front. Immunol.* **2019**, *10*, 3167. [CrossRef]
116. Bajwa, G.; DeBerardinis, R.J.; Shao, B.; Hall, B.; Farrar, J.D.; Gill, M.A. Cutting Edge: Critical Role of Glycolysis in Human Plasmacytoid Dendritic Cell Antiviral Responses. *J. Immunol.* **2016**, *196*, 2004–2009. [CrossRef] [PubMed]
117. Wu, D.; Sanin, D.E.; Everts, B.; Chen, Q.; Qiu, J.; Buck, M.D.; Patterson, A.; Smith, A.M.; Chang, C.-H.; Liu, Z.; et al. Type 1 Interferons Induce Changes in Core Metabolism that Are Critical for Immune Function. *Immunity* **2016**, *44*, 1325–1336. [CrossRef]
118. Hurley, H.J.; Dewald, H.; Rothkopf, Z.S.; Singh, S.; Jenkins, F.; Deb, P.; De, S.; Barnes, B.J.; Fitzgerald-Bocarsly, P. Frontline Science: AMPK regulates metabolic reprogramming necessary for interferon production in human plasmacytoid dendritic cells. *J. Leukoc. Biol.* **2021**, *109*, 299–308. [CrossRef]
119. Cao, W.; Manicassamy, S.; Tang, H.; Kasturi, S.P.; Pirani, A.; Murthy, N.; Pulendran, B. Toll-like receptor-mediated induction of type I interferon in plasmacytoid dendritic cells requires the rapamycin-sensitive PI(3)K-mTOR-p70S6K pathway. *Nat. Immunol.* **2008**, *9*, 1157–1164. [CrossRef]
120. Raychaudhuri, D.; Bhattacharya, R.; Sinha, B.P.; Liu, C.S.C.; Ghosh, A.R.; Rahaman, O.; Bandopadhyay, P.; Sarif, J.; D'Rozario, R.; Paul, S.; et al. Lactate Induces Pro-tumor Reprogramming in Intratumoral Plasmacytoid Dendritic Cells. *Front. Immunol.* **2019**, *10*, 1878. [CrossRef] [PubMed]
121. Chehimi, J.; Campbell, D.E.; Azzoni, L.; Bacheller, D.; Papasavvas, E.; Jerandi, G.; Mounzer, K.; Kostman, J.; Trinchieri, G.; Montaner, L.J. Persistent Decreases in Blood Plasmacytoid Dendritic Cell Number and Function Despite Effective Highly Active Antiretroviral Therapy and Increased Blood Myeloid Dendritic Cells in HIV-Infected Individuals. *J. Immunol.* **2002**, *168*, 4796–4801. [CrossRef]
122. Meyers, J.H.; Justement, J.S.; Hallahan, C.W.; Blair, E.T.; Sun, Y.A.; O'Shea, M.A.; Roby, G.; Kottilil, S.; Moir, S.; Kovacs, C.M.; et al. Impact of HIV on Cell Survival and Antiviral Activity of Plasmacytoid Dendritic Cells. *PLoS ONE* **2007**, *2*, e458. [CrossRef] [PubMed]
123. Anthony, D.D.; Yonkers, N.L.; Post, A.B.; Asaad, R.; Heinzel, F.P.; Lederman, M.M.; Lehmann, P.V.; Valdez, H. Selective Impairments in Dendritic Cell-Associated Function Distinguish Hepatitis C Virus and HIV Infection. *J. Immunol.* **2004**, *172*, 4907–4916. [CrossRef] [PubMed]

124. Wertheimer, A.M.; Bakke, A.; Rosen, H.R. Direct enumeration and functional assessment of circulating dendritic cells in patients with liver disease. *Hepatology* **2004**, *40*, 335–345. [CrossRef]
125. Szabo, G.; Dolganiuc, A. Subversion of plasmacytoid and myeloid dendritic cell functions in chronic HCV infection. *Immunobiology* **2005**, *210*, 237–247. [CrossRef]
126. Brown, K.N.; Wijewardana, V.; Liu, X.; Barratt-Boyes, S.M. Rapid Influx and Death of Plasmacytoid Dendritic Cells in Lymph Nodes Mediate Depletion in Acute Simian Immunodeficiency Virus Infection. *PLoS Pathog.* **2009**, *5*, e1000413. [CrossRef]
127. Li, H.; Evans, T.I.; Gillis, J.; Connole, M.; Reeves, R.K. Bone Marrow–Imprinted Gut-Homing of Plasmacytoid Dendritic Cells (pDCs) in Acute Simian Immunodeficiency Virus Infection Results in Massive Accumulation of Hyperfunctional CD4 + pDCs in the Mucosae. *J. Infect. Dis.* **2015**, *211*, 1717–1725. [CrossRef]
128. Zhan, Y.; Chow, K.V.; Soo, P.; Xu, Z.; Brady, J.L.; Lawlor, K.E.; Masters, S.L.; O'keeffe, M.; Shortman, K.; Zhang, J.-G.; et al. Plasmacytoid dendritic cells are short-lived: Reappraising the influence of migration, genetic factors and activation on estimation of lifespan. *Sci. Rep.* **2016**, *6*, 25060. [CrossRef] [PubMed]
129. Liu, K.; Waskow, C.; Liu, X.; Yao, K.; Hoh, J.; Nussenzweig, M. Origin of dendritic cells in peripheral lymphoid organs of mice. *Nat. Immunol.* **2007**, *8*, 578–583. [CrossRef]
130. Dress, R.J.; Dutertre, C.-A.; Giladi, A.; Schlitzer, A.; Low, I.; Shadan, N.B.; Tay, A.; Lum, J.; Kairi, M.F.B.M.; Hwang, Y.Y.; et al. Plasmacytoid dendritic cells develop from Ly6D+ lymphoid progenitors distinct from the myeloid lineage. *Nat. Immunol.* **2019**, *20*, 852–864. [CrossRef]
131. Lehmann, C.; Lafferty, M.; Garzino-Demo, A.; Jung, N.; Hartmann, P.; Fätkenheuer, G.; Wolf, J.S.; van Lunzen, J.; Romerio, F. Plasmacytoid Dendritic Cells Accumulate and Secrete Interferon Alpha in Lymph Nodes of HIV-1 Patients. *PLoS ONE* **2010**, *5*, e11110. [CrossRef] [PubMed]
132. George, M.M.; Bhangoo, A. Human immune deficiency virus (HIV) infection and the hypothalamic pituitary adrenal axis. *Rev. Endocr. Metab. Disord.* **2013**, *14*, 105–112. [CrossRef] [PubMed]
133. Miller, A.H.; Spencer, R.L.; Pearce, B.D.; Pisell, T.L.; Tanapat, P.; Leung, J.J.; Dhabhar, F.S.; McEwen, B.S.; Biron, C.A. Effects of viral infection on corticosterone secretion and glucocorticoid receptor binding in immune tissues. *Psychoneuroendocrinology* **1997**, *22*, 455–474. [CrossRef]
134. Boor, P.P.C.; Metselaar, H.J.; Mancham, S.; Tilanus, H.W.; Kusters, J.G.; Kwekkeboom, J. Prednisolone Suppresses the Function and Promotes Apoptosis of Plasmacytoid Dendritic Cells. *Am. J. Transplant.* **2006**, *6*, 2332–2341. [CrossRef] [PubMed]
135. Guiducci, C.; Gong, M.; Xu, Z.; Gill, M.; Chaussabel, D.; Meeker, T.; Chan, J.H.; Wright, T.; Punaro, M.; Bolland, S.; et al. TLR recognition of self nucleic acids hampers glucocorticoid activity in lupus. *Nature* **2010**, *465*, 937–941. [CrossRef]
136. Lepelletier, Y.; Zollinger, R.; Ghirelli, C.; Raynaud, F.; Hadj-Slimane, R.; Cappuccio, A.; Hermine, O.; Liu, Y.-J.; Soumelis, V. Toll-like receptor control of glucocorticoid-induced apoptosis in human plasmacytoid predendritic cells (pDCs). *Blood* **2010**, *116*, 3389–3397. [CrossRef]
137. Schlitzer, A.; Heiseke, A.F.; Einwächter, H.; Reindl, W.; Schiemann, M.; Manta, C.-P.; See, P.; Niess, J.-H.; Suter, T.; Ginhoux, F.; et al. Tissue-specific differentiation of a circulating CCR9− pDC-like common dendritic cell precursor. *Blood* **2012**, *119*, 6063–6071. [CrossRef]
138. Schlitzer, A.; Loschko, J.; Mair, K.; Vogelmann, R.; Henkel, L.; Einwächter, H.; Schiemann, M.; Niess, J.-H.; Reindl, W.; Krug, A. Identification of CCR9− murine plasmacytoid DC precursors with plasticity to differentiate into conventional DCs. *Blood* **2011**, *117*, 6562–6570. [CrossRef]
139. Zuniga, E.I.; McGavern, D.B.; Pruneda-Paz, J.L.; Teng, C.; Oldstone, M.B.A. Bone marrow plasmacytoid dendritic cells can differentiate into myeloid dendritic cells upon virus infection. *Nat. Immunol.* **2004**, *5*, 1227–1234. [CrossRef]
140. Liou, L.-Y.; Blasius, A.L.; Welch, M.J.; Colonna, M.; Oldstone, M.B.A.; Zuniga, E.I. In vivo conversion of BM plasmacytoid DC into CD11b+ conventional DC during virus infection. *Eur. J. Immunol.* **2008**, *38*, 3388–3394. [CrossRef]
141. Toma-Hirano, M.; Namiki, S.; Miyatake, S.; Arai, K.; Kamogawa-Schifter, Y. Type I interferon regulates pDC maturation and Ly49Q expression. *Eur. J. Immunol.* **2007**, *37*, 2707–2714. [CrossRef] [PubMed]
142. Alcántara-Hernández, M.; Leylek, R.; Wagar, L.E.; Engleman, E.G.; Keler, T.; Marinkovich, M.P.; Davis, M.M.; Nolan, G.P.; Idoyaga, J. High-Dimensional Phenotypic Mapping of Human Dendritic Cells Reveals Interindividual Variation and Tissue Specialization. *Immunity* **2017**, *47*, 1037–1050.e6. [CrossRef] [PubMed]
143. Leylek, R.; Alcántara-Hernández, M.; Lanzar, Z.; Lüdtke, A.; Perez, O.A.; Reizis, B.; Idoyaga, J. Integrated Cross-Species Analysis Identifies a Conserved Transitional Dendritic Cell Population. *Cell Rep.* **2019**, *29*, 3736–3750.e8. [CrossRef] [PubMed]
144. Leylek, R.; Alcántara-Hernández, M.; Granja, J.M.; Chavez, M.; Perez, K.; Diaz, O.R.; Li, R.; Satpathy, A.T.; Chang, H.Y.; Idoyaga, J. Chromatin Landscape Underpinning Human Dendritic Cell Heterogeneity. *Cell Rep.* **2020**, *32*, 108180. [CrossRef]
145. Villani, A.-C.; Satija, R.; Reynolds, G.; Sarkizova, S.; Shekhar, K.; Fletcher, J.; Griesbeck, M.; Butler, A.; Zheng, S.; Lazo, S.; et al. Single-cell RNA-seq reveals new types of human blood dendritic cells, monocytes, and progenitors. *Science* **2017**, *356*, eaah4573. [CrossRef]
146. Bar-On, L.; Birnberg, T.; Lewis, K.L.; Edelson, B.T.; Bruder, D.; Hildner, K.; Buer, J.; Murphy, K.M.; Reizis, B.; Jung, S. CX_3CR1^+ $CD8\alpha^+$ dendritic cells are a steady-state population related to plasmacytoid dendritic cells. *Proc. Natl. Acad. Sci. USA* **2010**, *107*, 14745–14750. [CrossRef]
147. Lau, C.M.; Nish, S.A.; Yogev, N.; Waisman, A.; Reiner, S.L.; Reizis, B. Leukemia-associated activating mutation of Flt3 expands dendritic cells and alters T cell responses. *J. Exp. Med.* **2016**, *213*, 415–431. [CrossRef]

148. Cervantes-Barragan, L.; Züst, R.; Weber, F.; Spiegel, M.; Lang, K.S.; Akira, S.; Thiel, V.; Ludewig, B. Control of coronavirus infection through plasmacytoid dendritic-cell- derived type I interferon. *Blood* **2007**, *109*, 1131–1137. [CrossRef]
149. Bastard, P.; Rosen, L.B.; Zhang, Q.; Michailidis, E.; Hoffmann, H.H.; Zhang, Y.; Dorgham, K.; Philippot, Q.; Rosain, J.; Béziat, V.; et al. Autoantibodies against type I IFNs in patients with life-threatening COVID-19. *Science* **2020**, *370*, eabd4585. [CrossRef]
150. Zhang, Q.; Liu, Z.; Moncada-Velez, M.; Chen, J.; Ogishi, M.; Bigio, B.; Yang, R.; Arias, A.A.; Zhou, Q.; Han, J.E.; et al. Inborn errors of type I IFN immunity in patients with life-threatening COVID-19. *Science* **2020**, *370*, eabd4570. [CrossRef]
151. Van Der Made, C.I.; Simons, A.; Schuurs-Hoeijmakers, J.; Van Den Heuvel, G.; Mantere, T.; Kersten, S.; Van Deuren, R.C.; Steehouwer, M.; Van Reijmersdal, S.V.; Jaeger, M.; et al. Presence of Genetic Variants among Young Men with Severe COVID-19. *JAMA—J. Am. Med. Assoc.* **2020**, *324*, 663–673. [CrossRef]
152. Bastard, P.; Gervais, A.; Le Voyer, T.; Rosain, J.; Philippot, Q.; Manry, J.; Michailidis, E.; Hoffmann, H.-H.; Eto, S.; Garcia-Prat, M.; et al. Autoantibodies neutralizing type I IFNs are present in ~4% of uninfected individuals over 70 years old and account for ~20% of COVID-19 deaths. *Sci. Immunol.* **2021**, *6*, eabl4340. [CrossRef]
153. Asano, T.; Boisson, B.; Onodi, F.; Matuozzo, D.; Moncada-Velez, M.; Renkilaraj, M.R.L.M.; Zhang, P.; Meertens, L.; Bolze, A.; Materna, M.; et al. X-linked recessive TLR7 deficiency in ~1% of men under 60 years old with life-threatening COVID-19. *Sci. Immunol.* **2021**, *6*, eabl4348. [CrossRef]
154. Klein, S.L.; Dhakal, S.; Ursin, R.L.; Deshpande, S.; Sandberg, K.; Mauvais-Jarvis, F. Biological sex impacts COVID-19 outcomes. *PLOS Pathog.* **2020**, *16*, e1008570. [CrossRef] [PubMed]
155. Consortium, W.S.T. Repurposed Antiviral Drugs for Covid-19—Interim WHO Solidarity Trial Results. *N. Engl. J. Med.* **2020**, *384*, 497–511. [CrossRef]
156. Monk, P.D.; Marsden, R.J.; Tear, V.J.; Brookes, J.; Batten, T.N.; Mankowski, M.; Gabbay, F.J.; Davies, D.E.; Holgate, S.T.; Ho, L.P.; et al. Safety and efficacy of inhaled nebulised interferon beta-1a (SNG001) for treatment of SARS-CoV-2 infection: A randomised, double-blind, placebo-controlled, phase 2 trial. *Lancet Respir. Med.* **2021**, *9*, 196–206. [CrossRef]
157. Hung, I.F.N.; Lung, K.C.; Tso, E.Y.K.; Liu, R.; Chung, T.W.H.; Chu, M.Y.; Ng, Y.Y.; Lo, J.; Chan, J.; Tam, A.R.; et al. Triple combination of interferon beta-1b, lopinavir–ritonavir, and ribavirin in the treatment of patients admitted to hospital with COVID-19: An open-label, randomised, phase 2 trial. *Lancet* **2020**, *395*, 1695–1704. [CrossRef]
158. Wang, N.; Zhan, Y.; Zhu, L.; Hou, Z.; Liu, F.; Song, P.; Qiu, F.; Wang, X.; Zou, X.; Wan, D.; et al. Retrospective Multicenter Cohort Study Shows Early Interferon Therapy Is Associated with Favorable Clinical Responses in COVID-19 Patients. *Cell Host Microbe* **2020**, *28*, 455–464.e2. [CrossRef] [PubMed]
159. Hadjadj, J.; Yatim, N.; Barnabei, L.; Corneau, A.; Boussier, J.; Smith, N.; Péré, H.; Charbit, B.; Bondet, V.; Chenevier-Gobeaux, C.; et al. Impaired type I interferon activity and inflammatory responses in severe COVID-19 patients. *Science* **2020**, *369*, 718–724. [CrossRef]
160. Lucas, C.; Wong, P.; Klein, J.; Castro, T.B.R.; Silva, J.; Sundaram, M.; Ellingson, M.K.; Mao, T.; Oh, J.E.; Israelow, B.; et al. Longitudinal analyses reveal immunological misfiring in severe COVID-19. *Nature* **2020**, *584*, 463–469. [CrossRef]
161. Galani, I.E.; Rovina, N.; Lampropoulou, V.; Triantafyllia, V.; Manioudaki, M.; Pavlos, E.; Koukaki, E.; Fragkou, P.C.; Panou, V.; Rapti, V.; et al. Untuned antiviral immunity in COVID-19 revealed by temporal type I/III interferon patterns and flu comparison. *Nat. Immunol.* **2021**, *22*, 32–40. [CrossRef] [PubMed]
162. Onodi, F.; Bonnet-Madin, L.; Meertens, L.; Karpf, L.; Poirot, J.; Zhang, S.-Y.; Picard, C.; Puel, A.; Jouanguy, E.; Zhang, Q.; et al. SARS-CoV-2 induces human plasmacytoid predendritic cell diversification via UNC93B and IRAK4. *J. Exp. Med.* **2021**, *218*, e20201387. [CrossRef] [PubMed]
163. Moreno-Eutimio, M.A.; López-Macías, C.; Pastelin-Palacios, R. Bioinformatic analysis and identification of single-stranded RNA sequences recognized by TLR7/8 in the SARS-CoV-2, SARS-CoV, and MERS-CoV genomes. *Microbes Infect.* **2020**, *22*, 226–229. [CrossRef] [PubMed]
164. Picard, C.; Puel, A.; Bonnet, M.; Ku, C.-L.; Bustamante, J.; Yang, K.; Soudais, C.; Dupuis, S.; Feinberg, J.; Fieschi, C.; et al. Pyogenic Bacterial Infections in Humans with IRAK-4 Deficiency. *Science.* **2003**, *299*, 2076–2079. [CrossRef]
165. Tabeta, K.; Hoebe, K.; Janssen, E.M.; Du, X.; Georgel, P.; Crozat, K.; Mudd, S.; Mann, N.; Sovath, S.; Goode, J.; et al. The Unc93b1 mutation 3d disrupts exogenous antigen presentation and signaling via Toll-like receptors 3, 7 and 9. *Nat. Immunol.* **2006**, *7*, 156–164. [CrossRef]
166. Fallerini, C.; Daga, S.; Mantovani, S.; Benetti, E.; Picchiotti, N.; Francisci, D.; Paciosi, F.; Schiaroli, E.; Baldassarri, M.; Fava, F.; et al. Association of toll-like receptor 7 variants with life-threatening COVID-19 disease in males: Findings from a nested case-control study. *eLife* **2021**, *10*. [CrossRef] [PubMed]
167. Huang, C.; Wang, Y.; Li, X.; Ren, L.; Zhao, J.; Hu, Y.; Zhang, L.; Fan, G.; Xu, J.; Gu, X.; et al. Clinical features of patients infected with 2019 novel coronavirus in Wuhan, China. *Lancet* **2020**, *395*, 497–506. [CrossRef]
168. Pauken, K.E.; Wherry, E.J. Overcoming T cell exhaustion in infection and cancer. *Trends Immunol.* **2015**, *36*, 265–276. [CrossRef]

Article

Prevention of CD8 T Cell Deletion during Chronic Viral Infection

David G. Brooks [1,2,3,*], Antoinette Tishon [1], Michael B. A. Oldstone [1] and Dorian B. McGavern [4,*]

1 Viral Immunobiology Laboratory, Department of Immunology and Microbial Science, The Scripps Research Institute, 10550 North Torrey Pines Rd., La Jolla, CA 92037, USA; atishon@scripps.edu (A.T.); mbaobo@scripps.edu (M.B.A.O.)
2 Princess Margaret Cancer Center, University Health Network, Toronto, ON M5G 2M9, Canada
3 Department of Immunology, University of Toronto, Toronto, ON M5S 1A8, Canada
4 Viral Immunology and Intravital Imaging Section, National Institute of Neurological Disorders and Stroke, The National Institutes of Health, 10 Center Drive, Bethesda, MD 20895, USA
* Correspondence: dbrooks@uhnresearch.ca (D.G.B.); mcgavernd@ninds.nih.gov (D.B.M.)

Abstract: During chronic viral infections, CD8 T cells rapidly lose antiviral and immune-stimulatory functions in a sustained program termed exhaustion. In addition to this loss of function, CD8 T cells with the highest affinity for viral antigen can be physically deleted. Consequently, treatments designed to restore function to exhausted cells and control chronic viral replication are limited from the onset by the decreased breadth of the antiviral T cell response. Yet, it remains unclear why certain populations of CD8 T cells are deleted while others are preserved in an exhausted state. We report that CD8 T cell deletion during chronic viral infection can be prevented by therapeutically lowering viral replication early after infection. The initial resistance to deletion enabled long-term maintenance of antiviral cytolytic activity of the otherwise deleted high-affinity CD8 T cells. In combination with decreased virus titers, CD4 T cell help and prolonged interactions with costimulatory molecules B7-1/B7-2 were required to prevent CD8 T cell deletion. Thus, therapeutic strategies to decrease early virus replication could enhance virus-specific CD8 T cell diversity and function during chronic infection.

Keywords: LCMV; chronic infection; CD8 T cells; CD4 T cell help; T cell deletion; exhaustion; costimulation; antiviral therapy; ribavirin

Citation: Brooks, D.G.; Tishon, A.; Oldstone, M.B.A.; McGavern, D.B. Prevention of CD8 T Cell Deletion during Chronic Viral Infection. *Viruses* **2021**, *13*, 1189. https://doi.org/10.3390/v13071189

Academic Editor: Juan C. De la Torre

Received: 19 May 2021
Accepted: 18 June 2021
Published: 22 June 2021

1. Introduction

At the beginning of what will become a chronic infection, heightened and sustained antigenic signaling and inflammatory factors enhance the expression of immunosuppressive cytokines and multiple inhibitory receptors/ligands (e.g., IL10, PD1, PDL1) that program and maintain T cell exhaustion and viral persistence [1–4]. The exhaustion of antiviral T cell function is observed during many chronic viral infections, including HIV and hepatitis B and C virus (HBV, HCV) infections of humans, lymphocytic choriomeningitis virus (LCMV) infection of rodents, and in response to many cancers [3], suggesting that shared mechanisms triggered by increased and/or sustained antigen regulate T cell inactivation. Although antiviral T cells with diminished function can remain in the immune repertoire, CD8 T cells with the highest affinity for viral antigens are sometimes physically deleted and thus absent throughout the infection [5–7]. Since virus-specific T cells continue to maintain some effector function and exert long-term pressure on virus replication, they often serve as targets for immune restorative therapies [3]. Unfortunately, the deletion of high affinity CD8 T cells can greatly hamper such therapeutic efforts and impede the host in its fight against a chronic virus. Understanding the mechanisms governing T cell exhaustion and deletion is important if we intend to eradicate persistent viral infections. We have already begun to identify distinct mechanisms that control of the functional states of antiviral T cells. For example, we demonstrated previously that loss of cytokine production and effector functions can be separated from the process of cellular expansion [8]. It therefore stands to reason the physical deletion of T cells might also be preventable despite functional exhaustion.

Deletion of high-affinity CD8 T cells has multiple potential consequences. First, it results in depletion of effector cells well equipped to fight infection, thereby limiting the breadth, magnitude, and efficacy of the antiviral response. Second, with high-affinity antiviral CD8 T cells removed, therapeutic interventions designed to control established chronic infections by resurrecting T cell responses are limited to restoring function to T cells of a low to moderate affinity. At the very least, preservation of otherwise deleted T cells has the potential to broaden the long-term antiviral response by providing more cells. Thus, methods that prevent the deletion of high-affinity CD8 T cells could be of great benefit to both the endogenous and therapeutic control of chronic viral infections.

The mechanisms that regulate deletion during viral infection are still unclear, although levels of antigen stimulation at the epitope level is higher for cells that are deleted [9], leading to the theory that cognate peptide-MHC density on the cell surface determines physical deletion versus maintenance of functionally exhausted T cells. Virus-specific CD4 T cells also participate in the fate of CD8 T cells. For example, CD4 T cell help is required to sustain virus-specific CD8 T cells during chronic infection [10,11], and re-establishment of CD4 T cell responses [12], specifically of CD4 Th1 cells [13], in the midst of an established chronic infection can prevent the progressive numerical and functional decay of virus-specific CD8 T cells. However, the early role of CD4 T cells in the priming of CD8 T cells, during what will eventually become a chronic infection, is less clear. In fact, early functional rescue of CD4 T cells during persistent viral infection by the administration of antiviral therapy still leads to CD8 T cell exhaustion [8].

In this study, we sought insights into the mechanisms that give rise CD8 T cell deletion during a persistent viral infection. We demonstrate that distinct from CD8 T cell functional exhaustion, the physical deletion of high-affinity LCMV-$NP_{396–404}$ specific CD8 T cells in chronic LCMV infection is due to increased virus infection and the loss of costimulatory signals. Therapeutic reduction of viral replication with the antiviral drug Ribavirin (Rb) prevented deletion of high affinity virus-specific CD8 T cells, which were then maintained in a functional state long after Rb treatment was ceased. Importantly, deletion of CD8 T cells was not linked solely to levels of virus replication. Instead, continued CD4 T cell help and prolonged costimulatory interactions with B7-1 (CD80) and B7-2 (CD86) were also needed to prevent deletion.

2. Materials and Methods

2.1. Mice and Virus

C57BL/6 (H-2^b, Thy1.2+) mice were from The Rodent Breeding Colony at the Scripps Research Institute. CD4−/− mice were bred from homozygous breeding pairs obtained from The Jackson Laboratory (Bar Harbor, ME, USA). The LCMV-$GP_{61–80}$-specific CD4+ T cell receptor (TCR) transgenic (SMARTA) mice have been described previously [14] and were gifts from Hans Hengartner and Rolf Zinkernagel (University Hospital of Zurich, Zurich, Switzerland). All mice were housed under specific pathogen-free conditions. Mouse handling conformed to the requirements of The Scripps Research Institute Animal Research Committee. Mice were infected intravenously with 2×10^6 plaque forming units (PFU) of LCMV-Arm or LCMV-Cl13. Virus stocks were prepared as described [14]. Viral titers were determined by plaque formation on Vero cells [14].

2.2. Flow Cytometry

Total splenocytes were stimulated for 5 h with 2 μg/mL of the MHC class I restricted LCMV-$NP_{396–404}$ peptide (>99% pure; Synpep, Dublin, CA, USA) in the presence of 50 U/mL recombinant murine IL-2 (R&D Systems, Minneapolis, MN, USA) and 1 mg/mL brefeldin A (Sigma, St. Louis, MO, USA). The absolute number of $NP_{396–404}$-specific CD8

T cells was determined by multiplying the frequency of IFNγ+ cells by the total number of splenocytes.

2.3. *Ribavirin Treatment*

Mice were treated with 100 mg/kg Rb (1-b-D-ribofuranosyl-1,2,4-triazole-3-carboxa mide, Virazole; ICN Pharmaceuticals, Costa Mesa, CA, USA) in PBS. We observed no toxicity with the use of Rb. Treatment was administered once daily intraperitoneally (i.p.) on days 1–8 after LCMV infection. Control mice were injected in parallel with PBS (vehicle-only control).

2.4. *Costimulation Blockade*

Mice received 250 μg CTLA4-Ig (Sigma, non-cytolytic) i.p. once daily or every other day on days 4–8 after infection. Control mice were treated in parallel with 250 μg matched mouse IgG2a isotype control antibody (BD Pharmingen, La Jolla, CA, USA).

2.5. *In Vitro and In Vivo Cytotoxicity Assays*

For in vitro ^{51}Cr release assays, splenocytes on day 9 after the indicated infection were incubated for 5 h with either unlabeled MC57 cells, LCMV-$GP_{33–41}$ peptide labelled MC57 cells, or $NP_{396–404}$ peptide labeled MC57 cells. For in vivo CTL assays, splenocytes from naïve C57BL/6 mice were either left unlabeled or were labeled with 2 μg/mL of the LCMV-$NP_{396–404}$ peptide. These target cell populations were then differentially labeled with CFSE (2.5 or 0.25 μg/mL) and transferred i.v. (5×10^6 cells of each population) into mice previously infected with LCMV. Fifteen hours post transfer, splenocytes were analyzed for CFSE expression by flow cytometry. Percent lysis = [1 − (ratio NP396 to unlabeled)] × 100.

2.6. *Statistical Analysis*

Statistical analyses were performed using GraphPad Prism 6 software (GraphPad Software, Inc., San Diego, CA, USA). In the line and bar graphs the error bars indicate standard deviation.

3. Results

3.1. *Lowering Virus Titers during Early Chronic Infection Preserves High-Affinity LCMV-$NP_{396–404}$ Dpecific CD8 T Cells*

To determine how early changes in virus replication affect T cell responses, we infected mice with LCMV-Clone13 (Cl13) to generate a chronic infection [15]. Mice were then treated daily with the antiviral drug Ribavirin (Rb) beginning one day after infection until day 8 [8]. As we demonstrated previously [8], early antiviral therapy with Rb rapidly lowered virus titers, leading to enhanced control of virus replication after the treatment was discontinued (Figure 1A). To assess whether the accelerated viral control required adaptive immunity or resulted solely in the initial lowering of viral titer, we infected Rag−/− mice (lacking B and T cells) with LCMV-Cl13. Even in the absence of B cells and T cells, Rb treatment led to a 4-fold decrease in viral replication (Figure 1B). However, virus titers remained similarly elevated in PBS vs. Rb treated Rag−/− mice following withdrawal of therapy (Figure 1B), indicating that the accelerated viral clearance conferred by Rb treatment in wild type (WT) mice was reliant on adaptive immune responses.

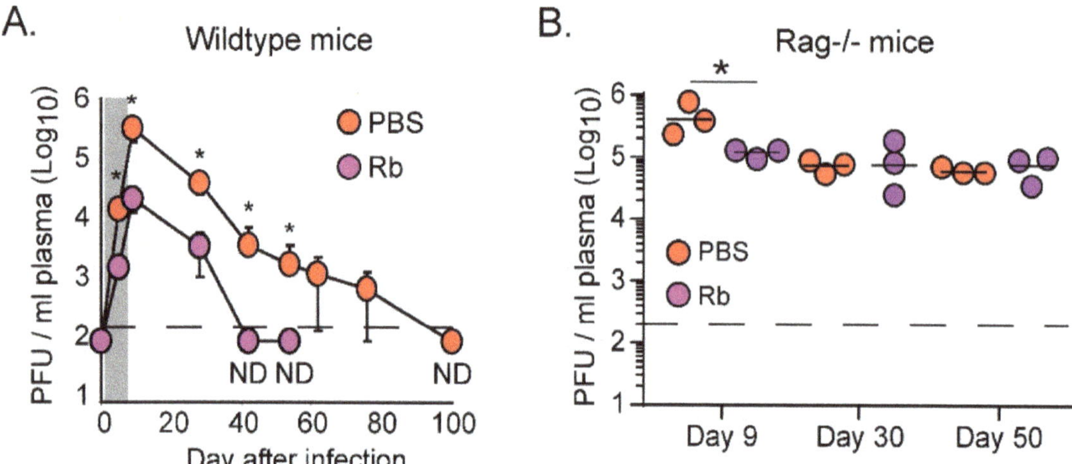

Figure 1. Accelerated control of chronic viremia following Rb therapy is dependent on adaptive immune responses. (**A**) Plasma viral titers from LCMV-Clone 13 (Cl13) infected C57BL/6 mice either PBS contro-treated (red circles) or treated with Ribavirin (Rb; purple circles) were determined at the indicated time point. Data are expressed as PFU per milliliter of plasma. Shaded area indicates time of Rb treatment (day 1–8 after infection). Each time point represents the average ± standard deviation (SD) of 4 mice per group. ND = not detected. (**B**) Infectious virus in the plasma of LCMV-Cl13 infected Rag−/− mice either PBS-treated or treated with Rb. Each circle represents a single animal. Data are representative of 2 experiments with 3–4 mice per group. * $p < 0.05$; one-way ANOVA.

We next sought to determine how the initial decrease in virus titers altered the virus-specific CD8 T cell response. We have shown previously that LCMV-specific CD4 T cells robustly resisted exhaustion when virus titers were decreased by early Rb treatment, with a smaller functional restoration of LCMV-$GP_{33–41}$-specific CD8 T cells [8]. Interestingly, there was an increase in both the frequency and number of the IFNγ-producing high-affinity LCMV-$NP_{396–404}$-specific CD8 T cells following Rb treatment (Figure 2A). Across experiments, the number of IFNγ-producing LCMV-$NP_{396–404}$-specific CD8 T cells was increased approximately 2–5-fold in Rb treated compared to untreated LCMV-Cl13 infected mice (Figure 2A). Rb treatment also increased the number of IFNγ-producing LCMV-$NP_{396–404}$-specific CD8 T cells following acute LCMV-Arm infection, albeit to a lesser extent than during chronic LCMV-Cl13 infection (Figure 2A). Despite the increased number of IFNγ-producing LCMV-$NP_{396–404}$ specific CD8 T cells in Rb treated chronic LCMV-Cl13 infected mice, the frequency of IFNγ/TNFα co-producing cells was not increased (Figure 2B), suggesting that preservation of the response was still accompanied by loss of polyfunctional cytokine potential. However, preservation of LCMV-$NP_{396–404}$ specific CD8 T cells following early Rb treatment resulted in increased $NP_{396–404}$-specific lysis (CTL) at day 9 post-infection (likely reflecting the increased number of NP396–404-specific cells), whereas this treatment had a minimal effect on the LCMV-$GP_{33–41}$ specific CTL activity (Figure 2C). These data indicate that Rb physical rescues high-affinity NP_{396}-specific CD8 T cells and preserves their CTL activity.

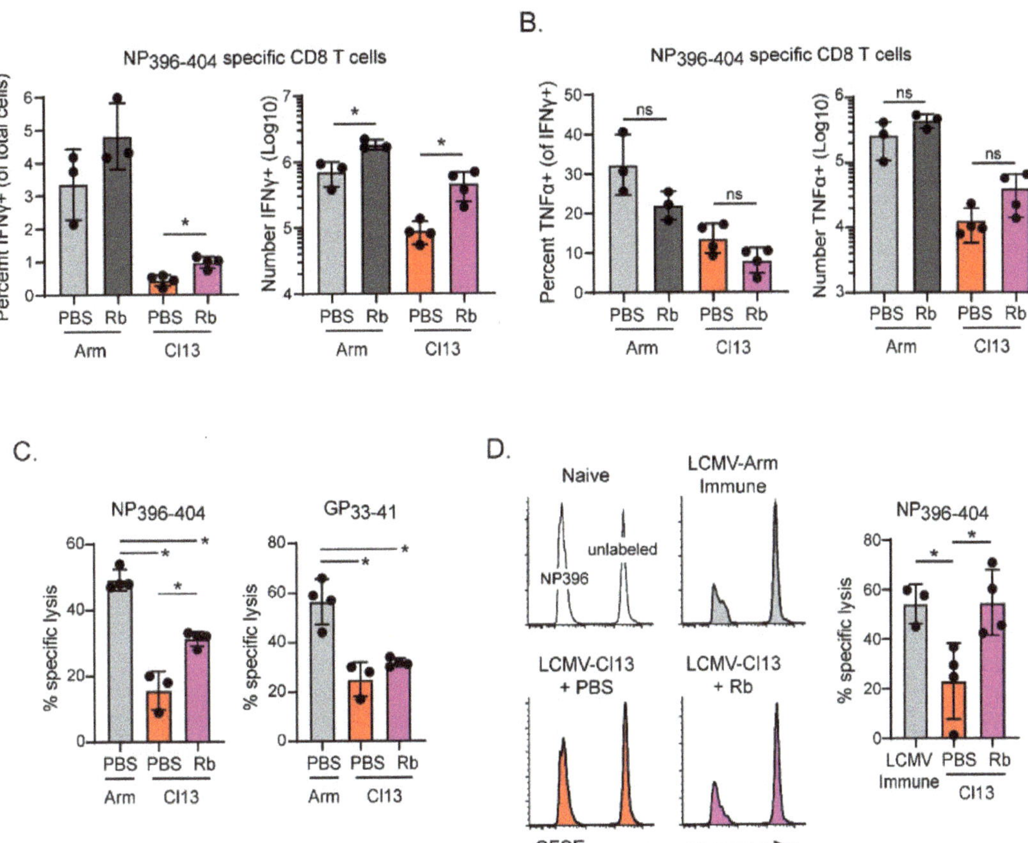

Figure 2. Early prevention and long-term functional rescue of high-affinity CD8 T cells. (**A**,**B**). Splenocytes from PBS-treated or Rb-treated LCMV-Arm or LCMV-Cl13 infected C57BL/6 mice were analyzed on day 9 after infection. Graphs illustrate (**A**) the frequency and number of IFNγ-producing, and (**B**) of TNFα and IFNγ double producing $NP_{396–404}$-specific CD8 T cells following ex vivo peptide restimulation. The bars represent the average ± SD of 4 mice (individual circles) in each group and represent multiple experiments. *, $p < 0.05$. (**C**) In vitro ^{51}Cr release activity against target cells coated with LCMV-$NP_{396–404}$ peptide or LCMV-$GP_{33–41}$ peptide (effector:target ratio of 50:1) on day 9 post infection. (**D**) In vivo CTL assay (day 50 after infection). Peptide-unlabeled or $NP_{396–404}$ peptide-labeled target cells (splenocytes) were co-transferred into LCMV-naïve mice, LCMV-Arm immune mice (day 50 after infection) or LCMV-Cl13 infected mice (day 50 after infection) that were either PBS-treated or had been treated with Rb on days 1 to 8 after infection. Histograms show a single mouse with each condition, and the bar graph indicates the percent $NP_{396–404}$-specific lysis. The bar graph shows the average ± SD of 3 (Arm infection) or 4 (Cl 13 and Cl 13 + Rb infections) mice per group. The circles on each bar represent individual mice. * $p < 0.05$; student's t-test (unpaired, two-tailed).

We next assessed whether the early preservation of LCMV-$NP_{396–404}$-specific CD8 T cells resulted in long-term CTL lytic activity by performing an in vivo CTL assay. Mice were treated with vehicle or Rb on days 1–8 after LCMV-Cl13 infection. In parallel, mice were left naïve to LCMV infection or infected with acutely resolved LCMV-Arm to generate LCMV-immune mice. Fifty days after infection, unlabeled or $NP_{396–404}$-peptide labeled splenocytes from naïve mice were transferred into the LCMV-naïve mice, LCMV-Arm 'immune' and LCMV-Cl13 infected mice. LCMV-Cl13 infected mice had diminished NP396–404 directed CTL lysis compared to LCMV-immune mice, although residual lysis was still observed in comparison to LCMV-naïve mice (Figure 2D). Importantly, LCMV-Cl13 infected mice initially treated with Rb exhibited sustained CTL lytic activity against the LCMV $NP_{396–404}$ epitope that was similar to the level observed in LCMV-Arm immune mice (Figure 2D), indicating that preventing the initial deletion of the NP396–404 CD8 T cell response resulted in preserved lytic activity of these cells.

3.2. *CD4 T Cells Are Required to Preserve High-Affinity LCMV-$NP_{396–404}$ Specific CD8 T Cells*

A main effect of early antiviral Rb treatment and diminished viral replication following LCMV-Cl13 infection was the rescue of CD4 T cell function [8]. To determine the role of CD4 T cells in preserving LCMV-$NP_{396–404}$-specific CD8 T cells, we used CD4-deficient mice. Consistent with a minimal impact of Rb treatment in enhancing $NP_{396–404}$-specific CD8 T cells following acute LCMV-Arm infection, no alteration in the number of IFNγ-producing $NP_{396–404}$-specific CD8 T cells was observed in Rb treated LCMV-Arm infected CD4−/− mice (Figure 3A). In contrast, the number of IFNγ-producing $NP_{396–404}$-specific CD8 T cells were similarly deleted in Rb-treated and PBS-treated CD4−/− mice following LCMV-Cl13 infection (Figure 3A). Further, deletion occurred in Rb treated CD4−/− mice despite a substantial reduction in virus titers (Figure 3A). These data suggest that CD4 T cells are required to prevent deletion of IFNγ-producing high-affinity $NP_{396–404}$-specific CD8 T cells in Rb treated mice secondary to the reduction in viral load.

3.3. *Prolonged Costimulatory Interactions Are Required to Sustain Virus-Specific CD4 T Cells and High-Affinity LCMV-$NP_{396–404}$ Specific CD8 T Cells*

Decreased LCMV-Cl13 titers following Rb treatment coincided with upregulation of the costimulatory molecules B7-1 and B7-2 on dendritic cells (DCs; Figure 3B) and other antigen presenting cell (APC) populations (B cells and macrophages) [8], suggesting that prolonged or repeated costimulatory interactions during chronic viral infection might sustain T cells when virus loads are reduced. To evaluate the effect of prolonged/repeated costimulation, mice were infected with LCMV-Cl13, treated with the antiviral compound Rb to diminish viral replication, and then treated with CTLA4-Ig to specifically block CD28/CTLA4:B7-1/B7-2 interactions. CTLA4-Ig or mouse IgG2a isotype control antibody was administered beginning on day 4 after infection to avoid inhibiting the initial priming interactions [14]. The CTLA4-Ig used did not deplete cells, as evidenced by the similar numbers of DCs in Rb-treated LCMV-Cl13 infected mice given isotype or CTLA4-Ig, that were both higher than LCMV-Cl13 infection alone (Figure 3C). The preservation of LCMV-$NP_{396–404}$-specific CD8 T cells afforded by early Rb treatment was now inhibited by blocking costimulation with CTLA4-Ig (Figure 3D). Importantly, the Rb+CTLA4-Ig treatment did not prevent the reduction in virus titers induced by Rb treatment alone (Figure 3E). These data suggest the decrease in virus titers alone was not sufficient to rescue NP396-specific CD8 T cells, but that prolonged costimulation was required after suppression of virus replication.

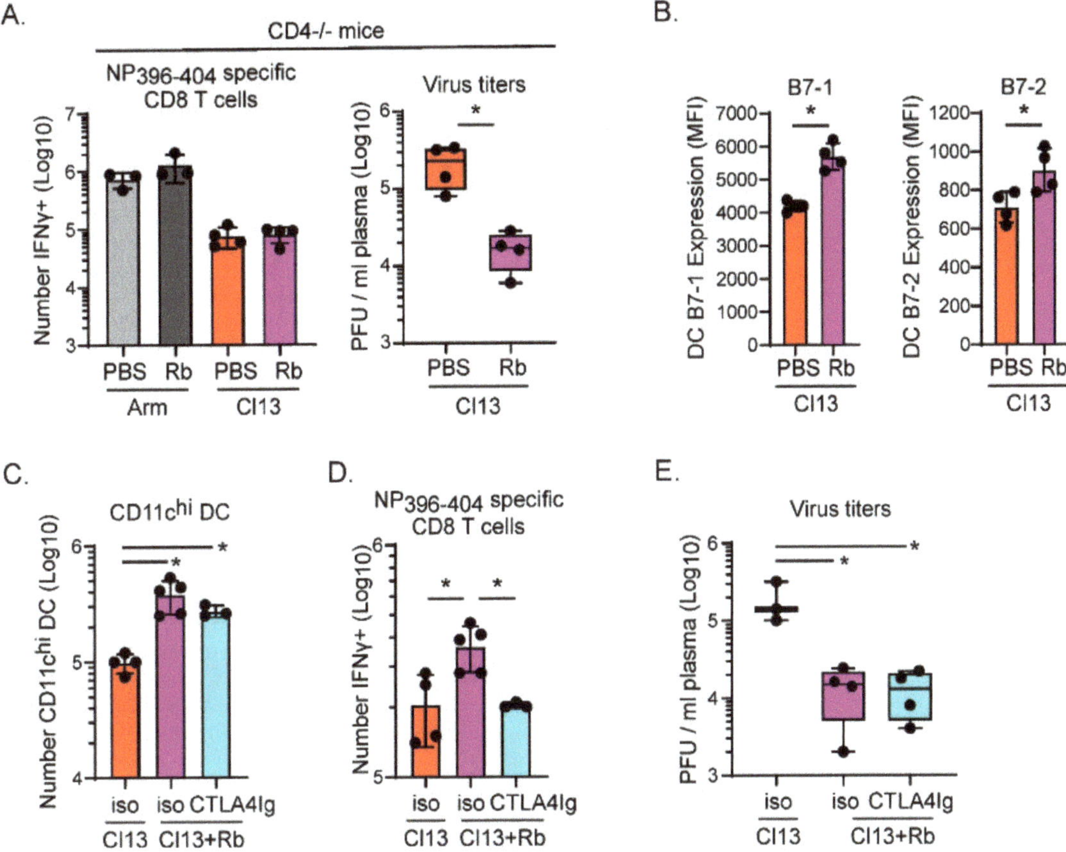

Figure 3. CD4 T cells and prolonged costimulation are required in combination with lower virus replication to prevent CD8 T cell deletion. (**A**). Splenocytes from PBS-treated or Rb-treated LCMV-Arm or LCMV-Cl13 infected CD4−/− mice were analyzed on day 9 after infection. The bar graph represents the number of IFNγ-producing $NP_{396–404}$ specific CD8 T cells and is the average ± SD of 4 mice per group. Box and whisker indicate day 9 virus titers in PBS-treated and Rb-treated CD4−/− mice infected with LCMV-Cl13. In the box and whisker plots, the midline represents the median and the error bars the minimum and maximum values (whiskers). Circles indicate individual mice. (**B**). Bars indicate the mean fluorescence intensity (MFI) of B7-1 and B7-2 expression on splenic $CD11c^{hi}$ dendritic cells in PBS treated or Rb-treated mice on day 9 after LCMV-Cl13 infection. Circles indicate the MFI in individual mice. (**C–E**). Splenocytes from PBS-treated or Rb-treated LCMV-Cl13 infected mice were analyzed on day 9 after infection. Mice were treated with PBS or Rb and then on days 4–8 after infection given CTLA4-Ig or an isotype antibody. Bars indicate (**C**) the number of $CD11c^{hi}$ dendritic cells, (**D**) the number of IFNγ-producing NP396-specific CD8 T cells, and (**E**) plasma virus titers. Data are representative of 2 or more experiments with 3–5 mice per group. * $p < 0.05$; student's *t*-test (unpaired, two-tailed).

4. Discussion

Our studies demonstrate that deletion of high-affinity CD8 T cells during chronic viral infection can be prevented by decreasing early virus titers and enabling costimulatory B7.1/B7.2 interactions. Unlike functional inactivation, after which cells are physically present, deleted cells are absent throughout infection and cannot be resurrected. Consequently, treatments designed to restore function to inactivated cells [1] may have limited efficacy due to the extinction of high-affinity cells and the decreased breadth of the T cell response. Preventing deletion of high-affinity CD8 T cells creates the opportunity to restore and stimulate a broader repertoire of antiviral T cells to a control persistent viral replication. A major thrust of current therapeutic vaccination strategies is to restore function to inactivated T cells with the hope that they will control the chronic viral infection. It is important to note that these data utilized $NP_{396-404}$ specific IFNγ production and not MCH I tetramer to quantify the $NP_{396-404}$-specific CD8 T cells. Thus, although the deletion of $NP_{396-404}$ reactive cells is prevented, it still remains to be determined whether this is due to survival of the cells and/or increased functionality. Importantly, our data indicate that if the early deletion of high-affinity CD8 T cells is abrogated, these cells can be functionally maintained and engaged throughout a state of viral persistence.

LCMV, like Lassa fever virus, is an Old-World arenavirus and is often used as a model of Lassa fever virus infection. In response to Lassa fever virus infection, survival versus death is associated with the amount of virus produced [16]. A recent study demonstrated that surviving individuals mounted a more robust Lassa virus-specific CD4 and CD8 T cell response relative to those that died from the disease [17,18]. Interestingly, Rb is effective in treating Lassa fever virus infection, and our data regarding the preservation of high affinity CTL in response to Rb provide another potential mechanism for how therapeutic control might be achieved in Lassa fever patients.

Functionally sustaining and enhancing CD4 T cell help during viral persistence (and particularly Th1 responses) maintains and enhances virus-specific CD8 T cells that are present throughout a chronic infection [12,13,19–21]. Nevertheless, virus-specific CD8 T cell exhaustion and deletion still occur at early time points post-infection despite the presence of CD4 T cells (albeit with attenuated function themselves) [10,11,14]. Importantly, our data demonstrate that CD4 T cells help prevent deletion of high-affinity NP396-specific CD8 T cells after virus titers are therapeutically decreased with Rb in the early stages of a chronic LCMV infection. Exactly how this early CD4 T cell help is delivered remains to be defined, but one possibility is increased IL-2 production by virus-specific CD4 T cells resulting from the diminished viral titers [8]. During HIV infection, deletion of high-affinity CD8 T cells inversely correlates with the benefit of early antiretroviral therapy and the maintenance of CD4 T cell activity [7]. Thus, the mechanisms leading to CD8 T cell deletion during states of heightened virus replication may be similar between rodent and human chronic infections.

Finally, our data suggest that mechanisms in addition to high levels of antigen exposure and prolonged/heightened TCR signaling result in T cell deletion [5,9]. Specifically, we observed that while deletion can be prevented by lowering virus titers with Rb, maintenance of these high-affinity CD8 T cells still requires additional interactions with costimulatory molecules. Thus, reduction of viral titers alone is insufficient when it comes to preventing deletion of high-affinity CD8 T cells. We identified sustained interactions with costimulatory molecules as one mechanism that helps prevent CTL deletion, but there are likely to be others. We postulate that decreasing viral replication with early antiviral therapy elevates costimulatory molecule expression on APCs, which provides CD8 T cells with the continued signals they need to survive despite viral persistence. The early preservation of these high-affinity CD8 T cells through decreased virus replication, augmented CD4 help, and enhanced costimulatory signals has important implications for virus control and immune restorative therapies during chronic viral infections.

Author Contributions: Conceptualization, D.G.B.; Investigation, A.T.; Supervision, M.B.A.O. and D.B.M. All authors have read and agreed to the published version of the manuscript.

Funding: This work was supported by NIH training grant AI07244-22 (D.G.B.), the Canadian Institutes of Health Research (CIHR) Foundation Grant FDN148386 (D.G.B.), the National Institutes of Health (NIH) grant AI085043 (D.G.B.), the Scotiabank Research Chair to (D.G.B.), NIH grants AI09484, AI45927 (M.B.A.O.), NS048866-01 (D.B.M.), a Dana Foundation grant (D.B.M.), and the intramural program of the National Institute of Neurological Disorders and Stroke (D.B.M.).

Institutional Review Board Statement: Mouse handling was approved by The Scripps Research Institute Animal Research Committee.

Informed Consent Statement: Not Applicable.

Data Availability Statement: Reagent requests can be made to D.G.B.

Conflicts of Interest: The authors declare no conflict of interest.

References

1. Barber, D.L.; Wherry, E.J.; Masopust, D.; Zhu, B.; Allison, J.P.; Sharpe, A.H.; Freeman, G.J.; Ahmed, R. Restoring function in exhausted CD8 T cells during chronic viral infection. *Nature* **2006**, *439*, 682–687. [CrossRef] [PubMed]
2. Brooks, D.G.; Trifilo, M.J.; Edelmann, K.H.; Teyton, L.; McGavern, D.B.; Oldstone, M.B. Interleukin-10 determines viral clearance or persistence in vivo. *Nat. Med.* **2006**, *12*, 1301–1309. [CrossRef]
3. McLane, L.M.; Abdel-Hakeem, M.S.; Wherry, E.J. CD8 T Cell Exhaustion During Chronic Viral Infection and Cancer. *Annu. Rev. Immunol.* **2019**, *37*, 457–495. [CrossRef]
4. Snell, L.M.; McGaha, T.L.; Brooks, D.G. Type I Interferon in Chronic Virus Infection and Cancer. *Trends Immunol.* **2017**, *38*, 542–557. [CrossRef] [PubMed]
5. Zajac, A.J.; Blattman, J.N.; Murali-Krishna, K.; Sourdive, D.J.; Suresh, M.; Altman, J.D.; Ahmed, R. Viral immune evasion due to persistence of activated T cells without effector function. *J. Exp. Med.* **1998**, *188*, 2205–2213. [CrossRef] [PubMed]
6. Gallimore, A.; Glithero, A.; Godkin, A.; Tissot, A.C.; Pluckthun, A.; Elliott, T.; Hengartner, H.; Zinkernagel, R. Induction and exhaustion of lymphocytic choriomeningitis virus-specific cytotoxic T lymphocytes visualized using soluble tetrameric major histocompatibility complex class I-peptide complexes. *J. Exp. Med.* **1998**, *187*, 1383–1393. [CrossRef]
7. Oxenius, A.; Price, D.A.; Easterbrook, P.J.; O'Callaghan, C.A.; Kelleher, A.D.; Whelan, J.A.; Sontag, G.; Sewell, A.K.; Phillips, R.E. Early highly active antiretroviral therapy for acute HIV-1 infection preserves immune function of CD8+ and CD4+ T lymphocytes. *Proc. Natl. Acad. Sci. USA* **2000**, *97*, 3382–3387. [CrossRef]
8. Brooks, D.G.; McGavern, D.B.; Oldstone, M.B. Reprogramming of antiviral T cells prevents inactivation and restores T cell activity during persistent viral infection. *J. Clin. Investig.* **2006**, *116*, 1675–1685. [CrossRef]
9. Wherry, E.J.; Blattman, J.N.; Murali-Krishna, K.; van der Most, R.; Ahmed, R. Viral persistence alters CD8 T-cell immunodominance and tissue distribution and results in distinct stages of functional impairment. *J. Virol.* **2003**, *77*, 4911–4927. [CrossRef]
10. Matloubian, M.; Concepcion, R.J.; Ahmed, R. CD4+ T cells are required to sustain CD8+ cytotoxic T-cell responses during chronic viral infection. *J. Virol.* **1994**, *68*, 8056–8063. [CrossRef] [PubMed]
11. Battegay, M.; Moskophidis, D.; Rahemtulla, A.; Hengartner, H.; Mak, T.W.; Zinkernagel, R.M. Enhanced establishment of a virus carrier state in adult CD4+ T-cell-deficient mice. *J. Virol.* **1994**, *68*, 4700–4704. [CrossRef] [PubMed]
12. Aubert, R.D.; Kamphorst, A.O.; Sarkar, S.; Vezys, V.; Ha, S.J.; Barber, D.L.; Ye, L.; Sharpe, A.H.; Freeman, G.J.; Ahmed, R. Antigen-specific CD4 T-cell help rescues exhausted CD8 T cells during chronic viral infection. *Proc. Natl. Acad. Sci. USA* **2011**, *108*, 21182–21187. [CrossRef]
13. Snell, L.M.; Osokine, I.; Yamada, D.H.; De la Fuente, J.R.; Elsaesser, H.J.; Brooks, D.G. Overcoming CD4 Th1 Cell Fate Restrictions to Sustain Antiviral CD8 T Cells and Control Persistent Virus Infection. *Cell Rep.* **2016**, *16*, 3286–3296. [CrossRef]
14. Brooks, D.G.; Teyton, L.; Oldstone, M.B.; McGavern, D.B. Intrinsic functional dysregulation of CD4 T cells occurs rapidly following persistent viral infection. *J. Virol.* **2005**, *79*, 10514–10527. [CrossRef] [PubMed]
15. Ahmed, R.; Salmi, A.; Butler, L.D.; Chiller, J.M.; Oldstone, M.B. Selection of genetic variants of lymphocytic choriomeningitis virus in spleens of persistently infected mice. Role in suppression of cytotoxic T lymphocyte response and viral persistence. *J. Exp. Med.* **1984**, *160*, 521–540. [CrossRef] [PubMed]
16. Oldstone, M.B.A. *Virus, Plagues, and History*, 2nd ed.; Oxford University Press: Oxford, UK, 2020.
17. Sakabe, S.; Hartnett, J.N.; Ngo, N.; Goba, A.; Momoh, M.; Sandi, J.D.; Kanneh, L.; Cubitt, B.; Garcia, S.D.; Ware, B.C.; et al. Identification of Common CD8(+) T Cell Epitopes from Lassa Fever Survivors in Nigeria and Sierra Leone. *J. Virol.* **2020**, *94*. [CrossRef] [PubMed]
18. Sullivan, B.M.; Sakabe, S.; Hartnett, J.N.; Ngo, N.; Goba, A.; Momoh, M.; Demby Sandi, J.; Kanneh, L.; Cubitt, B.; Garcia, S.D.; et al. High crossreactivity of human T cell responses between Lassa virus lineages. *PLoS Pathog.* **2020**, *16*, e1008352. [CrossRef]
19. Elsaesser, H.; Sauer, K.; Brooks, D.G. IL-21 is required to control chronic viral infection. *Science* **2009**, *324*, 1569–1572. [CrossRef]
20. Yi, J.S.; Du, M.; Zajac, A.J. A vital role for interleukin-21 in the control of a chronic viral infection. *Science* **2009**, *324*, 1572–1576. [CrossRef]
21. Frohlich, A.; Kisielow, J.; Schmitz, I.; Freigang, S.; Shamshiev, A.T.; Weber, J.; Marsland, B.J.; Oxenius, A.; Kopf, M. IL-21R on T cells is critical for sustained functionality and control of chronic viral infection. *Science* **2009**, *324*, 1576–1580. [CrossRef] [PubMed]

Review

Viral Control of Glioblastoma

Nicole Mihelson and Dorian B. McGavern *

Viral Immunology & Intravital Imaging Section, National Institute of Neurological Disorders and Stroke, National Institutes of Health, Building 10, Room 5N240C, Bethesda, MD 20892, USA; nicole.mihelson@nih.gov
* Correspondence: mcgavernd@mail.nih.gov; Tel.: +1-301-443-7949

Abstract: Glioblastoma multiforme (GBM) is a universally lethal cancer of the central nervous system. Patients with GBM have a median survival of 14 months and a 5-year survival of less than 5%, a grim statistic that has remained unchanged over the last 50 years. GBM is intransigent for a variety of reasons. The immune system has a difficult time mounting a response against glioblastomas because they reside in the brain (an immunologically dampened compartment) and generate few neoantigens relative to other cancers. Glioblastomas inhabit the brain like sand in the grass and display a high degree of intra- and inter-tumoral heterogeneity, impeding efforts to therapeutically target a single pathway. Of all potential therapeutic strategies to date, virotherapy offers the greatest chance of counteracting each of the obstacles mounted by GBM. Virotherapy can xenogenize a tumor that is deft at behaving like "self", triggering adaptive immune recognition in an otherwise immunologically quiet compartment. Viruses can also directly lyse tumor cells, creating damage and further stimulating secondary immune reactions that are detrimental to tumor growth. In this review, we summarize the basic immune mechanisms underpinning GBM immune evasion and the recent successes achieved using virotherapies.

Keywords: glioblastoma multiforme; immune evasion; virotherapies; immunotherapeutic strategies

Citation: Mihelson, N.; McGavern, D.B. Viral Control of Glioblastoma. *Viruses* **2021**, *13*, 1264. https://doi.org/10.3390/v13071264

Academic Editors: Michael B. A. Oldstone and Juan C. De la Torre

Received: 2 June 2021
Accepted: 24 June 2021
Published: 29 June 2021

1. Introduction

Glioblastoma multiforme (GBM) is intractable and among the most lethal human cancers. GBM is the most common tumor of the adult central nervous system (CNS) and is thought to arise from either neuroglial progenitors or de-differentiated glial cells [1–3]. GBM has a median survival of 14 months, and its 5-year survival rate has remained unchanged for nearly half a century, even in the face of new standards of care [4–6]. Many theories have been advanced for why GBM is universally lethal and intransigent. First, it harbors a bland mutagenomic profile relative to other cancers, due in part to a few canonical mutations driving oncogenesis as opposed to carcinogens [7]. Second, GBM exhibits vast intra- and inter-individual heterogeneity at both the level of the tumor and immune cells (phenotypically, spatially, temporally, etc.) [8,9] Third, GBM is an intrinsically infiltrative and migratory tumor akin to "sand in the grass," sitting behind both a blood–brain barrier and inside of a precious-real-estate organ that is unforgiving when it comes to supporting inflammation and immunity.

With the exception of tumor-treating fields (application of low-intensity alternating electric fields) that have increased median survival from 16 to 21 months as well increased 5-year survival from 5% to 13% in exploratory studies [10], virtually none of the many new therapies under investigation have materially prolonged GBM survival [11]. Viral therapy, if further optimized and developed, may be one exception [12]. The idea of using viruses (and other pathogens) to treat GBM was borne from the anecdotes of GBM patients contracting infections in the surgical site and surviving for years. In fact, a 2011 study of 200 GBM patients in Rome found that the 10 who contracted wound infections lived twice as long as their sterile counterparts [13]. One neurosurgeon went on record with *The New Yorker* to say, "If I ever get a GBM, put your finger in your keister and put it in the wound" [14]. Admittedly, this sounds less elegant than checkpoint blockade, but for reasons that will be advanced in this review, such an approach (or an adjacent—more regulated—approach in the vein of deliberate viral infection) is likely to bear more fruit than typical immunotherapeutic strategies.

2. Endogenous Viruses Found in GBM

GBM's vast intratumoral heterogeneity as well as its rank among the most lethal of all cancers has long prompted queries into its etiology. Viruses have captured the interests of investigators due to their role in the oncogenesis of other cancers, as well as the presence of viral nucleic acid and protein in GBM and the general state of immunosuppression in patients.

2.1. *Cytomegalovirus*

Human cytomegalovirus (CMV) is a beta herpes virus with a tropism for glial cells. CMV infects 50–90% of the adult population and seroprevalence increases with age [15]. In healthy individuals, CMV is a clinically irrelevant pathogen. In 2002, one of the first CMV studies in GBM showed that 27 of the 27 samples queried by immunohistochemistry and in situ hybridization expressed CMV genomic and protein material [16]. Since then, a plethora of studies have both confirmed and denied these original findings in tumor tissue and blood [17–20]. Much of the variability has been attributed to assay sensitivity and type, as well as the age of the samples queried [21]. Some groups have also shown that CMV expression was higher in GBM compared to low grade gliomas and epilepsy controls [17], and in newer samples compared with older ones [22]. In studies showing CMV gene products in tumor tissue, their absence in adjacent non-neoplastic tissue begged the question of why lytic phase CMV did not spread. Further, investigators wondered whether CMV's presence was a product of the chaos in the tumor bed rather than a cause. Indeed, CMV primarily infects quiescent cells, pushes them to G_1 to induce viral gene expression, and arrests the cell cycle prior to host DNA synthesis in S phase—behavior not entirely conducive with oncogenesis [23]. Furthermore, the prevalence of CMV in nearly 9 out of 10 adults does not match with the prevalence of GBM (1 out of 30,000). Thus, CMV has largely been ruled out as an oncogenic virus, though controversy remains as to whether it is an oncomodulator. According to this theory, CMV does not directly lead to malignant transformation but can perturb cellular processes in ways that promote oncogenic signaling and tumor growth. Treatment of GBM patients with the anti-viral drug Valganciclovir, in addition to standard of care GBM therapy, has been shown to improve median overall survival in newly diagnosed patients [24]. Reverse translational studies in syngeneic murine models revealed that tumor-bearing mice that had been perinatally infected with CMV fared worse than their uninfected counterparts, due in part to increased pericyte recruitment and enhanced angiogenesis in the tumor bed. Treating infected mice with antiviral therapy improved their survival by decreasing PDGF expression and impairing angiogenesis [25]. Collectively, these data support the notion that CMV might promote GBM growth in humans, but additional studies are required to prove this definitively.

2.2. *Epstein–Barr Virus*

Another oncomodulatory virus with questionable links to GBM is Epstein–Barr Virus (EBV). EBV is a ubiquitous gamma herpes virus that is present in approximately 90% of the adult population and manifests clinically as infectious mononucleosis during its acute lytic cycle. Following acute infection, EBV, like CMV, establishes life-long persistence. In its latency, EBV exists in one of three programs, each hallmarked by the expression of distinct RNA and proteins [26]. One such protein produced during latency programs II and III is LMP1, an oncoprotein that is critical for the carcinogenesis of various subsets of lymphomas and carcinomas [27]. This is in line with data that show EBV's main cellular reservoir is memory B cells [28,29]. The CNS is also gaining interest as a potential reservoir for EBV latency. Indeed, compliment receptor 2 (CR2), the primary receptor for

EBV, is expressed on astrocytes, and upon viral entry, astrocytes demonstrate increased proliferation [30,31]. EBV's association with a plethora of CNS indications including, but not limited to, ataxia, encephalitis, and demyelinating diseases, as well as its frequent presence in CNS lymphomas, has prompted inquiry into its role in gliomagenesis [32,33]. However, results thus far have been discordant and point to no clear role for EBV in GBM [34].

2.3. Human Endogenous Retroviruses

Studies linking human endogenous retroviruses (HERVs) to GBM are sparse, but there appears to be active interest in determining if HERVs contribute to human cancers. HERVs, which are remnants of retroviral germline infections, make up ~8% of the human genome [35]. The majority of HERVs are inactive and unable to replicate. However, certain HERVs, such as those in the HERV-K family, have retained open reading frames, and their viral products are translated during embryogenesis but typically silenced in healthy differentiated cells [36,37]. Due to genetic and epigenetic dysregulation in malignantly transformed cells, high levels of HERV products have been found in the following cancers: breast, lung, prostate, hepatocellular carcinoma, germ cell, melanoma, leukemia, lymphoma, and ovarian, among others [38–44]. Very recently, one study found that HERV transcripts (mostly from the ERV1 superfamily) were upregulated in GBM compared to adjacent tissue, though notably HERV-K was downregulated [45]. One other study reported that HERV-K could not be detected in GBM [46]. Given these conflicting reports and dearth of information, the role of HERVs in human GBM (if any) remains to be elucidated.

2.4. Exploiting Endogenous Viruses in GBM for Treatment

One reason why GBM is such an intractable tumor is because it can grow relatively unseen by the adaptive immune system until too late (i.e., when the tumor mass is large) [47]. This relative "invisibility" occurs at various steps of what should otherwise be the typical cancer immunity cycle: (1) release of tumor antigens, (2) antigen presentation by dendritic cells (DCs) and other professional antigen presenting cells (APCs), (3) priming and activation of T cells in secondary lymphoid tissues, (4) infiltration and engagement of the tumor by tumor-specific T cells [48,49]. Given that GBM is thought to be driven primarily by canonical mutations rather than carcinogens or other environmental perturbations, it harbors a far smaller neoantigen burden than other cancers [50–52]. Functionally, this means there are less altered self-antigens available for recognition by the immune system in steps 1 and 2 above. Furthermore, the brain lacks a population of migratory APCs to traffic tumor neoantigen to draining lymph nodes. There is also no evidence to date that lymphatic vessels directly contact and drain the brain parenchyma. Instead, parenchymal antigen is thought to drain along the 150 nm wide basement membranes of capillaries and arteries, as well as along exiting neural sheaths [53–55]. The little parenchymal antigen that does escape these tight barrier systems would enter the dura mater and potentially drain via its dedicated lymphatics [56,57]. However, we have observed in a murine model of GBM that the drainage of tumor antigen into the cervical lymph nodes is negligible, which impinges on step 3 in the cancer immunity cycle, unless drainage can be enhanced [47]. Lastly, the blood–brain and CSF–brain barriers are designed to limit leukocyte migration into the brain to prevent immunopathology [58,59]. In fact, CNS endothelium only supports the migration of activated leukocytes, and even then, those that extravasate into the perivascular space are sequestered there unless they recognize cognate antigen on perivascular APCs [60–62]. The latter would only occur if appropriate tumor antigen drained in the first place. These features impede step 4 of the cancer immunity cycle. In concert, the unique anatomy and immunology of the CNS afford glioblastomas with the ability to remain well ensconced and hidden in the brain. Therefore, it might be incredibly useful to leverage endogenous viral infections against GBM.

CMV antigens in infected tumor cells could potentially be exploited as neoantigens worth targeting (Figure 1A). Indeed, given the well-documented subclinical CMV viremia observed in GBM patients, there have been several adoptive T cell transfer and vaccine strategies employing CMV antigens as proxies for GBM-specific antigens and CMV-specific in place of tumor-specific T cells. A notable hallmark of CMV infection is its maintenance of outsized pools of oligoclonal memory T cells, which accumulate over time in a process called "memory inflation". Impressively, up to one quarter of the total CD8 T cell population of elderly CMV seropositive individuals is CMV-specific [63,64], and in contrast to other latent infections, CMV-specific T cells remain highly functional [65]. Such frequencies of oligoclonal T cells are rarely achieved even with vaccination. This, coupled with the high prevalence of CMV seropositivity in the adult population, makes co-opting this pre-existing pool of "tumor-specific" T cells attractive and pragmatic as a therapeutic.

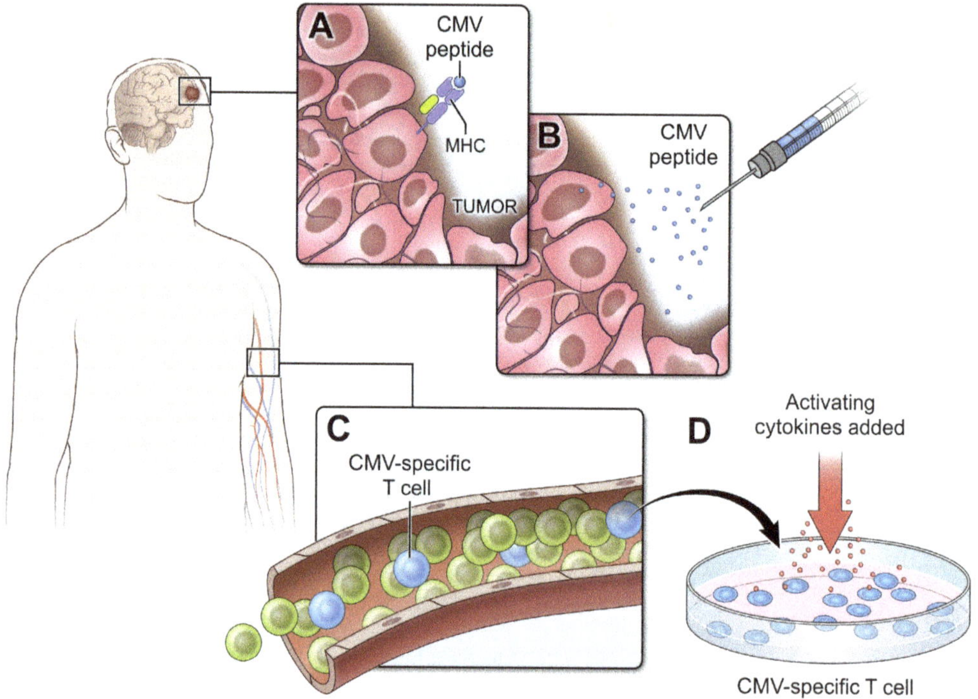

Figure 1. CMV xenogenization of glioblastoma. (**A**) GBM presents CMV peptide on the surface of MHC molecules. (**B**) CMV peptide injected either systemically or intratumorally can be used to activate CMV-specific T cells and route them to the tumor. (**C**,**D**) CMV-specific T cells can comprise up to 10% of the human T cell repertoire and represent a significant contingent of immune cells that can be harnessed for anti-GBM therapy. Many strategies currently involve transferring autologous T cells that have been stimulated ex vivo. Major histocompatibility complex (MHC), cytomegalovirus (CMV).

The entire approach of rendering GBM antigenically foreign to an otherwise unperturbed host through endogenous viruses predicates on intact viral antigen presentation on the tumor cell surface (Figure 1A). Though CMV has been detected in GBM, these assays do not test nor guarantee that CMV peptides are presented by major histocompatibility complexes (MHC), and there is adequate reason to doubt the extent to which appropriate antigen presentation occurs. Over half of human GBM specimens show loss of HLA class I antigens, and of the remaining fraction, half show faint staining, leaving only 20% of tumors with robust, intact membrane-bound MHC I expression [66,67]. The assessment of MHC I expression should thus be routine in GBM care and should occur prior to the start of CD8 T cell-based immunotherapies. The absence of MHC can be irreversible, as in the case of loss of heterozygosity (LOH) mutations at 6p21 and 15q21, loci corresponding to HLA class I and β2-microglobulin (β2m), respectively, which are critical components of the MHC molecule [68]. This occurs in almost half of GBM cases, and therapeutic strategies for these patients should be independent of MHC class I antigen presentation and CD8+ T cell engagement [68]. In other cases, the absence of surface MHC class I can be reversed with appropriate cytokine stimulation (e.g., type I and II interferons) provided the HLA-A, HLA-B, and HLA-C gene loci are intact [69].

Patients with intact MHC loci might benefit from efforts underway to activate CMV-specific T cells and promote their subsequent engagement of brain tumors by injecting CMV peptides, and early reports demonstrate the safety and feasibility of this approach (Figure 1B) [70,71]. Other methods to activate CMV-specific T cells (Figure 1C) have included the transfer of autologous DCs pulsed ex vivo with CMV lower matrix protein pp65 as an adjuvant to standard of care chemoradiation. Across three sequential, independently conducted clinical trials, nearly one third of GBM patients receiving pp65 DCs became "exceptional long-term survivors" [72,73]. More recent efforts to treat patients with autologous CMV-specific T cells activated and expanded ex vivo (Figure 1D) demonstrated safety as an adjuvant therapy for primary and recurrent GBM and the potential to extend survival in part via epitope spreading [74–76]. The latter is particularly important given how "quiet" the adaptive immune response would otherwise be. GBM also has a well-documented history of antigenic escape following therapeutic pressure [77]. Therefore, harnessing anti-viral immunity to wage an effective GBM response will likely require polyclonal T cell responses as well as epitope spreading to other antigens. The functional attributes of CMV-specific CD8 T cells are also important. To improve the polyfunctionality of therapeutic T cells, investigators recently combined the transfer of autologous CMV pp65-specific CD8+ T cells with pp65 RNA-loaded DCs. This combination improved frequencies of IFNγ, TNFα and CCL3 positive, polyfunctional CMV-specific T cells, which correlated with an increased overall survival [78]. However, it remains to be determined how these functional attributes influenced the subsequent antitumor response.

The strategies summarized in Figure 1 aim to combat the low immunogenicity of GBM and mount an adaptive antitumor immune response by harnessing CMV xenogenization of the tumor [79]. The efficacy of these strategies depends on there being sufficient cognate peptide-MHC I complexes on the surface of tumor cells for subsequent CD8+ T cell engagement and killing. However, MHC I loss is irreversible in nearly half of GBM patients [68], and in the majority of the remaining half, MHC is significantly downregulated [66,67,69]. In patients with a loss of MHC I expression, immunotherapeutic strategies must rely on leukocyte–tumor interactions that are independent of MHC (e.g., NK cells or antibodies). On the other hand, patients with inducible MHC I expression should be treated with dual approaches that first upregulate tumor antigen presentation (e.g., with type I or II interferons) and then promote cytolytic engagement of the tumor by anti-tumor (viral) CD8+ T cells.

3. Oncolytic Viral Therapy

Oncolytic viruses (OVs) are naturally occurring or genetically engineered viruses that selectively infect and lyse tumor cells, disseminating to adjacent malignant cells to repeat the process. OVs exploit the inherent dysregulation within the cancer cell—from cell signaling abnormalities to incapacitated antiviral machinery—as well as the cancer's own immune evasion pathways to preferentially infect and reproduce in neoplastic cells [80]. For this reason, oncolytic viruses do not necessarily rely on neoantigens at the tumor cell surface for their selective infection of tumor cells. OVs orchestrate antitumor activity through two distinct but related mechanisms. The first is their direct lysis of tumor cells, and the second is the concomitant innate and adaptive response mounted in response to damaged-associated molecular patterns (DAMPs) and viral pathogen-associated molecular patterns (PAMPs) released from dying tumor cells (Figure 2). The latter is thought to break immune tolerance against GBM. In fact, the molecular signals and cytokines induced by infection with OVs activate and attract a variety of APCs, as well as innate and adaptive immune cell subsets. Remarkably, the induction of a pleiotropic anti-GBM response by OVs has promoted survival in a subset of GBM patients.

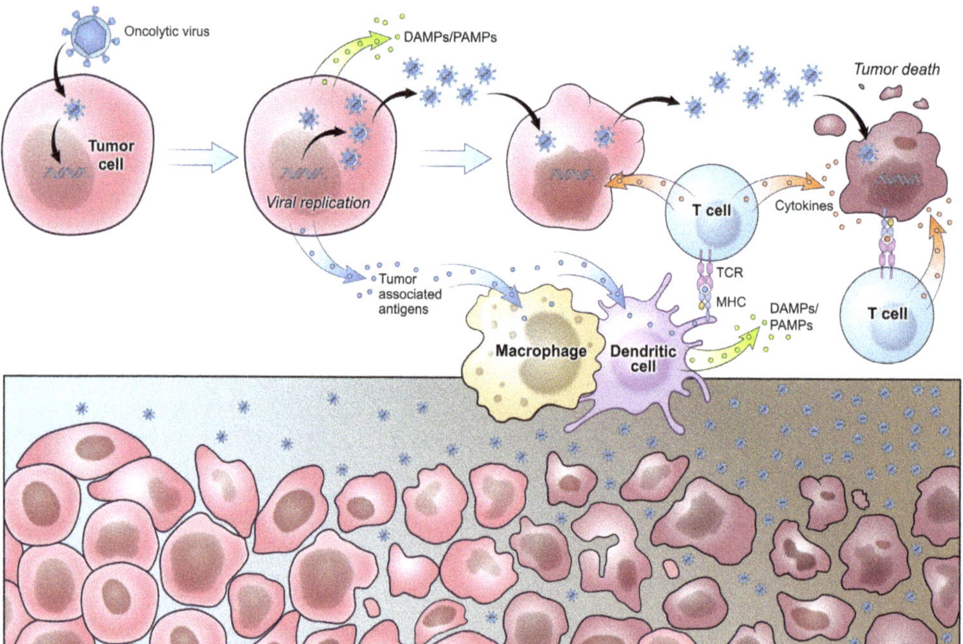

Figure 2. Viral oncolysis of glioblastoma. Oncolytic viruses are either naturally selected or engineered to preferentially infect tumor cells. Upon entry, viruses co-opt tumor cell machinery to support their own replication and lyse tumor cells in the process. Viral progeny are then released and continue infecting surrounding tumor cells. Secondary immune reactions develop in response to damage- and pathogen-associated molecular patterns (DAMPs and PAMPs), breaking immune tolerance against GBM. Increased antigen presentation facilitates the generation of both anti-viral and anti-tumor T cells that help destroy tumor cells. T cell receptor (TCR), major histocompatibility complex (MHC).

Among the most promising candidates is a prototype oncolytic poliovirus (PV) recombinant, PVSRIPO—a modified poliovirus lacking neurovirulence [81]. The tropism of PVSRIPO is based on CD155, which is a type I transmembrane glycoprotein broadly upregulated on a variety of tumor cells, but not on normal adult neural cells, and on APCs [82]. The infection of DCs by PVSRIPO does not lyse them. Rather, it induces a strong type I interferon response, enhances antigen presentation, and promotes the ability of DCs to prime tumor-specific T cells that traffic to the tumor microenvironment [83,84]. In a recent Phase 2 dosing study, PVSRIPO delivered intratumorally via catheter modestly increased survival relative to a historical control group from 11.3 to 12.5 months. However, more impressive was the fact that survival plateaued at 21% in PVSRIP patients at 24 months and 36 months compared to 14% and 4%, respectively, in historical controls [12]. These data suggest that a robust anti-tumor immune response was sustained in patients who responded to treatment, although additional studies into the HLA status and anti-tumor immune response in surviving patients are warranted. It will also be important to determine whether PVSRIPO promoted the release of hidden tumor neoantigens. Though most patients experienced adverse events related to brain inflammation, a concern with OVs, edema was successfully managed in all cases. Some potential confounding variables in this exciting trial include the younger age of study participants as well their higher initial gross total resection rates and lower rates of prior bevacizumab (anti-VEGF) therapy than historical controls.

In other OV trials, subsets of patients have similarly experienced durable clinical responses. Following intratumoral inoculation of a tumor-specific adenovirus, DNX-2041, 20% of patients were alive at 36 months [85]. Radiographic data in responders suggested immunogenic cell death, which is hypothesized to be the mechanism of action with OVs. G47Δ, a third-generation type I Herpes Simplex Virus (HSV-1) developed in Japan, received "Sakigake" breakthrough status and is expected to receive fast-track drug approval soon. Recent preliminary data from a Phase II study following six intratumoral injections of G47Δ showed a one year survival rate of 92.3% relative to the historical control rate of 15% [86]. HSV-1 possesses a few advantageous properties. First, it is highly proliferative and thus can achieve high titers. Second, it does not integrate into the host genome. Third, anti-HSV-1 agents to which G47Δ is hypersensitive are readily available, providing a safety break if needed [87]. Despite the capacity of HSV-1 to infect a variety of cells, the deletion of ICP6, a ribonucleotide reductase (RR) required for viral DNA synthesis, in G47Δ renders the virus dependent on abundant RR in the cells it infects. Tumor cells, but not normal cells, express sufficient compensatory levels of RR to support G47Δ replication [88].

Another virus that has attracted attention for its neurotropism is Zika Virus (ZIKV). It was recently reported that ZIKV preferentially lyses patient-derived glioma stem cells (GSCs) over differentiated GBM cells or human neural precursors, due in part to increased surface integrin alpha chain V expression [89,90]. GSCs are thought to drive tumor growth, immune evasion, and recurrence given their self-renewing multipotent nature, selective therapeutic resistance, and capacity to co-opt angiogenesis to self-sustain [1]. The use of ZIKV in GBM is still in the preclinical phase.

OVs offer the promise of not only warming an otherwise "cold" tumor niche, but also of delivering therapeutic cargo encoded in the virus. This is of particular importance in GBM since the various barrier systems of the CNS passively and actively (via efflux pumps) work to prevent the infiltration of delivered drugs. Indeed, drug concentrations in the tumor microenvironment can be less than 20% of their concentration in the plasma (Figure 3A) [91]. Arming OVs with immunostimulants provides another opportunity for this new class of therapeutics to break immune tolerance against GBM and control cargo delivery both spatially and temporally (Figure 3B). In pre-clinical studies using a fatal genetic model of GBM, an oncolytic HSV encoding IL-12 resulted in an unprecedented degree of protection and a 30% cure [92].

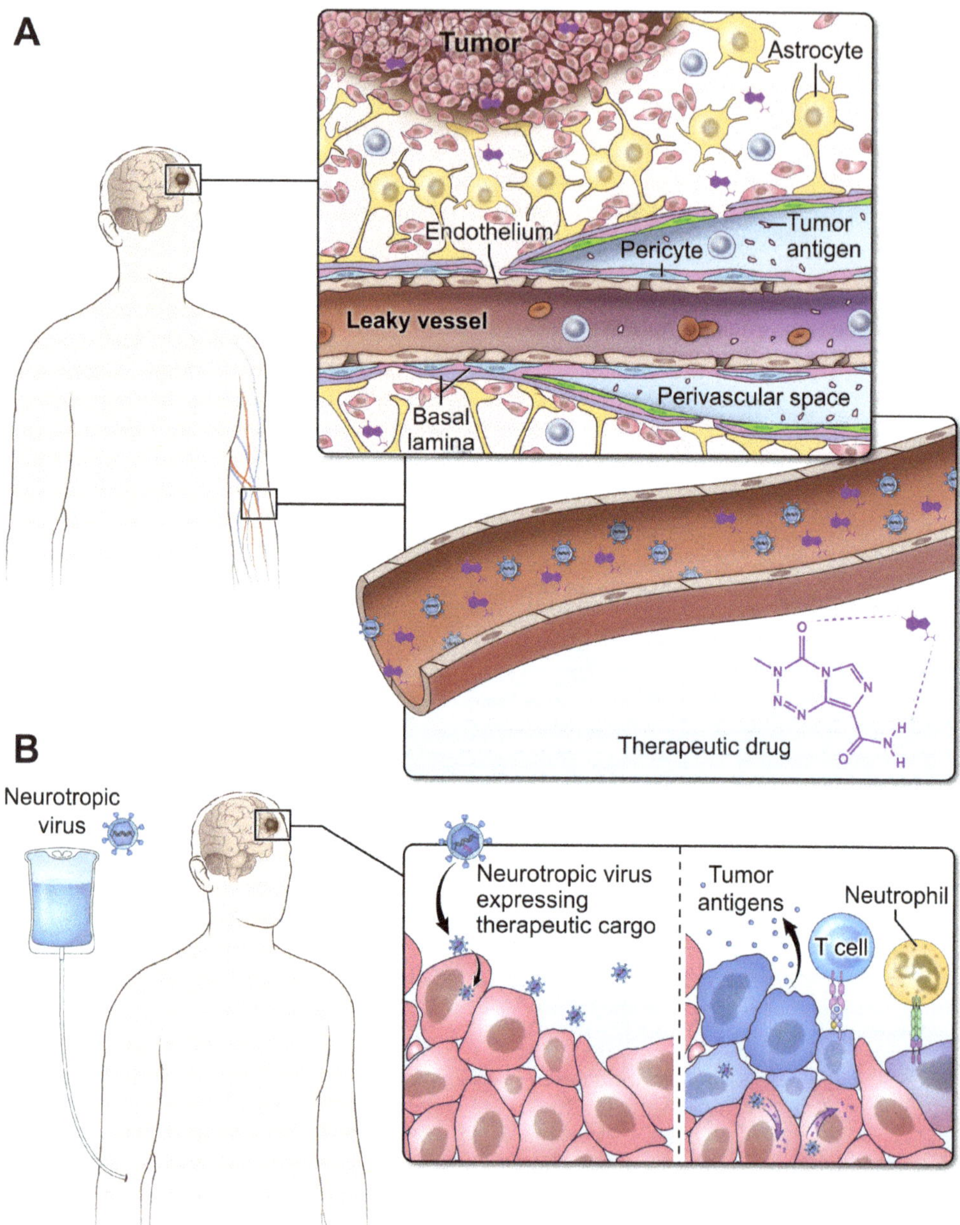

Figure 3. The blood–brain barrier (BBB) makes it challenging to deliver drugs to GBM. (**A**) Approximately 20% of the delivered concentration of anticancer therapeutics reaches the tumor bed despite the presence of leaky vessels. The BBB limits the degree to which therapeutics can access GBM. (**B**) Neurotropic viruses are naturally selected biologic agents that can traverse the BBB and infect the tumor bed. These viruses can be engineered to carry therapeutic cargo that is delivered directly to tumor cells, stimulating innate and adaptive immune responses.

4. Concluding Remarks

GBM remains one of the deadliest cancers, and its long-term survival rate has not changed in the last 50 years. There is a plethora of reasons for GBM's recalcitrance and lethality, a few of which include its striking heterogeneity, subset of stem cells, and capacity to induce systemic immune suppression. There are also significant limitations in the capacity of the brain to detect GBM antigens and support afferent as well as efferent immune responses. Few interventions can overcome all these hurdles except for virotherapy. Further studies are warranted to better understand the mechanisms by which these virotherapies work and improve upon their efficacy in all GBM patients. It will be important to map out determinants of long-term durable anti-tumor immune responses and identify other immunomodulatory therapies that provide synergistic combinatorial effects. Nonetheless, the rapid development of viral therapies for GBM starting at the turn of this century may ultimately have the greatest impact on improving life expectancy for GBM patients.

Funding: This research was supported the intramural program at the National Institute of Neurological Disorders and Stroke, National Institutes of Health (NIH).

Institutional Review Board Statement: Not applicable.

Informed Consent Statement: Not applicable.

Data Availability Statement: Not applicable.

Acknowledgments: We thank Alan Hoofring and Ethan Tyler in the NIH Medical Arts Design Section for generating the illustrations shown in Figures 1–3.

Conflicts of Interest: The authors declare no conflict of interest.

References

1. Gimple, R.C.; Bhargava, S.; Dixit, D.; Rich, J.N. Glioblastoma stem cells: Lessons from the tumor hierarchy in a lethal cancer. *Genes Dev.* **2019**, *33*, 591–609. [CrossRef]
2. Ostrom, Q.T.; Gittleman, H.; Fulop, J.; Liu, M.; Blanda, R.; Kromer, C.; Wolinsky, Y.; Kruchko, C.; Barnholtz-Sloan, J.S. CBTRUS Statistical Report: Primary Brain and Central Nervous System Tumors Diagnosed in the United States in 2008–2012. *Neuro-Oncology* **2015**, *17* (Suppl. 4), iv1–iv62. [CrossRef]
3. Lee, J.H.; Lee, J.E.; Kahng, J.Y.; Kim, S.H.; Park, J.S.; Yoon, S.J.; Um, J.Y.; Kim, W.K.; Lee, J.K.; Park, J.; et al. Human glioblastoma arises from subventricular zone cells with low-level driver mutations. *Nature* **2018**, *560*, 243–247. [CrossRef]
4. Stupp, R.; Hegi, M.E.; Mason, W.P.; van den Bent, M.J.; Taphoorn, M.J.; Janzer, R.C.; Ludwin, S.K.; Allgeier, A.; Fisher, B.; Belanger, K.; et al. Effects of radiotherapy with concomitant and adjuvant temozolomide versus radiotherapy alone on survival in glioblastoma in a randomised phase III study: 5-year analysis of the EORTC-NCIC trial. *Lancet Oncol.* **2009**, *10*, 459–466. [CrossRef]
5. Poon, M.T.C.; Sudlow, C.L.M.; Figueroa, J.D.; Brennan, P.M. Longer-term ($\geq$ 2 years) survival in patients with glioblastoma in population-based studies pre- and post-2005: A systematic review and meta-analysis. *Sci. Rep.* **2020**, *10*, 11622. [CrossRef]
6. Fukushima, S.; Narita, Y.; Miyakita, Y.; Ohno, M.; Takizawa, T.; Takusagawa, Y.; Mori, M.; Ichimura, K.; Tsuda, H.; Shibui, S. A case of more than 20 years survival with glioblastoma, and development of cavernous angioma as a delayed complication of radiotherapy. *Neuropathology* **2013**, *33*, 576–581. [CrossRef]
7. Kim, J.; Lee, I.H.; Cho, H.J.; Park, C.K.; Jung, Y.S.; Kim, Y.; Nam, S.H.; Kim, B.S.; Johnson, M.D.; Kong, D.S.; et al. Spatiotemporal Evolution of the Primary Glioblastoma Genome. *Cancer Cell* **2015**, *28*, 318–328. [CrossRef] [PubMed]
8. Wang, Q.; Hu, B.; Hu, X.; Kim, H.; Squatrito, M.; Scarpace, L.; deCarvalho, A.C.; Lyu, S.; Li, P.; Li, Y.; et al. Tumor Evolution of Glioma-Intrinsic Gene Expression Subtypes Associates with Immunological Changes in the Microenvironment. *Cancer Cell* **2017**, *32*, 42–56.e6. [CrossRef] [PubMed]
9. Darmanis, S.; Sloan, S.A.; Croote, D.; Mignardi, M.; Chernikova, S.; Samghababi, P.; Zhang, Y.; Neff, N.; Kowarsky, M.; Caneda, C.; et al. Single-Cell RNA-Seq Analysis of Infiltrating Neoplastic Cells at the Migrating Front of Human Glioblastoma. *Cell Rep.* **2017**, *21*, 1399–1410. [CrossRef] [PubMed]
10. Stupp, R.; Taillibert, S.; Kanner, A.; Read, W.; Steinberg, D.; Lhermitte, B.; Toms, S.; Idbaih, A.; Ahluwalia, M.S.; Fink, K.; et al. Effect of Tumor-Treating Fields Plus Maintenance Temozolomide vs Maintenance Temozolomide Alone on Survival in Patients With Glioblastoma: A Randomized Clinical Trial. *JAMA* **2017**, *318*, 2306–2316. [CrossRef]
11. Fisher, J.P.; Adamson, D.C. Current FDA-Approved Therapies for High-Grade Malignant Gliomas. *Biomedicines* **2021**, *9*, 324. [CrossRef]

12. Desjardins, A.; Gromeier, M.; Herndon, J.E., 2nd; Beaubier, N.; Bolognesi, D.P.; Friedman, A.H.; Friedman, H.S.; McSherry, F.; Muscat, A.M.; Nair, S.; et al. Recurrent Glioblastoma Treated with Recombinant Poliovirus. *N. Engl. J. Med.* **2018**, *379*, 150–161. [CrossRef]
13. De Bonis, P.; Albanese, A.; Lofrese, G.; de Waure, C.; Mangiola, A.; Pettorini, B.L.; Pompucci, A.; Balducci, M.; Fiorentino, A.; Lauriola, L.; et al. Postoperative infection may influence survival in patients with glioblastoma: Simply a myth? *Neurosurgery* **2011**, *69*, 864–868. [CrossRef]
14. Eakin, E. Bacteria on the Brain. *The New Yorker*, 7 December 2015.
15. Cannon, M.J.; Schmid, D.S.; Hyde, T.B. Review of cytomegalovirus seroprevalence and demographic characteristics associated with infection. *Rev. Med. Virol.* **2010**, *20*, 202–213. [CrossRef]
16. Cobbs, C.S.; Harkins, L.; Samanta, M.; Gillespie, G.Y.; Bharara, S.; King, P.H.; Nabors, L.B.; Cobbs, C.G.; Britt, W.J. Human cytomegalovirus infection and expression in human malignant glioma. *Cancer Res.* **2002**, *62*, 3347–3350. [PubMed]
17. Scheurer, M.E.; Bondy, M.L.; Aldape, K.D.; Albrecht, T.; El-Zein, R. Detection of human cytomegalovirus in different histological types of gliomas. *Acta Neuropathol.* **2008**, *116*, 79–86. [CrossRef] [PubMed]
18. Poltermann, S.; Schlehofer, B.; Steindorf, K.; Schnitzler, P.; Geletneky, K.; Schlehofer, J.R. Lack of association of herpesviruses with brain tumors. *J. Neurovirol.* **2006**, *12*, 90–99. [CrossRef] [PubMed]
19. Mitchell, D.A.; Xie, W.; Schmittling, R.; Learn, C.; Friedman, A.; McLendon, R.E.; Sampson, J.H. Sensitive detection of human cytomegalovirus in tumors and peripheral blood of patients diagnosed with glioblastoma. *Neuro-Oncology* **2008**, *10*, 10–18. [CrossRef] [PubMed]
20. Sabatier, J.; Uro-Coste, E.; Pommepuy, I.; Labrousse, F.; Allart, S.; Trémoulet, M.; Delisle, M.B.; Brousset, P. Detection of human cytomegalovirus genome and gene products in central nervous system tumours. *Br. J. Cancer* **2005**, *92*, 747–750. [CrossRef] [PubMed]
21. Dziurzynski, K.; Chang, S.M.; Heimberger, A.B.; Kalejta, R.F.; McGregor Dallas, S.R.; Smit, M.; Soroceanu, L.; Cobbs, C.S. Consensus on the role of human cytomegalovirus in glioblastoma. *Neuro-Oncology* **2012**, *14*, 246–255. [CrossRef]
22. Ranganathan, P.; Clark, P.A.; Kuo, J.S.; Salamat, M.S.; Kalejta, R.F. Significant association of multiple human cytomegalovirus genomic Loci with glioblastoma multiforme samples. *J. Virol.* **2012**, *86*, 854–864. [CrossRef] [PubMed]
23. Spector, D.H. Human cytomegalovirus riding the cell cycle. *Med. Microbiol. Immunol.* **2015**, *204*, 409–419. [CrossRef]
24. Stragliotto, G.; Pantalone, M.R.; Rahbar, A.; Bartek, J.; Söderberg-Naucler, C. Valganciclovir as Add-on to Standard Therapy in Glioblastoma Patients. *Clin. Cancer Res.* **2020**, *26*, 4031–4039. [CrossRef] [PubMed]
25. Krenzlin, H.; Behera, P.; Lorenz, V.; Passaro, C.; Zdioruk, M.; Nowicki, M.O.; Grauwet, K.; Zhang, H.; Skubal, M.; Ito, H.; et al. Cytomegalovirus promotes murine glioblastoma growth via pericyte recruitment and angiogenesis. *J. Clin. Investig.* **2019**, *129*, 1671–1683. [CrossRef]
26. Münz, C. Latency and lytic replication in Epstein-Barr virus-associated oncogenesis. *Nat. Rev. Microbiol.* **2019**, *17*, 691–700. [CrossRef] [PubMed]
27. Salahuddin, S.; Fath, E.K.; Biel, N.; Ray, A.; Moss, C.R.; Patel, A.; Patel, S.; Hilding, L.; Varn, M.; Ross, T.; et al. Epstein-Barr Virus Latent Membrane Protein-1 Induces the Expression of SUMO-1 and SUMO-2/3 in LMP1-positive Lymphomas and Cells. *Sci. Rep.* **2019**, *9*, 208. [CrossRef] [PubMed]
28. Heath, E.; Begue-Pastor, N.; Chaganti, S.; Croom-Carter, D.; Shannon-Lowe, C.; Kube, D.; Feederle, R.; Delecluse, H.J.; Rickinson, A.B.; Bell, A.I. Epstein-Barr virus infection of naïve B cells in vitro frequently selects clones with mutated immunoglobulin genotypes: Implications for virus biology. *PLoS Pathog.* **2012**, *8*, e1002697. [CrossRef] [PubMed]
29. Hatton, O.L.; Harris-Arnold, A.; Schaffert, S.; Krams, S.M.; Martinez, O.M. The interplay between Epstein-Barr virus and B lymphocytes: Implications for infection, immunity, and disease. *Immunol. Res.* **2014**, *58*, 268–276. [CrossRef]
30. Menet, A.; Speth, C.; Larcher, C.; Prodinger, W.M.; Schwendinger, M.G.; Chan, P.; Jäger, M.; Schwarzmann, F.; Recheis, H.; Fontaine, M.; et al. Epstein-Barr virus infection of human astrocyte cell lines. *J. Virol.* **1999**, *73*, 7722–7733. [CrossRef]
31. Słońska, A.; Cymerys, J.; Chodkowski, M.; Bąska, P.; Krzyżowska, M.; Bańbura, M.W. Human herpesvirus type 2 infection of primary murine astrocytes causes disruption of the mitochondrial network and remodeling of the actin cytoskeleton: An in vitro morphological study. *Arch. Virol.* **2021**, *166*, 1371–1383. [CrossRef]
32. Fujimoto, H.; Asaoka, K.; Imaizumi, T.; Ayabe, M.; Shoji, H.; Kaji, M. Epstein-Barr virus infections of the central nervous system. *Intern. Med.* **2003**, *42*, 33–40. [CrossRef] [PubMed]
33. Soldan, S.S.; Lieberman, P.M. Epstein-Barr Virus Infection in the Development of Neurological Disorders. *Drug Discov. Today Dis. Models* **2020**, *32 Pt A*, 35–52. [CrossRef]
34. Strong, M.J.; Blanchard, E.t.; Lin, Z.; Morris, C.A.; Baddoo, M.; Taylor, C.M.; Ware, M.L.; Flemington, E.K. A comprehensive next generation sequencing-based virome assessment in brain tissue suggests no major virus–tumor association. *Acta Neuropathol. Commun.* **2016**, *4*, 71. [CrossRef] [PubMed]
35. Markovitz, D.M. "Reverse genomics" and human endogenous retroviruses. *Trans. Am. Clin. Climatol. Assoc.* **2014**, *125*, 57–62. [PubMed]
36. Grow, E.J.; Flynn, R.A.; Chavez, S.L.; Bayless, N.L.; Wossidlo, M.; Wesche, D.J.; Martin, L.; Ware, C.B.; Blish, C.A.; Chang, H.Y.; et al. Intrinsic retroviral reactivation in human preimplantation embryos and pluripotent cells. *Nature* **2015**, *522*, 221–225. [CrossRef]

37. Mayer, J.; Sauter, M.; Rácz, A.; Scherer, D.; Mueller-Lantzsch, N.; Meese, E. An almost-intact human endogenous retrovirus K on human chromosome 7. *Nat. Genet.* **1999**, *21*, 257–258. [CrossRef]
38. Serafino, A.; Balestrieri, E.; Pierimarchi, P.; Matteucci, C.; Moroni, G.; Oricchio, E.; Rasi, G.; Mastino, A.; Spadafora, C.; Garaci, E.; et al. The activation of human endogenous retrovirus K (HERV-K) is implicated in melanoma cell malignant transformation. *Exp. Cell Res.* **2009**, *315*, 849–862. [CrossRef]
39. Kleiman, A.; Senyuta, N.; Tryakin, A.; Sauter, M.; Karseladze, A.; Tjulandin, S.; Gurtsevitch, V.; Mueller-Lantzsch, N. HERV-K(HML-2) GAG/ENV antibodies as indicator for therapy effect in patients with germ cell tumors. *Int. J. Cancer* **2004**, *110*, 459–461. [CrossRef]
40. Ma, W.; Hong, Z.; Liu, H.; Chen, X.; Ding, L.; Liu, Z.; Zhou, F.; Yuan, Y. Human Endogenous Retroviruses-K (HML-2) Expression Is Correlated with Prognosis and Progress of Hepatocellular Carcinoma. *Biomed. Res. Int.* **2016**, *2016*, 8201642. [CrossRef]
41. Ishida, T.; Obata, Y.; Ohara, N.; Matsushita, H.; Sato, S.; Uenaka, A.; Saika, T.; Miyamura, T.; Chayama, K.; Nakamura, Y.; et al. Identification of the HERV-K gag antigen in prostate cancer by SEREX using autologous patient serum and its immunogenicity. *Cancer Immun.* **2008**, *8*, 15.
42. Kahyo, T.; Tao, H.; Shinmura, K.; Yamada, H.; Mori, H.; Funai, K.; Kurabe, N.; Suzuki, M.; Tanahashi, M.; Niwa, H.; et al. Identification and association study with lung cancer for novel insertion polymorphisms of human endogenous retrovirus. *Carcinogenesis* **2013**, *34*, 2531–2538. [CrossRef] [PubMed]
43. Wang-Johanning, F.; Radvanyi, L.; Rycaj, K.; Plummer, J.B.; Yan, P.; Sastry, K.J.; Piyathilake, C.J.; Hunt, K.K.; Johanning, G.L. Human endogenous retrovirus K triggers an antigen-specific immune response in breast cancer patients. *Cancer Res.* **2008**, *68*, 5869–5877. [CrossRef]
44. Wang-Johanning, F.; Liu, J.; Rycaj, K.; Huang, M.; Tsai, K.; Rosen, D.G.; Chen, D.T.; Lu, D.W.; Barnhart, K.F.; Johanning, G.L. Expression of multiple human endogenous retrovirus surface envelope proteins in ovarian cancer. *Int. J. Cancer* **2007**, *120*, 81–90. [CrossRef] [PubMed]
45. Yuan, Z.; Yang, Y.; Zhang, N.; Soto, C.; Jiang, X.; An, Z.; Zheng, W.J. Human Endogenous Retroviruses in Glioblastoma Multiforme. *Microorganisms* **2021**, *9*, 764. [CrossRef] [PubMed]
46. Kessler, A.F.; Wiesner, M.; Denner, J.; Kämmerer, U.; Vince, G.H.; Linsenmann, T.; Löhr, M.; Ernestus, R.I.; Hagemann, C. Expression-analysis of the human endogenous retrovirus HERV-K in human astrocytic tumors. *BMC Res. Notes* **2014**, *7*, 159. [CrossRef] [PubMed]
47. Song, E.; Mao, T.; Dong, H.; Boisserand, L.S.B.; Antila, S.; Bosenberg, M.; Alitalo, K.; Thomas, J.L.; Iwasaki, A. VEGF-C-driven lymphatic drainage enables immunosurveillance of brain tumours. *Nature* **2020**, *577*, 689–694. [CrossRef] [PubMed]
48. Chen, D.S.; Mellman, I. Oncology Meets Immunology: The Cancer-Immunity Cycle. *Immunity* **2013**, *39*, 1–10. [CrossRef]
49. Ozga, A.J.; Chow, M.T.; Luster, A.D. Chemokines and the immune response to cancer. *Immunity* **2021**, *54*, 859–874. [CrossRef]
50. Alexandrov, L.B.; Nik-Zainal, S.; Wedge, D.C.; Aparicio, S.A.; Behjati, S.; Biankin, A.V.; Bignell, G.R.; Bolli, N.; Borg, A.; Børresen-Dale, A.L.; et al. Signatures of mutational processes in human cancer. *Nature* **2013**, *500*, 415–421. [CrossRef]
51. Neftel, C.; Laffy, J.; Filbin, M.G.; Hara, T.; Shore, M.E.; Rahme, G.J.; Richman, A.R.; Silverbush, D.; Shaw, M.L.; Hebert, C.M.; et al. An Integrative Model of Cellular States, Plasticity, and Genetics for Glioblastoma. *Cell* **2019**, *178*, 835–849.e21. [CrossRef]
52. Lawrence, M.S.; Stojanov, P.; Polak, P.; Kryukov, G.V.; Cibulskis, K.; Sivachenko, A.; Carter, S.L.; Stewart, C.; Mermel, C.H.; Roberts, S.A.; et al. Mutational heterogeneity in cancer and the search for new cancer-associated genes. *Nature* **2013**, *499*, 214–218. [CrossRef] [PubMed]
53. Carare, R.O.; Bernardes-Silva, M.; Newman, T.A.; Page, A.M.; Nicoll, J.A.; Perry, V.H.; Weller, R.O. Solutes, but not cells, drain from the brain parenchyma along basement membranes of capillaries and arteries: Significance for cerebral amyloid angiopathy and neuroimmunology. *Neuropathol. Appl. Neurobiol.* **2008**, *34*, 131–144. [CrossRef] [PubMed]
54. Ma, Q.; Ineichen, B.V.; Detmar, M.; Proulx, S.T. Outflow of cerebrospinal fluid is predominantly through lymphatic vessels and is reduced in aged mice. *Nat. Commun.* **2017**, *8*, 1434. [CrossRef] [PubMed]
55. Kida, S.; Pantazis, A.; Weller, R.O. CSF drains directly from the subarachnoid space into nasal lymphatics in the rat. Anatomy, histology and immunological significance. *Neuropathol. Appl. Neurobiol.* **1993**, *19*, 480–488. [CrossRef]
56. Ahn, J.H.; Cho, H.; Kim, J.H.; Kim, S.H.; Ham, J.S.; Park, I.; Suh, S.H.; Hong, S.P.; Song, J.H.; Hong, Y.K.; et al. Meningeal lymphatic vessels at the skull base drain cerebrospinal fluid. *Nature* **2019**, *572*, 62–66. [CrossRef]
57. Louveau, A.; Smirnov, I.; Keyes, T.J.; Eccles, J.D.; Rouhani, S.J.; Peske, J.D.; Derecki, N.C.; Castle, D.; Mandell, J.W.; Lee, K.S.; et al. Structural and functional features of central nervous system lymphatic vessels. *Nature* **2015**, *523*, 337–341. [CrossRef]
58. Engelhardt, B.; Ransohoff, R.M. Capture, crawl, cross: The T cell code to breach the blood-brain barriers. *Trends Immunol.* **2012**, *33*, 579–589. [CrossRef]
59. Mastorakos, P.; McGavern, D. The anatomy and immunology of vasculature in the central nervous system. *Sci. Immunol.* **2019**, *4*, eaav0492. [CrossRef]
60. Greter, M.; Heppner, F.L.; Lemos, M.P.; Odermatt, B.M.; Goebels, N.; Laufer, T.; Noelle, R.J.; Becher, B. Dendritic cells permit immune invasion of the CNS in an animal model of multiple sclerosis. *Nat. Med.* **2005**, *11*, 328–334. [CrossRef]
61. Lodygin, D.; Odoardi, F.; Schläger, C.; Körner, H.; Kitz, A.; Nosov, M.; van den Brandt, J.; Reichardt, H.M.; Haberl, M.; Flügel, A. A combination of fluorescent NFAT and H2B sensors uncovers dynamics of T cell activation in real time during CNS autoimmunity. *Nat. Med.* **2013**, *19*, 784–790. [CrossRef]

62. Kawakami, N.; Nägerl, U.V.; Odoardi, F.; Bonhoeffer, T.; Wekerle, H.; Flügel, A. Live imaging of effector cell trafficking and autoantigen recognition within the unfolding autoimmune encephalomyelitis lesion. *J. Exp. Med.* **2005**, *201*, 1805–1814. [CrossRef]
63. Khan, N.; Shariff, N.; Cobbold, M.; Bruton, R.; Ainsworth, J.A.; Sinclair, A.J.; Nayak, L.; Moss, P.A. Cytomegalovirus seropositivity drives the CD8 T cell repertoire toward greater clonality in healthy elderly individuals. *J. Immunol.* **2002**, *169*, 1984–1992. [CrossRef]
64. Alejenef, A.; Pachnio, A.; Halawi, M.; Christmas, S.E.; Moss, P.A.; Khan, N. Cytomegalovirus drives Vδ2neg γδ T cell inflation in many healthy virus carriers with increasing age. *Clin. Exp. Immunol.* **2014**, *176*, 418–428. [CrossRef] [PubMed]
65. Kim, J.; Kim, A.R.; Shin, E.C. Cytomegalovirus Infection and Memory T Cell Inflation. *Immune Netw.* **2015**, *15*, 186–190. [CrossRef]
66. Facoetti, A.; Nano, R.; Zelini, P.; Morbini, P.; Benericetti, E.; Ceroni, M.; Campoli, M.; Ferrone, S. Human leukocyte antigen and antigen processing machinery component defects in astrocytic tumors. *Clin. Cancer Res.* **2005**, *11*, 8304–8311. [CrossRef] [PubMed]
67. Castro, M.; Sipos, B.; Pieper, N.; Biskup, S. Major histocompatibility complex class 1 (MHC1) loss among patients with glioblastoma (GBM). *J. Clin. Oncol.* **2020**, *38*, e14523. [CrossRef]
68. Yeung, J.T.; Hamilton, R.L.; Ohnishi, K.; Ikeura, M.; Potter, D.M.; Nikiforova, M.N.; Ferrone, S.; Jakacki, R.I.; Pollack, I.F.; Okada, H. LOH in the HLA class I region at 6p21 is associated with shorter survival in newly diagnosed adult glioblastoma. *Clin. Cancer Res.* **2013**, *19*, 1816–1826. [CrossRef]
69. Satoh, E.; Mabuchi, T.; Satoh, H.; Asahara, T.; Nukui, H.; Naganuma, H. Reduced expression of the transporter associated with antigen processing 1 molecule in malignant glioma cells, and its restoration by interferon-gamma and -beta. *J. Neurosurg.* **2006**, *104*, 264–271. [CrossRef]
70. Cuburu, N.; Bialkowski, L.; Pontejo, S.M.; Bell, A.; Kim, R.; Thompson, C.D.; Lowy, D.R.; Schiller, J.T. Abstract 5563: Harnessing pre-existing viral immunity for development of a broadly applicable tumor immunotherapy. *Cancer Res.* **2020**, *80*, 5563.
71. Thompson, E.; Landi, D.; Thompson, E.; Lipp, E.; Balajonda, B.; Herndon, J., II; Buckley, E.; Flahiff, C.; Jaggers, D.; Schroeder, K.; et al. CTIM-21. Peptide vaccine directed to CMV pp65 for treatment of recurrent malignant glioma and medulloblastoma in children and young adults: Preliminary results of a phase I trial. *Neuro-Oncology* **2020**, *22* (Suppl. 2), ii37. [CrossRef]
72. Batich, K.A.; Mitchell, D.A.; Healy, P.; Herndon, J.E.; Sampson, J.H. Once, Twice, Three Times a Finding: Reproducibility of Dendritic Cell Vaccine Trials Targeting Cytomegalovirus in Glioblastoma. *Clin. Cancer Res.* **2020**, *26*, 5297. [CrossRef] [PubMed]
73. Mitchell, D.A.; Batich, K.A.; Gunn, M.D.; Huang, M.N.; Sanchez-Perez, L.; Nair, S.K.; Congdon, K.L.; Reap, E.A.; Archer, G.E.; Desjardins, A.; et al. Tetanus toxoid and CCL3 improve dendritic cell vaccines in mice and glioblastoma patients. *Nature* **2015**, *519*, 366–369. [CrossRef] [PubMed]
74. Smith, C.; Lineburg, K.E.; Martins, J.P.; Ambalathingal, G.R.; Neller, M.A.; Morrison, B.; Matthews, K.K.; Rehan, S.; Crooks, P.; Panikkar, A.; et al. Autologous CMV-specific T cells are a safe adjuvant immunotherapy for primary glioblastoma multiforme. *J. Clin. Investig.* **2020**, *130*, 6041–6053. [CrossRef] [PubMed]
75. Schuessler, A.; Smith, C.; Beagley, L.; Boyle, G.M.; Rehan, S.; Matthews, K.; Jones, L.; Crough, T.; Dasari, V.; Klein, K.; et al. Autologous T-cell therapy for cytomegalovirus as a consolidative treatment for recurrent glioblastoma. *Cancer Res.* **2014**, *74*, 3466–3476. [CrossRef] [PubMed]
76. Ghazi, A.; Ashoori, A.; Hanley, P.J.; Brawley, V.S.; Shaffer, D.R.; Kew, Y.; Powell, S.Z.; Grossman, R.; Grada, Z.; Scheurer, M.E.; et al. Generation of polyclonal CMV-specific T cells for the adoptive immunotherapy of glioblastoma. *J. Immunother.* **2012**, *35*, 159–168. [CrossRef]
77. Brown, C.E.; Alizadeh, D.; Starr, R.; Weng, L.; Wagner, J.R.; Naranjo, A.; Ostberg, J.R.; Blanchard, M.S.; Kilpatrick, J.; Simpson, J.; et al. Regression of Glioblastoma after Chimeric Antigen Receptor T-Cell Therapy. *N. Engl. J. Med.* **2016**, *375*, 2561–2569. [CrossRef]
78. Reap, E.A.; Suryadevara, C.M.; Batich, K.A.; Sanchez-Perez, L.; Archer, G.E.; Schmittling, R.J.; Norberg, P.K.; Herndon, J.E.; Healy, P.; Congdon, K.L.; et al. Dendritic Cells Enhance Polyfunctionality of Adoptively Transferred T Cells That Target Cytomegalovirus in Glioblastoma. *Cancer Res.* **2018**, *78*, 256. [CrossRef]
79. Kobayashi, M. Viral Xenogenization of Intact Tumor Cells. In *Advances in Cancer Research*; Klein, G., Weinhouse, S., Eds.; Academic Press: Cambridge, MA, USA, 1979; Volume 30, pp. 279–299.
80. Kaufman, H.L.; Kohlhapp, F.J.; Zloza, A. Oncolytic viruses: A new class of immunotherapy drugs. *Nat. Rev. Drug Discov.* **2015**, *14*, 642–662. [CrossRef] [PubMed]
81. Dobrikova, E.Y.; Goetz, C.; Walters, R.W.; Lawson, S.K.; Peggins, J.O.; Muszynski, K.; Ruppel, S.; Poole, K.; Giardina, S.L.; Vela, E.M.; et al. Attenuation of neurovirulence, biodistribution, and shedding of a poliovirus:rhinovirus chimera after intrathalamic inoculation in Macaca fascicularis. *J. Virol.* **2012**, *86*, 2750–2759. [CrossRef]
82. Chandramohan, V.; Bryant, J.D.; Piao, H.; Keir, S.T.; Lipp, E.S.; Lefaivre, M.; Perkinson, K.; Bigner, D.D.; Gromeier, M.; McLendon, R.E. Validation of an Immunohistochemistry Assay for Detection of CD155, the Poliovirus Receptor, in Malignant Gliomas. *Arch. Pathol. Lab. Med.* **2017**, *141*, 1697–1704. [CrossRef]
83. Brown, M.C.; Holl, E.K.; Boczkowski, D.; Dobrikova, E.; Mosaheb, M.; Chandramohan, V.; Bigner, D.D.; Gromeier, M.; Nair, S.K. Cancer immunotherapy with recombinant poliovirus induces IFN-dominant activation of dendritic cells and tumor antigen-specific CTLs. *Sci. Transl. Med.* **2017**, *9*, eaan4220. [CrossRef]
84. Mosaheb, M.M.; Dobrikova, E.Y.; Brown, M.C.; Yang, Y.; Cable, J.; Okada, H.; Nair, S.K.; Bigner, D.D.; Ashley, D.M.; Gromeier, M. Genetically stable poliovirus vectors activate dendritic cells and prime antitumor CD8 T cell immunity. *Nat. Commun.* **2020**, *11*, 524. [CrossRef]

85. Lang, F.F.; Conrad, C.; Gomez-Manzano, C.; Yung, W.K.A.; Sawaya, R.; Weinberg, J.S.; Prabhu, S.S.; Rao, G.; Fuller, G.N.; Aldape, K.D.; et al. Phase I Study of DNX-2401 (Delta-24-RGD) Oncolytic Adenovirus: Replication and Immunotherapeutic Effects in Recurrent Malignant Glioma. *J. Clin. Oncol.* **2018**, *36*, 1419–1427. [CrossRef] [PubMed]
86. Todo, T. ATIM-14. Results of phase II clinical trial of oncolytic herpes virus G47Δ in patients with glioblastoma. *Neuro-Oncology* **2019**, *21* (Suppl. 6), vi4. [CrossRef]
87. Agarwalla, P.K.; Aghi, M.K. Oncolytic herpes simplex virus engineering and preparation. *Methods Mol. Biol.* **2012**, *797*, 1–19.
88. Mineta, T.; Rabkin, S.D.; Yazaki, T.; Hunter, W.D.; Martuza, R.L. Attenuated multi-mutated herpes simplex virus-1 for the treatment of malignant gliomas. *Nat. Med.* **1995**, *1*, 938–943. [CrossRef]
89. Zhu, Z.; Gorman, M.J.; McKenzie, L.D.; Chai, J.N.; Hubert, C.G.; Prager, B.C.; Fernandez, E.; Richner, J.M.; Zhang, R.; Shan, C.; et al. Zika virus has oncolytic activity against glioblastoma stem cells. *J. Exp. Med.* **2017**, *214*, 2843–2857. [CrossRef] [PubMed]
90. Zhu, Z.; Mesci, P.; Bernatchez, J.A.; Gimple, R.C.; Wang, X.; Schafer, S.T.; Wettersten, H.I.; Beck, S.; Clark, A.E.; Wu, Q.; et al. Zika Virus Targets Glioblastoma Stem Cells through a SOX2-Integrin αvβ5 Axis. *Cell Stem Cell* **2020**, *26*, 187–204.e10. [CrossRef] [PubMed]
91. Zhou, Q.; Guo, P.; Kruh, G.D.; Vicini, P.; Wang, X.; Gallo, J.M. Predicting human tumor drug concentrations from a preclinical pharmacokinetic model of temozolomide brain disposition. *Clin. Cancer Res.* **2007**, *13*, 4271–4279. [CrossRef]
92. Alessandrini, F.; Menotti, L.; Avitabile, E.; Appolloni, I.; Ceresa, D.; Marubbi, D.; Campadelli-Fiume, G.; Malatesta, P. Eradication of glioblastoma by immuno-virotherapy with a retargeted oncolytic HSV in a preclinical model. *Oncogene* **2019**, *38*, 4467–4479. [CrossRef]

Article

Screening and Identification of Lujo Virus Inhibitors Using a Recombinant Reporter Virus Platform

Stephen R. Welch , Jessica R. Spengler , Sarah C. Genzer, Payel Chatterjee, Mike Flint , Éric Bergeron , Joel M. Montgomery, Stuart T. Nichol, César G. Albariño and Christina F. Spiropoulou *

Viral Special Pathogens Branch, Centers for Disease Control and Prevention, U.S. Department of Health and Human Services, Atlanta, GA 30329, USA; yos6@cdc.gov (S.R.W.); wsk7@cdc.gov (J.R.S.); muz8@cdc.gov (S.C.G.); mwe9@cdc.gov (P.C.); vfa3@cdc.gov (M.F.); exj8@cdc.gov (É.B.); ztq9@cdc.gov (J.M.M.); stn1@cdc.gov (S.T.N.); bwu4@cdc.gov (C.G.A.)

* Correspondence: ccs8@cdc.gov

Abstract: Lujo virus (LUJV), a highly pathogenic arenavirus, was first identified in 2008 in Zambia. To aid the identification of effective therapeutics for LUJV, we developed a recombinant reporter virus system, confirming reporter LUJV comparability with wild-type virus and its utility in high-throughput antiviral screening assays. Using this system, we evaluated compounds with known and unknown efficacy against related arenaviruses, with the aim of identifying LUJV-specific and potential new pan-arenavirus antivirals. We identified six compounds demonstrating robust anti-LUJV activity, including several compounds with previously reported activity against other arenaviruses. These data provide critical evidence for developing broad-spectrum antivirals against high-consequence arenaviruses.

Keywords: Lujo virus; viral hemorrhagic fever; arenavirus; reporter virus; reverse genetics; antiviral screen; therapeutic; emerging viruses

Citation: Welch, S.R.; Spengler, J.R.; Genzer, S.C.; Chatterjee, P.; Flint, M.; Bergeron, É.; Montgomery, J.M.; Nichol, S.T.; Albariño, C.G.; Spiropoulou, C.F. Screening and Identification of Lujo Virus Inhibitors Using a Recombinant Reporter Virus Platform. *Viruses* **2021**, *13*, 1255. https://doi.org/10.3390/v13071255

Academic Editors: Michael B.A. Oldstone and Juan C. De la Torre

Received: 1 June 2021
Accepted: 23 June 2021
Published: 28 June 2021

1. Introduction

Lujo virus (LUJV), the first highly pathogenic arenavirus identified in Africa in over 40 years, was recognized after a cluster of severe viral hemorrhagic fever (VHF) cases in Southern Africa in September and October of 2008 [1,2]. The index case was identified in Zambia, and four subsequent nosocomial cases were detected in associated healthcare workers in South Africa. The infections initially presented as a non-specific febrile illness, but progressed in severity over 10–13 days to respiratory distress, neurological signs, and circulatory collapse [1]. To date, no additional cases of LUJV infection have been reported. However, the lack of a defined etiology or source of exposure for the index case; the apparent ease with which primary, secondary, and tertiary contacts were infected; and the unusually high case fatality rate (80%) caused significant alarm.

The family Arenaviridae encompasses three genera: *Mammarenavirus*, *Reptarenavirus*, and *Hartmanivirus* [3]. With the exception of the bat-borne Tacaribe virus, rodents are the natural hosts of the mammalian-specific mammarenaviruses, with each viral species closely associated with a single rodent species reservoir [4,5]. Infected rodents develop a persistent infection and continually shed the virus in body secretions such as urine and feces; human transmission occurs via contact with these secretions, either directly or in contaminated food products [6]. Mammarenaviruses are divided into two groups based on the geographic, genetic, and epidemiological relationship with their respective rodent hosts: Old World (or lymphocyte choriomeningitis/Lassa complex) species and New World (or Tacaribe complex) species [7]. Several *Mammarenavirus* species in both these groups can cause severe VHF disease in humans. Highly pathogenic Old World complex viruses such as LUJV and Lassa virus (LASV) are found in Africa, whereas the New World pathogens such as Junin (JUNV), Guanarito (GTOV), Machupo, and Chapare viruses are found in

South America [8]. Treatment options for these highly pathogenic mammarenaviruses remain limited. The only therapeutic agent currently in clinical use is convalescent plasma therapy for JUNV patients [9], although broad-spectrum antiviral nucleoside analogs such as ribavirin and favirpiravir (T-705) have demonstrated some clinical success in treating LASV if treatment is started early and when clinical signs are still mild [10]. While currently no FDA-approved drugs are licensed to specifically treat these pathogens, advances in drug discovery targeting specific areas of the viral life cycle, such as entry, trafficking, RNA replication, viral and host cell protein interactions, and budding are uncovering several promising avenues of research [11].

Here, we describe the development and rescue of a novel recombinant ZsGreen1 (ZsG)-expressing reporter LUJV, termed rLUJV/ZsG. By incorporating rLUJV/ZsG into a high-throughput antiviral compound screening assay, we were able to rapidly evaluate potential therapeutic candidates that were effective against LUJV. A panel of compound candidates was selected based on: (1) demonstrated activity against other arenaviruses; (2) demonstrated activity against other negative-sense RNA viruses; and (3) preliminary data from large in-house screening assays against multiple virus species. Of the 83 compounds tested, 20 demonstrated robust anti-LUJV activity; 6 were further confirmed to potently inhibit wild-type (wt) LUJV infection. Using LUJV in these assays provides critical support for antiviral activity of previously identified and novel compounds against re-emerging and emerging highly pathogenic arenaviruses.

2. Materials and Methods

2.1. Biosafety

All work with the infectious virus was conducted in a biosafety level 4 (BSL-4) laboratory at the Centers for Disease Control and Prevention (CDC, Atlanta, GA, USA) following established BSL-4 standard operating procedures approved by the Institutional Biosafety Committee. All recombinant virus work was approved by the Centers for Disease Control and Prevention Institutional Biosafety Committee.

2.2. Cells

Vero-E6, Huh7 (from Apath LLC, New York, NY, USA), and BSR-T7/5 cells were cultured in Dulbecco's modified Eagle's medium supplemented with 5% (v/v) fetal calf serum, non-essential amino acids, 100 U/mL penicillin, and 100 µg/mL streptomycin.

2.3. Rescue of Recombinant LUJV

Recombinant reporter LUJV expressing the green fluorescent protein ZsGreen1 (rLUJV/ZsG) was generated using a previously described reverse genetics system [12]. Briefly, to rescue recombinant LUJV, T7 promoter-based plasmids containing full-length antigenomic sense LUJV L and S genome segments (prototypic Zambian strain; Genbank NC_012777.1 and NC_012776.1, respectively) were transfected into BSR-T7/5 cells. The reporter rLUJV/ZsG was generated by substituting the full-length antigenomic sense S segment with a modified version encoding ZsG. At 5 days post-transfection, cell culture supernatants were harvested, clarified by low-speed centrifugation, and used to infect Vero-E6 cells. Virus stocks harvested 3–4 days post-infection were quantified in Vero-E6 cells by tissue culture infective dose 50 ($TCID_{50}$) assays, using either immunofluorescence (wt virus) or ZsG expression (reporter virus) [13].

2.4. Next-Generation Sequencing and Bioinformatics

RNA was extracted from cell culture supernatants (clarified by low-speed centrifugation) using the MagMAX-96 Total RNA Isolation Kit (Thermo Fisher Scientific, Waltham, MA, USA) on a 96-well Roche MagMAX extraction platform with a DNase-I treatment step according to manufacturer's instructions. RNA sequencing was performed using KAPA RNA HyperPrep kit (Roche, Basel, Switzerland) for library preparation and analyzed on the MiniSeq system (Illumina, San Diego, CA, USA).

2.5. Antiviral Compound Screening

Compounds were obtained from Selleckchem (Pittsburgh, PA, USA), Tocris (Bristol, UK), or MedChemExpress (Monmouth Junction, NJ, USA), with the exception of favipiravir, which was obtained from BOC Sciences (Shirley, NJ, USA). Huh7 cells were seeded at 1×10^4 cells/well in a 96-well plate 16–20 h prior to treatment with compounds diluted in dimethyl sulfoxide (DMSO; final DMSO concentration 0.5%). For experiments measuring fluorescence reduction, 1 h post-treatment, cells were infected with rLUJV/ZsG (MOI 0.1; passage 1 stock from Vero-E6 cells); 72 h post-infection (hpi), ZsG fluorescence was determined using a BioTek Synergy reader (height 6 mm; 100 gain/sensitivity). For experiments measuring the reduction in wt virus titers, 1 h post-treatment, cells were infected with LUJV (MOI 0.1; prototypic Zambian strain; Genbank NC_012777.1 and NC_012776.1; passaged twice in Vero-E6 cells from original isolate); 72 hpi, supernatants were collected and viral titers determined by $TCID_{50}$ in Vero-E6 cells. Cell viability (ATP content) was determined in parallel in compound-treated, mock-infected cells using CellTiter-Glo 2.0 (Promega, Madison, WI, USA). All experiments were performed in quadruplicate and repeated at least three times.

2.6. Microscopy

ZsG fluorescence in infected cells was imaged directly using an EVOS digital inverted microscope (Thermo-Fisher, Waltham, MA, USA). For immunofluorescence, cells were fixed in 10% formalin for 30 min, permeabilized in phosphate-buffered saline (PBS) containing 0.1% Triton-X100 (*v*/*v*) for 10 min, and blocked with PBS with 5% bovine serum albumin (*v*/*v*) for 30 min. Primary antibodies were α-LUJV HMAF (generated in-house, CDC, Atlanta, GA, USA, ref# 703712), and secondary antibodies were labeled with AlexaFluor488 (Thermo-Fisher, Waltham, MA, USA, #A11034).

2.7. Data Analysis

ZsG fluorescence values for compound-treated infected cells were normalized to mock-treated (DMSO only), infected cells and used to fit a 4-parameter equation to semi-log plots of the concentration-response data. From this, the compound concentrations inhibiting ZsG expression by 50% (EC_{50}) were interpolated. Cell viability was similarly calculated in compound-treated, mock-infected cells to determine the 50% cell cytotoxicity concentration (CC_{50}) of each compound. The selectivity index was calculated by dividing CC_{50} by EC_{50}. Data analysis was performed using GraphPad Prism v9 (GraphPad, San Diego, CA, USA). Suitability for high-throughput screening was determined using the Z prime (Z') score, a measure of statistical effect size, calculated as $Z' = 1 - (3 \times (\sigma_P + \sigma_N)/ \mid \mu_P - \mu_N \mid$, where σ_P = standard deviation of positive control, σ_N = standard deviation of negative control, μ_P = mean of positive control, and μ_N = mean of negative control. Z' values between 0.5 and 1.0 were considered acceptable. Signal-to-noise ratio was determined as $(\mu_P - \mu_B)/\sigma_B$, where μ_P = mean signal of positive control, μ_B = mean background signal, and σ_B = standard deviation of background signal. Significance was calculated using a one-sample *t*-test.

3. Results

3.1. Generation of a Recombinant LUJV Expressing ZsGreen1 Fluorescent Protein

LUJV contains two ambisense genome segments termed large (L) and small (S) with respect to their relative nucleotide length (Figure 1A). Each genome segment encodes two genes: RNA-dependent RNA polymerase (RDRP) and Z protein on the L, and the nucleoprotein (N) and glycoprotein precursor (GPC) on the S (Figure 1B). Transcription of monocistronic viral mRNA is initiated at the untranslated region (UTR) and terminates at the intergenomic region (IGR).

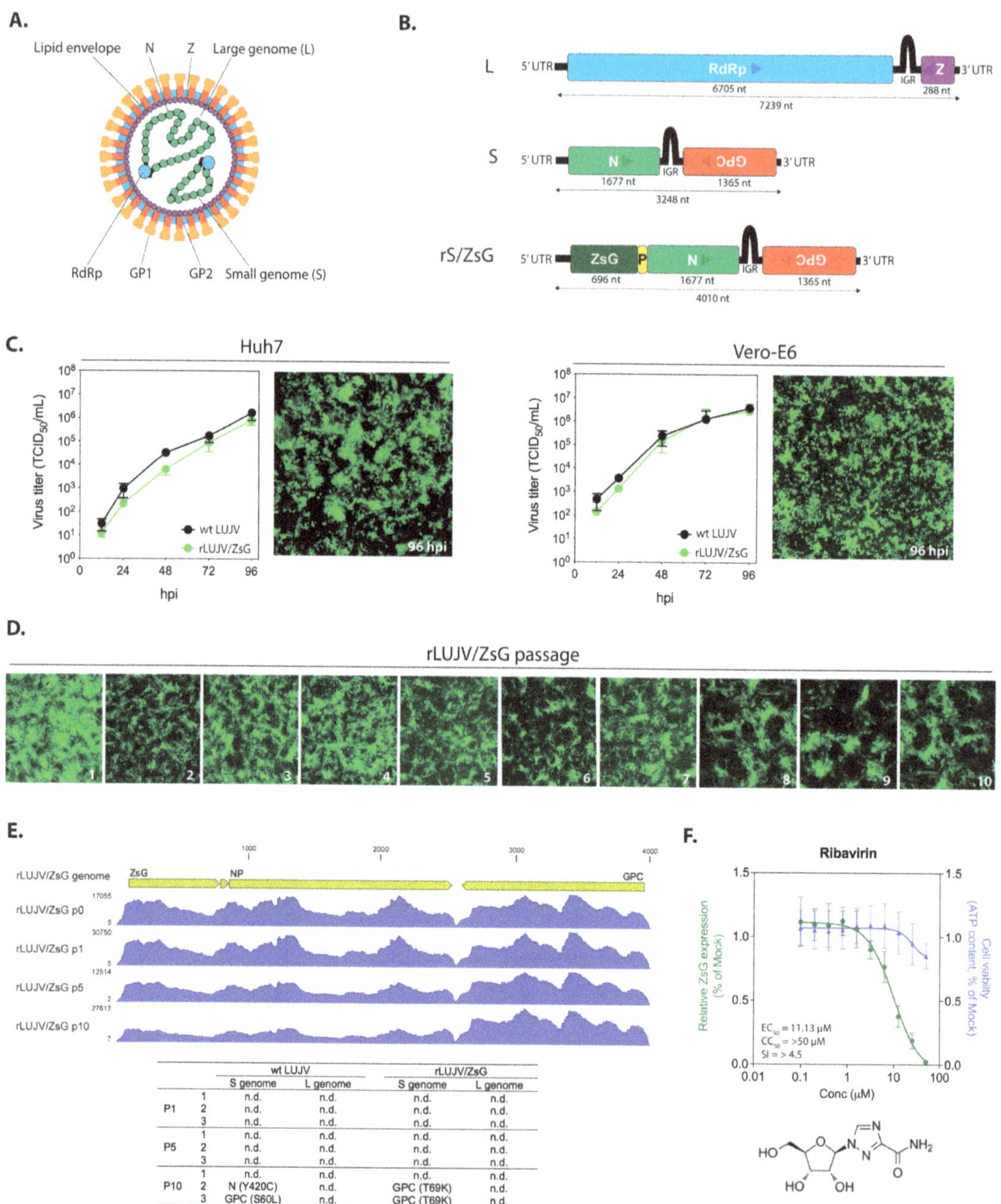

		wt LUJV		rLUJV/ZsG	
		S genome	L genome	S genome	L genome
P1	1	n.d.	n.d.	n.d.	n.d.
	2	n.d.	n.d.	n.d.	n.d.
	3	n.d.	n.d.	n.d.	n.d.
P5	1	n.d.	n.d.	n.d.	n.d.
	2	n.d.	n.d.	n.d.	n.d.
	3	n.d.	n.d.	n.d.	n.d.
P10	1	n.d.	n.d.	n.d.	n.d.
	2	N (Y420C)	n.d.	GPC (T69K)	n.d.
	3	GPC (S60L)	n.d.	GPC (T69K)	n.d.

Figure 1. Design and characterization of rLUJV/ZsG. (**A**) Schematic representation of the Lujo virus (LUJV) virion; and (**B**) the LUJV large (L) and small (S) genome segments (antigenomic sense). Boxes represent the viral protein coding sequences (CDS) for RNA-dependent RNA polymerase (RdRp), Z matrix protein (Z), glycoprotein precursor (GPC), and nucleoprotein (N). Arrows represent the coding direction. The recombinant S genome segment (rS/ZsG) contains the ZsGreen1 (ZsG) CDS attached to the N CDS via a porcine teshovirus-1 2A (P2A; P in the figure) nucleotide sequence. GP1,

glycoprotein 1; GP2, glycoprotein 2; UTR, untranslated region; IGR, intergenic region; nt, nucleotides. **(C)** Viral growth curves in Vero-E6 and Huh7 cells infected with either wild-type (wt) LUJV or rLUJV/ZsG at MOI 0.1. $TCID_{50}$ titers were determined at 12, 24, 48, 72, and 96 h post-infection (hpi). Representative fluorescent microscopy images (10× magnification) of rLUJV/ZsG-infected cell monolayers at 96 hpi are presented. **(D)** rLUJV/ZsG was passaged 10 times in Vero-E6 cells, with representative fluorescent microscopy images presented (10 × magnification) of infected monolayers taken at time of passage (3–4 days post-infection). **(E)** NGS analysis of wt LUJV and rLUJV/ZsG passaged 10 times in Vero-E6 cells. Viral RNA analyzed at passages 1, 5, and 10 showed no changes in S genome segment coverage over the passages. The associated table indicates the non-synonymous mutations detected in the S genome segments of each of the 3 replicates at passages 1, 5, and 10 (n.d., none detected). **(F)** The rLUJV/ZsG fluorescence reduction screening assay was evaluated with the antiviral compound ribavirin. Reduction in ZsG fluorescence in Huh7 cells treated with a range of ribavirin concentrations 1 h prior to infection with rLUJV/ZsG at MOI 0.1 was determined, with fluorescence values normalized to mock-treated cells (DMSO only). Cell viability was assessed at each ribavirin concentration by measuring ATP content (green), and values were normalized to those of mock-treated cells. Each point represents the mean of quadruplicate wells, with error bars indicating standard deviation; graphs are representative of 2 independent experiments. Associated EC_{50}, CC_{50}, and SI values are provided, along with the chemical structure of ribavirin.

Previous attempts to generate mammarenaviruses expressing reporter plasmids have either incorporated additional genome segments, as done with tri-segmented recombinant lymphocytic choriomeningitis virus (LCMV) [14], or introduced additional coding sequences (CDS) into the S genome segment, as in a recombinant reporter LASV [15,16]. Here, to generate a recombinant LUJV expressing the fluorescent protein ZsG, we constructed a modified S genome segment in which the ZsG coding sequence was inserted upstream of the N coding sequence, with a porcine teshovirus-1 2A peptide linker sequence (P2A) inserted between ZsG and NP (Figure 1B). This genetic format transcribes a single viral mRNA, but the two separate proteins are expressed via a ribosomal skipping event during P2A translation [17]. The modified S genome segment was incorporated into the previously described reverse genetics system for LUJV [12]. The recombinant reporter LUJV, able to express all parental viral proteins alongside the inserted ZsG reporter, was successfully rescued using this system and was termed rLUJV/ZsG.

To establish the suitability of rLUJV/ZsG as a surrogate for wt LUJV, we first investigated viral growth kinetics. Huh7 or Vero-E6 cells were infected with either wt LUJV or rLUJV/ZsG at MOI 0.1, and titers were determined at 12, 24, 48, 72, and 96 hpi. In both cell lines, growth kinetics of wt LUJV and rLUJV/ZsG were comparable at each time point assessed. In cells infected with rLUJV/ZsG, strong ZsG expression was observed in monolayers of both cell types at 96 hpi (Figure 1C). To assess the stability of the ZsG insert, we passaged rLUJV/ZsG 10 times in Vero-E6 cells (in triplicate). At each passage point, strong ZsG expression was detected in the monolayer, indicating that the ZsG-P2A-NP modification in the S genome segment was stable (Figure 1D). To further assess the genome stability, we used next-generation sequencing (NGS) to analyze the viral RNA from passages 1, 5, and 10, and compared the sequences to those of wt LUJV passaged concurrently (Figure 1E). The ZsG CDS was retained over 10 passages and no mutations were seen in any of the replicates. Similarly, no mutations were detected in the L genome segment over 10 passages. In the S genome segment, no mutations were seen over 5 passages in either wt LUJV or rLUJV/ZsG. At passage 10, 2 of the 3 replicate samples of both viruses had accumulated one non-synonymous, non-identical mutation.

3.2. Suitability of the LUJV Reporter Virus For Use in an In Vitro Screening Assay

To determine the suitability of rLUJV/ZsG for in vitro screening assays, potential LUJV inhibitors were screened using rLUJV/ZsG in Huh7 cells, which have been previously used in the discovery and evaluation of antiviral compounds for other VHF pathogens [16,18,19]. The assays were optimized using ribavirin, a well-documented broad-spectrum antiviral. The Z' score (0.89) and signal-to-noise ratio (190:1) were calculated using 50 μM ribavirin as the positive control and DMSO vehicle only as the negative control and indicated a robust assay. Ribavirin dose-response curves demonstrated a concentration-dependent reduction in ZsG fluorescence, with a calculated EC_{50} value of 11.13 ± 1.4 μM and CC_{50} of >50 μM (Figure 1F). These values are broadly similar to those reported for other arenaviruses, including LASV (2.47 μM) and JUNV (18.3 μM) [16,20], indicating that rLUJV/ZsG is suitable for a screening assay that uses reduction in ZsG fluorescence to indicate inhibition of viral replication.

3.3. Antiviral Compound Screening

To identify anti-LUJV therapeutics, a wide range of compounds was identified from published data on arenavirus inhibitors, broad-spectrum antivirals, and preliminary in-house screens using an FDA-approved library of compounds. Virus inhibition was initially determined for each of 83 candidate compounds at three concentrations (5000, 500, and 50 nM), with cell viability confirmed concurrently (Supplementary Table S1). Twenty compounds demonstrated robust anti-LUJV activity (ZsG fluorescence reduced to <30% of control) while maintaining >80% cell viability. These compounds were further evaluated using an extended concentration curve (Table 1). Six of the twenty compounds had SI $\geq$ 10, demonstrating promising efficacy and applicability for therapeutic use; efficacy was further confirmed against wt LUJV.

The 20 compounds demonstrating robust anti-LUJV activity can be broadly grouped based on known or predicted modes of antiviral action as: (1) inhibitors of viral replication; (2) inhibitors of viral entry, intracellular trafficking, or virus egress; (3) protein kinase inhibitors; (4) selective estrogen receptor modulators (SERMs); or (5) miscellaneous compounds not falling into the four previous groups.

Table 1. 50% effective concentration (EC_{50}), 50% cytotoxic concentration (CC_{50}), and selectivity indices (SI) of select compounds identified by rLUJV/ZsG reporter virus screening assay.

Inhibition Group	Compound	EC_{50} (μM)	CC_{50} (μM)	SI
Viral replication	2′-deoxy-2′-fluorocytidine (2-dFC)	0.533 ± 0.205	>50	>94
	AVN-944	0.201 ± 0.089	>12.5	>62
	Brequinar	0.109 ± 0.015	>12.5	>115
	Mycophenolic acid (MPA)	0.238 ± 0.097	2.035 ± 0.380	8.6
	Favipiravir (T-705)	2.951 ± 0.693	>50	>16.9
Viral entry, uncoating, and egress	Amiodarone	4.896 ± 0.513	13.63 ± 1.514	2.8
	Apilimod	1.438 ± 0.637	14.41 ± 3.245	10
	Isavucanazole	6.107 ± 0.926	12.69 ± 3.970	2.1
	Niclosamide	0.085 ± 0.029	0.272 ± 0.034	3.2
Protein kinase	AR-12	1.225 ± 0.444	5.522 ± 1.670	4.5
	BX-795	0.598 ± 0.274	9.913 ± 1.091	16.6
	Afatinib	2.256 ± 0.734	7.639 ± 0.699	3.4
SERMs	Bazedoxifene HCL	1.719 ± 0.481	6.679 ± 2.077	3.9
	Raloxifene (Evista)	1.500 ± 0.793	9.295 ± 1.314	6.2
	Tamoxifene citrate	3.594 ± 0.724	7.371 ± 1.036	2.1
	Toremifene citrate	4.217 ± 0.778	10.86 ± 3.144	2.6
Additional mechanisms	Benztropine mesylate	5.141 ± 1.515	>25	>4.9
	Clemastine fumarate	3.750 ± 1.790	13.12 ± 2.328	3.5
	Loperamide HCL	3.669 ± 0.986	12.29 ± 3.574	3.3
	Obatoclax	0.355 ± 0.105	1.502 ± 0.332	4.2

3.3.1. Inhibitors of Viral Replication

Some of the most potent anti-LUJV compounds and those with the highest SI in our screens were compounds predicted to target viral replication (Figure 2). The nucleoside analogs 2′-deoxy-2′-fluorocytidine (2-dFC) and favipiravir both had CC_{50} values > 50 μM, and EC_{50} values of 0.54 ± 0.21 μM and 2.95 ± 0.69 μM, respectively. 2-dFC has shown similar potency against LASV [16], the nairovirus Crimean-Congo hemorrhagic fever virus

(CCHFV) [19], and the paramyxovirus Sosuga virus [21] in previous studies, further demonstrating this compound's broad-species antiviral potential. The inosine-5′-monophosphate dehydrogenase inhibitors AVN-944 and mycophenolic acid (MPA) both demonstrated sub-micromolar EC_{50} values of 0.20 ± 0.09 μM and 0.24 ± 0.1 μM, respectively, although MPA's CC_{50} value of 2.0 ± 0.4 μM resulted in a reduced SI of 8.6. MPA has previously demonstrated antiviral effects against a wide range of RNA viruses including influenza [22], dengue virus [23], Zika virus [24], rotavirus [25], CCHFV [19], and hantavirus [26]. AVN-944 was previously shown to have antiviral activity in vitro against JUNV, reducing titers by >90% at concentrations of 7.5 μM and above [27]. The best overall performing compound was brequinar (EC_{50} 0.109 ± 0.02 μM, CC_{50} > 12.5 μM, SI > 115), a selective inhibitor of the enzyme dihydroorotate dehydrogenase. This compound is believed to act by blocking de novo pyrimidine biosynthesis, and recent studies have shown brequinar inhibits other arenaviruses such as LASV, JUNV, and LCMV [28], as well as other RNA virus species such as influenza strains A and B [29] and HIV-1 [30].

3.3.2. Inhibitors of Viral Entry, Trafficking, and Egress

The compounds apilimod, niclosamide, and isavucanazole have all been shown to inhibit LASV entry in vitro [31,32]. Of these, the most potent inhibitor of rLUJV/ZsG was niclosamide (EC_{50} 0.085 ± 0.029 μM, CC_{50} 0.272 ± 0.034 μM, SI = 3.2) with a sub-micromolar EC_{50} value (Figure 3A). Of the other two compounds, apilimod (EC_{50} 1.44 ± 0.6 μM, CC_{50} 14.41 ± 3.2 μM, SI = 10) demonstrated greater inhibition than isavucanazole (EC_{50} 6.1 ± 0.9 μM, CC_{50} 12.69 ± 3.9 μM, SI = 2.1). Interestingly, the niclosamide values for rLUJV/ZsG inhibition are very similar to the reported EC_{50} of 0.08 μM for LASV inhibition in a pseudovirus screening assay; however, the same study reported an EC_{50} of 0.05 μM for apilimod against LASV, which is considerably lower than EC = 1.44 μM for rLUJV/ZsG [31]. Isavucanazole was shown to inhibit LASV (EC_{50} 1.2 μM, CC_{50} > 30 μM) in an HIV-luc pseudovirus screen, but did not inhibit LUJV in the same screen [32]. Here, however, using rLUJV/ZsG, which better represents authentic viral replication, we observed a degree of inhibition using isavucanazole, which may indicate a mode of action different from blocking viral entry. Finally, amiodarone (EC_{50} 4.90 ± 0.51 μM, CC_{50} 13.63 ± 1.51 μM, SI = 2.8) is a cationic amphiphilic drug (CAD) that can accumulate within acidic intracellular vesicles, inhibiting membrane fusion and ribonucleoprotein release [33]. It can inhibit Ebola virus (EBOV) in vitro and in vivo [34,35], and has been shown to inhibit in vitro the new world arenavirus GTOV [36], SARS-CoV-1 [37], and hepatitis C virus [38]. While these four compounds all inhibited rLUJV/ZsG, the degree of cytotoxicity associated with all except apilimod reduced their respective SI values and resulted in a very narrow window of effectiveness.

3.3.3. Kinase Inhibitors

Kinase inhibitors are increasingly being examined as potential antiviral drugs. They make attractive options for study since large numbers of these compounds are continually being developed and approved for clinical use to treat cancers and inflammatory conditions, and because of increasing knowledge of host kinase use by viruses during infection [39–41]. AR-12 is a celecoxib derivative kinase inhibitor that downregulates the PI3K/Akt pathway, which is known to contribute to arenavirus budding [42]. BX-795 inhibits 3-phosphoinositide-dependent kinase 1, IKK-related kinase, TANK-binding kinase 1, and IKKε; afatanib is a tyrosine kinase inhibitor. AR-12 has previously been shown to inhibit both LASV [18] and JUNV [43], and BX-795 has shown promise as a potent inhibitor of herpes simplex virus 1 (HSV-1) and HSV-2 [44]. Against rLUJV/ZsG, BX-795 (EC_{50} 0.60 ± 0.27 μM, CC_{50} 9.91 ± 1.09 μM) performed the best, with an SI value of 16.6. AR-12 (EC_{50} 1.3 ± 0.4 μM, CC_{50} 5.5 ± 1.7 μM, SI = 4.5) and afatinib (EC_{50} 2.26 ± 0.73 μM, CC_{50} 7.64 ± 0.7 μM, SI = 3.4) both inhibited rLUJV/ZsG, but their cytotoxicity reduced their SI values compared to BX-795 (Figure 3B).

3.3.4. Selective Estrogen Receptor Modulators

Four of the compounds in the screen were SERMs, a class of compounds that are also CADs and have antiviral effects against several viral species, including EBOV [45]. The SERM performing best against rLUJV/ZsG was raloxifene (EC_{50} 1.5 ± 0.79 μM, CC_{50} 9.3 ± 1.3 μM, SI = 6.2) (Figure 4A). Bazedoxifene HCl (EC_{50} 1.72 ± 0.79 μM, CC_{50} 6.68 ± 2.08 μM, SI = 3.9) demonstrated a similar EC_{50} value, but was associated with significant cytotoxicity, which reduced the SI values. Both tamoxifene citrate (EC_{50} 3.59 ± 0.72 μM, CC_{50} 7.37 ± 1.04 μM, SI = 2.1) and toremifene citrate (EC_{50} 4.22 ± 0.78 μM, CC_{50} 10.86 ± 3.14 μM, SI = 2.6) inhibited rLUJV/ZsG, but the associated cytotoxicity again led to a very narrow window of antiviral effect.

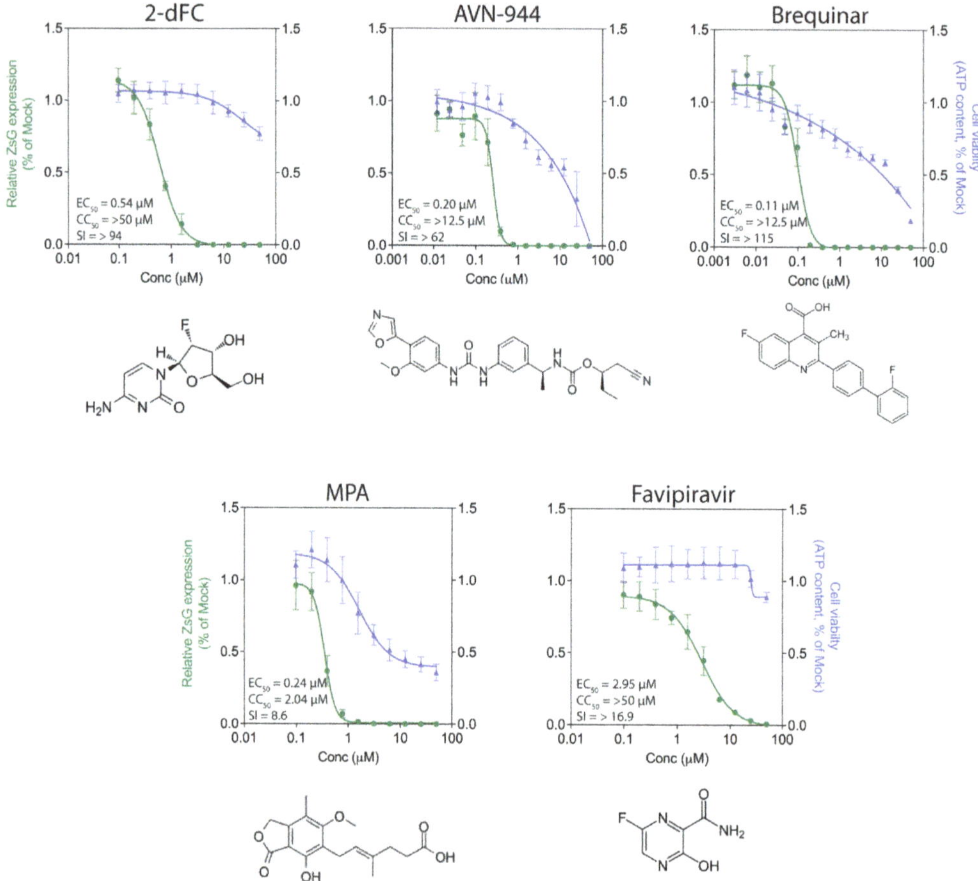

Figure 2. Antiviral activity of select compounds targeting viral replication processes. Concentration-response curves in Huh7 cells infected with rLUJV/ZsG (MOI 0.1) after being treated 1 h prior with 2′-deoxy-2′-fluorocytidine (2-dFC), AVN-944, brequinar, mycophenolic acid (MPA), or T-705. Relative ZsG expression is shown in green and cell viability in blue. Values were normalized to mock-treated (DMSO only) infected cells. Each point represents the mean of quadruplicate wells, with error bars indicating standard deviation; graphs are representative of at least 3 independent experiments. Associated EC_{50}, CC_{50}, and SI values are given, along with the chemical structure of each compound.

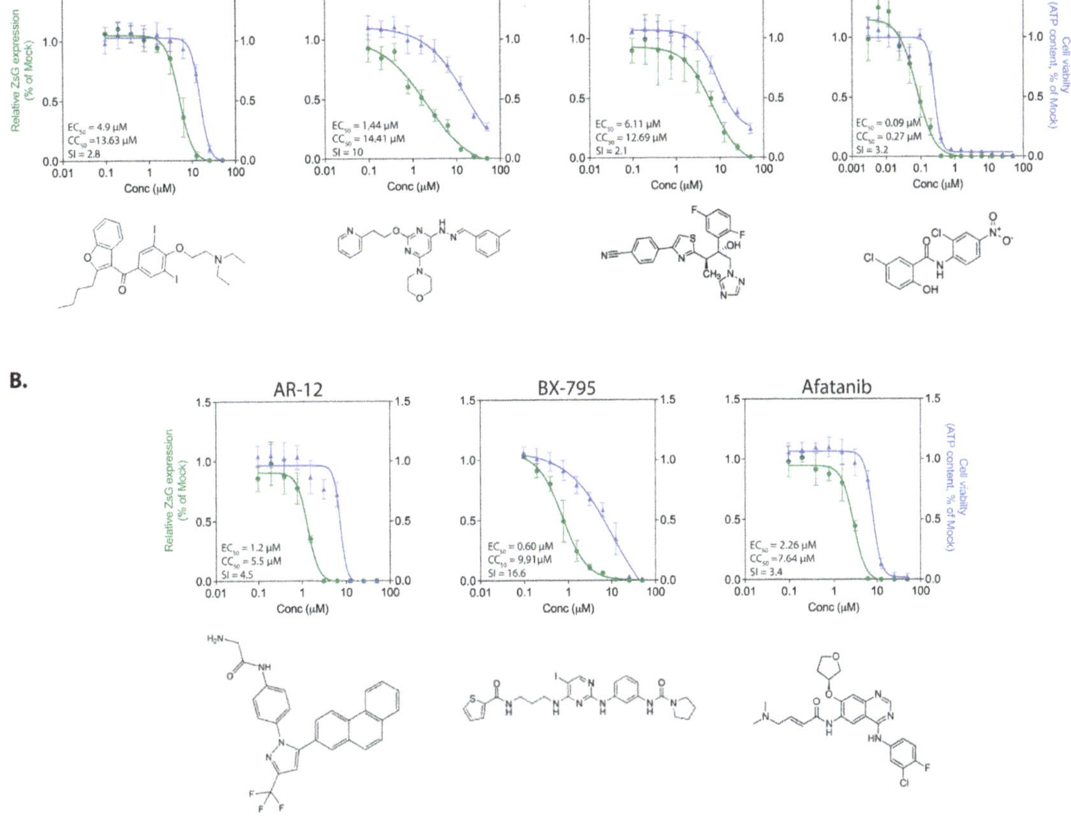

Figure 3. Antiviral activity of select protein kinase inhibitors and known entry inhibitors. Concentration-response curves in Huh7 cells infected with rLUJV/ZsG (MOI 0.1) after being treated 1 h prior with (**A**) compounds known to block arenavirus entry, egress, or intracellular tracking processes; or (**B**) protein kinase inhibitors. Relative ZsG expression is shown in green and cell viability is shown in blue. Values were normalized to mock-treated (DMSO only) infected cells. Each point represents the mean of quadruplicate wells, with error bars indicating standard deviation; graphs are representative of at least 3 independent experiments. Associated EC_{50}, CC_{50}, and SI values are provided along with the chemical structure of each compound.

3.3.5. Additional Compounds with Anti-rLUJV/ZsG Properties

Four of the compounds selected from the initial screen did not fit into the four previous groups based on predicted mode of action (Figure 4B). Three of these compounds are FDA-approved compounds: benztropine mesylate (EC_{50} 5.14 ± 1.52 μM, CC_{50} > 25, SI > 4.9) is a neurotransmitter inhibitor (anticholinergic); clemastine fumarate (EC_{50} 3.75 ± 1.79 μM, CC_{50} 13.12 ± 2.33 μM, SI = 3.5) acts primarily as an H1 histamine antagonist; and loperamide HCl (EC_{50} 3.67 ± 0.99 μM, CC_{50} 12.29 ± 3.57 μM, SI = 3.3) is an opioid receptor that targets μ-opioid receptors in the large intestine and acts as an antidiarrheal agent by decreasing intestinal movement. All three were able to moderately inhibit rLUJV/ZsG and had SI values between 3 and 5. The final compound, obatoclax (EC_{50} 0.36 ± 0.1 μM, CC_{50} 1.5 ± 0.33 μM, SI = 4.2) inhibits the Bcl-2 family of proteins and has been investigated as an experimental anti-cancer drug. It inhibited rLUJV/ZsG at sub-micromolar EC_{50} value, but was cytotoxic at a relatively low concentrations, which reduced SI to 4.2.

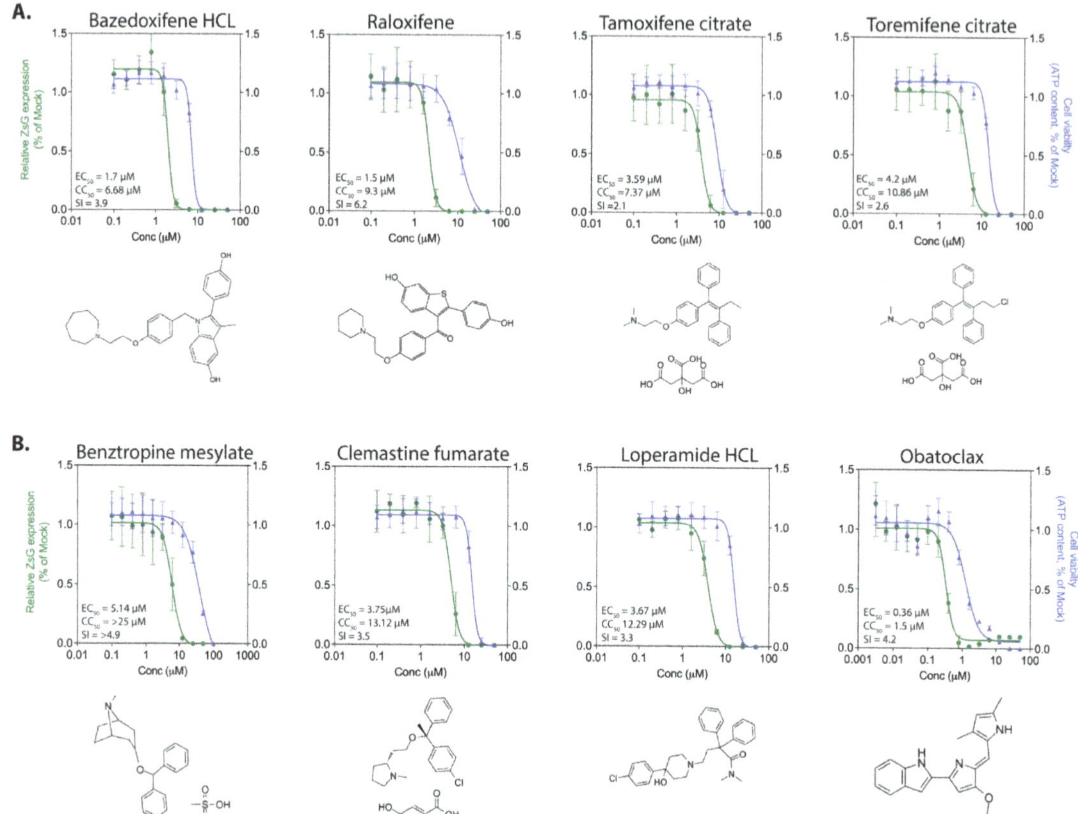

Figure 4. Antiviral activity of selective estrogen receptor modulators and additional compounds demonstrating antiviral effects. Concentration-response curves in Huh7 cells infected with rLUJV/ZsG (MOI 0.1) after being treated 1 h prior with: (**A**) selective estrogen receptor modulators (SERMs), or (**B**) protein kinase inhibitors. Relative ZsG expression is shown in green and cell viability is shown in blue. Values were normalized to mock-treated (DMSO only) infected cells. Each point represents the mean of quadruplicate wells, with error bars indicating standard deviation; graphs are representative of at least 3 independent experiments. Associated EC_{50}, CC_{50}, and SI values are provided, along with the chemical structure of each compound.

3.4. Confirmatory Screening with wt LUJV

To confirm that the antiviral activity of the compounds was not specific to the ZsG-expressing recombinant virus, the top six candidates (SI $\geq$ 10) were selected to be screened using wt LUJV; these compounds were 2-dFC, apilimod, AVN-944, brequinar, BX-795, and favipiravir. These six drugs were tested using a shortened concentration curve (5 dilutions centered around their calculated EC_{50} value), with cell viability determined concurrently. All compounds evaluated inhibited wt LUJV titers in a concentration-dependent manner, and all compounds reduced viral titers by approximately 4 logs while retaining >75% cell viability (Figure 5). These results further confirm the suitability of using a reduction of ZsG fluorescence in rLUJV/ZsG-infected cells as a convenient readout in place of wt virus infection.

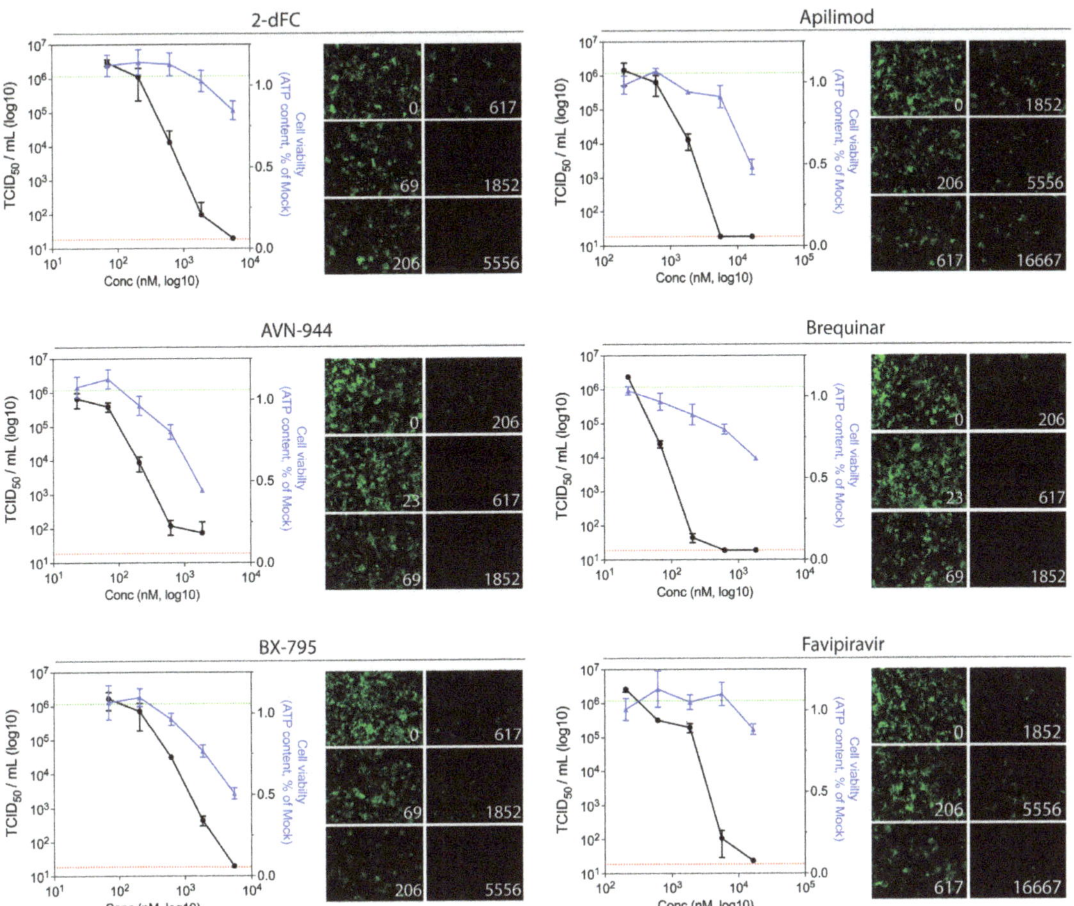

Figure 5. Ability of the most potent anti-LUJV compounds to inhibit wild-type virus replication. Titer reduction assays using wt LUJV were performed in Huh7 cells treated with varying concentrations of indicated compound 1 h prior to infection (MOI 0.1). For each compound concentration, inhibition of rLUJV/ZsG and cell viability were concurrently assessed. Four days post-infection, cell culture supernatants were collected and LUJV titers ($TCID_{50}$/mL) were determined in Vero-E6 cells by immunofluorescence assays. For each concentration, LUJV titers (black) and cell viability (blue) are shown. Dotted lines represent the limit of detection for the $TCID_{50}$ assay (red) and the mean wt LUJV titer in mock-treated (DMSO only) cells (green). Each point represents the mean of quadruplicate wells, with error bars indicating standard deviation; graphs are representative of at least 3 independent experiments. ZsG fluorescence in rLUJV/ZsG-infected Huh7 cells at each compound concentration is also shown (10× magnification).

4. Discussion

Only one outbreak of LUJV has been recorded to date, with four out of five patients succumbing to fatal disease. The effectiveness of existing antivirals in treating LUJV patients is unknown. Ribavirin was administered to the sole surviving patient; however, based on this single treated case, it is not possible to determine the effect of treatment on the outcome. Furthermore, given our knowledge of ribavirin treatment in other arenavirus infections, in which it is only partially effective and associated with significant side effects, investigations into uncovering effective and safe arenavirus therapeutic options are badly needed [7]. High-throughput screening assays utilizing recombinant reporter viruses as surrogates for the wild-type parental virus have previously been successful in identifying several anti-mammarenavirus compounds [15,16,46,47]. Here, using a novel recombinant reporter LUJV, we were able to identify six compounds with high efficacy against LUJV that warrant further evaluation.

Given the genetic diversity among species, the large number of different arenavirus host reservoirs, and the geographic range of these pathogens, the emergence of new arenaviruses pathogenic to humans remains an ongoing public health concern [7]. Importantly, several of the compounds identified here as effective against LUJV (i.e., favipiravir, 2-dFC, and brequinar) have also been shown to inhibit other arenavirus species. The nucleoside analogs favipiravir and 2-dFC are both known to inhibit other arenaviruses including LASV, JUNV, Machupo virus, and GTOV [16,48,49]. The most promising inhibitor identified here was the dihydroorotate dehydrogenase inhibitor brequinar (EC_{50} 0.11 $\pm$ 0.02 μM, CC_{50} > 12.5 μM, SI > 115), which has previously been shown to inhibit LASV, JUNV, and LCMV at similar sub-micromolar EC_{50} values [50]. These broad-spectrum antivirals inhibit a wide range of highly pathogenic arenavirus species, presumably by targeting conserved viral processes, and are thus promising candidates for treating current and future emerging arenaviruses. It should be noted that inhibitors of pyrimidine biosynthesis can exhibit potent antiviral activity in cell culture, but be ineffective in animal models, which is likely due to pyrimidines in the animals' diet overcoming the antiviral effect [51].

There is notable value in identifying efficacious compounds that have previously been approved for clinical use. Several of the compounds with anti-LUJV efficacy are already approved for other indications. Benztropine (Congentin) is used as an anti-tremor medication to treat Parkinson's disease, and has also been shown to have mild antiviral effects against EBOV (EC_{50} 8.1 μM) [52], MERS-CoV (EC_{50} 16.6 μM), and SARS-CoV (EC_{50} 21.6 μM) [53]. Clemastine, an approved anti-histamine, is also effective against EBOV (EC_{50} 5.4 μM) [52], SARS-CoV-2 (EC_{50} 0.95 μM), and SARS-CoV-1 (EC_{50} 6.6 $\pm$ 1.48 μM) [54]. While not yet approved, obatoclax is an inhibitor of the Bcl-2 family of proteins and an experimental anti-cancer drug that has undergone several phase II clinical trials; it has been shown to inhibit LCMV [28] as well as multiple other virus species including Rift Valley fever virus and HSV-2 [30]. The ability to rapidly screen large libraries of FDA-approved compounds or small molecules for antiviral efficacy is a huge advantage of our reporter virus systems, since comparable studies using wt virus are difficult to adapt for high-throughput screening and are prohibitively time-consuming, costly, and labor-intensive.

The six most promising candidates with both high efficacy and low cytotoxicity fell within three broadly defined groups based on modes of action: inhibiting viral replication; viral entry and trafficking; or protein kinases. The majority (four out of six) of these with the highest SI values (and therefore the most promising therapeutic candidates) were inhibitors of viral replication. Several other compounds that were identified demonstrated potent anti-LUJV activity, but had associated cytotoxicity, leading to low SI values and a narrow therapeutic window. Although this would likely make these compounds unattractive as single-drug treatment options, combination therapy using compounds with distinct modes of action may synergistically inhibit LUJV [19,27,55,56]. This would allow these more cytotoxic compounds to be used at far lower concentrations while maximizing their antiviral potential.

Here we describe the development of a recombinant LUJV engineered to express the fluorescent protein ZsGreen1 and validate its use in high-throughput screening assays for large or complex screens containing multiple compounds at various concentrations. We identified several highly efficacious anti-LUJV compounds with multiple predicted modes of action. Future studies are planned to investigate if potential synergistic actions increase the effectiveness of these compounds, as well as to advance the most promising candidates to studies using the available animal models of disease to evaluate in vivo effectiveness. Given the genetic diversity of mammarenaviruses and the high-consequence nature of their infections, it is promising to report that newly emerging species such as LUJV are similarly inhibited by several previously reported compounds. Identifying pan-arenavirus inhibiting compounds serves as a key step in the preparedness against new and emerging pathogenic threats.

Supplementary Materials: The following are available online at https://www.mdpi.com/article/10.3390/v13071255/s1. Table S1: Screening a panel of selected compounds for inhibition of rLUJV/ZsG fluorescence.

Author Contributions: Conceptualization, S.R.W., S.T.N., C.G.A. and C.F.S.; Data curation, S.R.W., J.R.S., S.C.G., P.C. and M.F.; Formal analysis, S.R.W., J.R.S., S.C.G.; Funding acquisition, J.M.M., S.T.N. and C.F.S.; Investigation, S.R.W., J.R.S., S.C.G., P.C., M.F., and C.G.A.; Methodology, S.R.W., J.R.S., S.C.G., P.C., M.F., É.B. and C.G.A.; Project administration, C.F.S.; Resources, É.B., C.G.A., C.F.S.; Supervision, M.F., J.M.M., S.T.N., C.F.S.; Visualization, S.R.W.; Writing—original draft, S.R.W., J.R.S., C.F.S.; Writing—review & editing, S.R.W., J.R.S., S.C.G., M.F., É.B., S.T.N., and C.F.S. All authors have read and agreed to the published version of the manuscript.

Funding: This work was partially supported by an appointment to the Research Participation Program at the Centers for Disease Control and Prevention administered by the Oak Ridge Institute for Science and Education through an interagency agreement between the U.S. Department of Energy and CDC [S.R.W.], and by CDC Emerging Infectious Disease Research Core Funds. The findings and conclusions in this report are those of the authors and do not necessarily represent the official position of the Centers for Disease Control and Prevention.

Institutional Review Board Statement: Not applicable.

Informed Consent Statement: Not applicable.

Data Availability Statement: Not applicable.

Acknowledgments: We thank Tatyana Klimova for assistance with editing the manuscript.

Conflicts of Interest: The authors declare no conflict of interest.

References

1. Paweska, J.T.; Sewlall, N.H.; Ksiazek, T.G.; Blumberg, L.H.; Hale, M.J.; Lipkin, W.I.; Weyer, J.; Nichol, S.T.; Rollin, P.E.; McMullan, L.K.; et al. Nosocomial outbreak of novel arenavirus infection, southern Africa. *Emerg. Infect. Dis.* **2009**, *15*, 1598–1602. [CrossRef] [PubMed]
2. Briese, T.; Paweska, J.T.; McMullan, L.K.; Hutchison, S.K.; Street, C.; Palacios, G.; Khristova, M.L.; Weyer, J.; Swanepoel, R.; Egholm, M.; et al. Genetic detection and characterization of Lujo virus, a new hemorrhagic fever-associated arenavirus from southern Africa. *PLoS Pathog.* **2009**, *5*, e1000455. [CrossRef]
3. Radoshitzky, S.R.; Buchmeier, M.J.; Charrel, R.N.; Clegg, J.C.S.; Gonzalez, J.-P.J.; Günther, S.; Hepojoki, J.; Kuhn, J.H.; Lukashevich, I.S.; Romanowski, V.; et al. ICTV Virus Taxonomy Profile: Arenaviridae. *J. Gen. Virol.* **2019**, *100*, 1200–1201. [CrossRef] [PubMed]
4. Gonzalez, J.P.; Emonet, S.; de Lamballerie, X.; Charrel, R. Arenaviruses. *Curr. Top. Microbiol. Immunol.* **2007**, *315*, 253–288.
5. Hallam, S.J.; Koma, T.; Maruyama, J.; Paessler, S. Review of Mammarenavirus Biology and Replication. *Front. Microbiol.* **2018**, *9*, 1751. [CrossRef] [PubMed]
6. Charrel, R.N.; de Lamballerie, X. Zoonotic aspects of arenavirus infections. *Veter. Microbiol.* **2010**, *140*, 213–220. [CrossRef] [PubMed]
7. Emonet, S.F.; de la Torre, J.C.; Domingo, E.; Sevilla, N. Arenavirus genetic diversity and its biological implications. *Infect. Genet. Evol.* **2009**, *9*, 417–429. [CrossRef]
8. Radoshitzky, S.R.; Bào, Y.; Buchmeier, M.J.; Charrel, R.N.; Clawson, A.N.; Clegg, C.S.; DeRisi, J.L.; Emonet, S.; Gonzalez, J.-P.; Kuhn, J.H.; et al. Past, present, and future of arenavirus taxonomy. *Arch. Virol.* **2015**, *160*, 1851–1874. [CrossRef]
9. Maiztegui, J.I.; Fernandez, N.J.; de Damilano, A.J. Efficacy of immune plasma in treatment of Argentine haemorrhagic fever and association between treatment and a late neurological syndrome. *Lancet* **1979**, *314*, 1216–1217. [CrossRef]
10. Raabe, V.N.; Kann, G.; Ribner, B.S.; Morales, A.; Varkey, J.B.; Mehta, A.K.; Lyon, G.M.; Vanairsdale, S.; Faber, K.; Becker, S.; et al. Favipiravir and Ribavirin Treatment of Epidemiologically Linked Cases of Lassa Fever. *Clin. Infect. Dis.* **2017**, *65*, 855–859. [CrossRef]
11. Brisse, M.E.; Ly, H. Hemorrhagic Fever-Causing Arenaviruses: Lethal Pathogens and Potent Immune Suppressors. *Front. Immunol.* **2019**, *10*, 372. [CrossRef]
12. Bergeron, É.; Chakrabarti, A.K.; Bird, B.H.; Dodd, K.A.; McMullan, L.K.; Spiropoulou, C.F.; Nichol, S.T.; Albariño, C.G. Reverse genetics recovery of Lujo virus and role of virus RNA secondary structures in efficient virus growth. *J. Virol.* **2012**, *86*, 10759–10765. [CrossRef]
13. Reed, L.J.; Muench, H. A simple method for estimating fifty percent endpoints. *Am. J. Epidemiol.* **1938**, *27*, 493–497. [CrossRef]

14. Emonet, S.F.; Garidou, L.; McGavern, D.B.; de la Torre, J.C. Generation of recombinant lymphocytic choriomeningitis viruses with trisegmented genomes stably expressing two additional genes of interest. *Proc. Natl. Acad. Sci. USA* **2009**, *106*, 3473–3478. [CrossRef]
15. Caì, Y.; Iwasaki, M.; Beitzel, B.F.; Yú, S.; Postnikova, E.N.; Cubitt, B.; DeWald, L.E.; Radoshitzky, S.R.; Bollinger, L.; Jahrling, P.B.; et al. Recombinant Lassa Virus Expressing Green Fluorescent Protein as a Tool for High-Throughput Drug Screens and Neutralizing Antibody Assays. *Viruses* **2018**, *10*, 655. [CrossRef]
16. Welch, S.R.; Guerrero, L.W.; Chakrabarti, A.K.; McMullan, L.K.; Flint, M.; Bluemling, G.R.; Painter, G.R.; Nichol, S.T.; Spiropoulou, C.F.; Albariño, C.G. Lassa and Ebola virus inhibitors identified using minigenome and recombinant virus reporter systems. *Antivir. Res.* **2016**, *136*, 9–18. [CrossRef]
17. Kim, J.H.; Lee, S.R.; Li, L.H.; Park, H.J.; Park, J.H.; Lee, K.Y.; Kim, M.K.; Shin, B.A.; Choi, S.Y. High cleavage efficiency of a 2A peptide derived from porcine teschovirus-1 in human cell lines, zebrafish and mice. *PLoS ONE* **2011**, *6*, e18556. [CrossRef] [PubMed]
18. Mohr, E.L.; McMullan, L.K.; Lo, M.K.; Spengler, J.R.; Bergeron, É.; Albariño, C.G.; Shrivastava-Ranjan, P.; Chiang, C.-F.F.; Nichol, S.T.; Spiropoulou, C.F.; et al. Inhibitors of cellular kinases with broad-spectrum antiviral activity for hemorrhagic fever viruses. *Antivir. Res.* **2015**, *120*, 40–47. [CrossRef] [PubMed]
19. Welch, S.R.; Scholte, F.E.M.M.; Flint, M.; Chatterjee, P.; Nichol, S.T.; Bergeron, É.; Spiropoulou, C.F. Identification of 2′-deoxy-2′-fluorocytidine as a potent inhibitor of Crimean-Congo hemorrhagic fever virus replication using a recombinant fluorescent reporter virus. *Antivir. Res.* **2017**, *147*, 91–99. [CrossRef]
20. García, C.C.; Candurra, N.A.; Damonte, E.B. Antiviral and virucidal activities against arenaviruses of zinc-finger active compounds. *Antivir. Chem. Chemother.* **2000**, *11*, 231–237. [CrossRef] [PubMed]
21. Welch, S.R.; Chakrabarti, A.K.; Wiggleton Guerrero, L.; Jenks, H.M.; Lo, M.K.; Nichol, S.T.; Spiropoulou, C.F.; Albariño, C.G. Development of a reverse genetics system for Sosuga virus allows rapid screening of antiviral compounds. *PLoS Negl. Trop. Dis.* **2018**, *12*, e0006326. [CrossRef] [PubMed]
22. To, K.K.W.; Mok, K.-Y.; Chan, A.S.F.; Cheung, N.N.; Wang, P.; Lui, Y.-M.; Chan, J.F.W.; Chen, H.; Chan, K.-H.; Kao, R.Y.T.; et al. Mycophenolic acid, an immunomodulator, has potent and broad-spectrum in vitro antiviral activity against pandemic, seasonal and avian influenza viruses affecting humans. *J. Gen. Virol.* **2016**, *97*, 1807–1817. [CrossRef]
23. Diamond, M.S.; Zachariah, M.; Harris, E. Mycophenolic Acid Inhibits Dengue Virus Infection by Preventing Replication of Viral RNA. *Virology* **2002**, *304*, 211–221. [CrossRef]
24. Barrows, N.J.; Campos, R.K.; Powell, S.T.; Prasanth, K.R.; Schott-Lerner, G.; Soto-Acosta, R.; Galarza-Muñoz, G.; McGrath, E.L.; Urrabaz-Garza, R.; Gao, J.; et al. A Screen of FDA-Approved Drugs for Inhibitors of Zika Virus Infection. *Cell Host Microbe* **2016**, *20*, 259–270. [CrossRef] [PubMed]
25. Yin, Y.; Wang, Y.; Dang, W.; Xu, L.; Su, J.; Zhou, X.; Wang, W.; Felczak, K.; van der Laan, L.J.W.; Pankiewicz, K.W.; et al. Mycophenolic acid potently inhibits rotavirus infection with a high barrier to resistance development. *Antivir. Res.* **2016**, *133*, 41–49. [CrossRef] [PubMed]
26. Sun, Y.; Chung, D.-H.; Chu, Y.-K.; Jonsson, C.B.; Parker, W.B. Activity of ribavirin against Hantaan virus correlates with production of ribavirin-5′-triphosphate, not with inhibition of IMP dehydrogenase. *Antimicrob. Agents Chemother.* **2007**, *51*, 84–88. [CrossRef] [PubMed]
27. Dunham, E.C.; Leske, A.; Shifflett, K.; Watt, A.; Feldmann, H.; Hoenen, T.; Groseth, A. Lifecycle modelling systems support inosine monophosphate dehydrogenase (IMPDH) as a pro-viral factor and antiviral target for New World arenaviruses. *Antivir. Res.* **2018**, *157*, 140–150. [CrossRef]
28. Kim, Y.-J.; Cubitt, B.; Chen, E.; Hull, M.V.; Chatterjee, A.K.; Cai, Y.; Kuhn, J.H.; de la Torre, J.C. The ReFRAME library as a comprehensive drug repurposing library to identify mammarenavirus inhibitors. *Antivir. Res.* **2019**, *169*, 104558. [CrossRef]
29. Park, J.-G.; Ávila-Pérez, G.; Nogales, A.; Blanco-Lobo, P.; de la Torre, J.C.; Martínez-Sobrido, L. Identification and Characterization of Novel Compounds with Broad-Spectrum Antiviral Activity against Influenza A and B Viruses. *J. Virol.* **2020**, *94*. [CrossRef]
30. Andersen, P.I.; Krpina, K.; Ianevski, A.; Shtaida, N.; Jo, E.; Yang, J.; Koit, S.; Tenson, T.; Hukkanen, V.; Anthonsen, M.W.; et al. Novel Antiviral Activities of Obatoclax, Emetine, Niclosamide, Brequinar, and Homoharringtonine. *Viruses* **2019**, *11*, 964. [CrossRef]
31. Hulseberg, C.E.; Fénéant, L.; Szymańska-de Wijs, K.M.; Kessler, N.P.; Nelson, E.A.; Shoemaker, C.J.; Schmaljohn, C.S.; Polyak, S.J.; White, J.M. Arbidol and Other Low-Molecular-Weight Drugs That Inhibit Lassa and Ebola Viruses. *J. Virol.* **2019**, *93*. [CrossRef]
32. Zhang, X.; Tang, K.; Guo, Y. The antifungal isavuconazole inhibits the entry of lassa virus by targeting the stable signal peptide-GP2 subunit interface of lassa virus glycoprotein. *Antivir. Res.* **2020**, *174*, 104701. [CrossRef]
33. Salata, C.; Calistri, A.; Parolin, C.; Baritussio, A.; Palù, G. Antiviral activity of cationic amphiphilic drugs. *Expert Rev. Anti. Infect. Ther.* **2017**, *15*, 483–492. [CrossRef]
34. Salata, C.; Baritussio, A.; Munegato, D.; Calistri, A.; Ha, H.R.; Bigler, L.; Fabris, F.; Parolin, C.; Palù, G.; Mirazimi, A. Amiodarone and metabolite MDEA inhibit Ebola virus infection by interfering with the viral entry process. *Pathog. Dis.* **2015**, *73*. [CrossRef]
35. Madrid, P.B.; Panchal, R.G.; Warren, T.K.; Shurtleff, A.C.; Endsley, A.N.; Green, C.E.; Kolokoltsov, A.; Davey, R.; Manger, I.D.; Gilfillan, L.; et al. Evaluation of Ebola Virus Inhibitors for Drug Repurposing. *ACS Infect. Dis.* **2015**, *1*, 317–326. [CrossRef] [PubMed]

36. Gehring, G.; Rohrmann, K.; Atenchong, N.; Mittler, E.; Becker, S.; Dahlmann, F.; Pöhlmann, S.; Vondran, F.W.R.; David, S.; Manns, M.P.; et al. The clinically approved drugs amiodarone, dronedarone and verapamil inhibit filovirus cell entry. *J. Antimicrob. Chemother.* **2014**, *69*, 2123–2131. [CrossRef]
37. Stadler, K.; Ha, H.R.; Ciminale, V.; Spirli, C.; Saletti, G.; Schiavon, M.; Bruttomesso, D.; Bigler, L.; Follath, F.; Pettenazzo, A.; et al. Amiodarone alters late endosomes and inhibits SARS coronavirus infection at a post-endosomal level. *Am. J. Respir. Cell Mol. Biol.* **2008**, *39*, 142–149. [CrossRef] [PubMed]
38. Cheng, Y.-L.; Lan, K.-H.; Lee, W.-P.; Tseng, S.-H.; Hung, L.-R.; Lin, H.-C.; Lee, F.-Y.; Lee, S.-D.; Lan, K.-H. Amiodarone inhibits the entry and assembly steps of hepatitis C virus life cycle. *Clin. Sci.* **2013**, *125*, 439–448. [CrossRef]
39. Schor, S.; Einav, S. Repurposing of Kinase Inhibitors as Broad-Spectrum Antiviral Drugs. *DNA Cell Biol.* **2018**, *37*, 63–69. [CrossRef] [PubMed]
40. Ott, P.A.; Adams, S. Small-molecule protein kinase inhibitors and their effects on the immune system: Implications for cancer treatment. *Immunotherapy* **2011**, *3*, 213–227. [CrossRef]
41. Gross, S.; Rahal, R.; Stransky, N.; Lengauer, C.; Hoeflich, K.P. Targeting cancer with kinase inhibitors. *J. Clin. Investig.* **2015**, *125*, 1780–1789. [CrossRef]
42. Urata, S.; Ngo, N.; de la Torre, J.C. The PI3K/Akt pathway contributes to arenavirus budding. *J. Virol.* **2012**, *86*, 4578–4585. [CrossRef]
43. Booth, L.; Roberts, J.L.; Ecroyd, H.; Tritsch, S.R.; Bavari, S.; Reid, S.P.; Proniuk, S.; Zukiwski, A.; Jacob, A.; Sepúlveda, C.S.; et al. AR-12 Inhibits Multiple Chaperones Concomitant With Stimulating Autophagosome Formation Collectively Preventing Virus Replication. *J. Cell. Physiol.* **2016**, *231*, 2286–2302. [CrossRef] [PubMed]
44. Su, A.-R.; Qiu, M.; Li, Y.-L.; Xu, W.-T.; Song, S.-W.; Wang, X.-H.; Song, H.-Y.; Zheng, N.; Wu, Z.-W. BX-795 inhibits HSV-1 and HSV-2 replication by blocking the JNK/p38 pathways without interfering with PDK1 activity in host cells. *Acta Pharmacol. Sin.* **2017**, *38*, 402–414. [CrossRef]
45. Johansen, L.M.; Brannan, J.M.; Delos, S.E.; Shoemaker, C.J.; Stossel, A.; Lear, C.; Hoffstrom, B.G.; Dewald, L.E.; Schornberg, K.L.; Scully, C.; et al. FDA-approved selective estrogen receptor modulators inhibit Ebola virus infection. *Sci. Transl. Med.* **2013**, *5*, 190ra79. [CrossRef]
46. Lee, A.M.; Rojek, J.M.; Gundersen, A.; Ströher, U.; Juteau, J.-M.; Vaillant, A.; Kunz, S. Inhibition of cellular entry of lymphocytic choriomeningitis virus by amphipathic DNA polymers. *Virology* **2008**, *372*, 107–117. [CrossRef] [PubMed]
47. Herring, S.; Oda, J.M.; Wagoner, J.; Kirchmeier, D.; O'Connor, A.; Nelson, E.A.; Huang, Q.; Liang, Y.; DeWald, L.E.; Johansen, L.M.; et al. Inhibition of Arenaviruses by Combinations of Orally Available Approved Drugs. *Antimicrob. Agents Chemother.* **2021**, *65*. [CrossRef]
48. Gowen, B.B.; Juelich, T.L.; Sefing, E.J.; Brasel, T.; Smith, J.K.; Zhang, L.; Tigabu, B.; Hill, T.E.; Yun, T.; Pietzsch, C.; et al. Favipiravir (T-705) inhibits Junín virus infection and reduces mortality in a guinea pig model of Argentine hemorrhagic fever. *PLoS Negl. Trop. Dis.* **2013**, *7*, e2614. [CrossRef]
49. Mendenhall, M.; Russell, A.; Juelich, T.; Messina, E.L.; Smee, D.F.; Freiberg, A.N.; Holbrook, M.R.; Furuta, Y.; de la Torre, J.-C.; Nunberg, J.H.; et al. T-705 (favipiravir) inhibition of arenavirus replication in cell culture. *Antimicrob. Agents Chemother.* **2011**, *55*, 782–787. [CrossRef] [PubMed]
50. Kim, Y.-J.; Cubitt, B.; Cai, Y.; Kuhn, J.H.; Vitt, D.; Kohlhof, H.; de la Torre, J.C. Novel Dihydroorotate Dehydrogenase Inhibitors with Potent Interferon-Independent Antiviral Activity against Mammarenaviruses In Vitro. *Viruses* **2020**, *12*, 821. [CrossRef] [PubMed]
51. Wang, Q.-Y.; Bushell, S.; Qing, M.; Xu, H.Y.; Bonavia, A.; Nunes, S.; Zhou, J.; Poh, M.K.; Florez de Sessions, P.; Niyomrattanakit, P.; et al. Inhibition of dengue virus through suppression of host pyrimidine biosynthesis. *J. Virol.* **2011**, *85*, 6548–6556. [CrossRef]
52. Johansen, L.M.; DeWald, L.E.; Shoemaker, C.J.; Hoffstrom, B.G.; Lear-Rooney, C.M.; Stossel, A.; Nelson, E.; Delos, S.E.; Simmons, J.A.; Grenier, J.M.; et al. A screen of approved drugs and molecular probes identifies therapeutics with anti-Ebola virus activity. *Sci. Transl. Med.* **2015**, *7*, 290ra89. [CrossRef] [PubMed]
53. Dyall, J.; Coleman, C.M.; Hart, B.J.; Venkataraman, T.; Holbrook, M.R.; Kindrachuk, J.; Johnson, R.F.; Olinger, G.G.; Jahrling, P.B.; Laidlaw, M.; et al. Repurposing of clinically developed drugs for treatment of Middle East respiratory syndrome coronavirus infection. *Antimicrob. Agents Chemother.* **2014**, *58*, 4885–4893. [CrossRef] [PubMed]
54. Yang, L.; Pei, R.; Li, H.; Ma, X.; Zhou, Y.; Zhu, F.; He, P.; Tang, W.; Zhang, Y.; Xiong, J.; et al. Identification of SARS-CoV-2 entry inhibitors among already approved drugs. *Acta Pharmacol. Sin.* **2020**. [CrossRef]
55. Delang, L.; Abdelnabi, R.; Neyts, J. Favipiravir as a potential countermeasure against neglected and emerging RNA viruses. *Antivir. Res.* **2018**, *153*, 85–94. [CrossRef] [PubMed]
56. Pomeroy, J.J.; Drusano, G.L.; Rodriquez, J.L.; Brown, A.N. Searching for synergy: Identifying optimal antiviral combination therapy using Hepatitis C virus (HCV) agents in a replicon system. *Antivir. Res.* **2017**, *146*, 149–152. [CrossRef]

Article

The Antiviral Effect of the Chemical Compounds Targeting DED/EDh Motifs of the Viral Proteins on Lymphocytic Choriomeningitis Virus and SARS-CoV-2

Mya Myat Ngwe Tun [1], Kouichi Morita [1], Takeshi Ishikawa [2,*] and Shuzo Urata [3,*]

1 Department of Virology, Institute of Tropical Medicine and Leading Program, Graduate School of Biomedical Science, Nagasaki University, 1-12-4 Sakamoto, Nagasaki 852-8523, Japan; myamyat@tm.nagasaki-u.ac.jp (M.M.N.T.); moritak@nagasaki-u.ac.jp (K.M.)
2 Department of Chemistry, Biotechnology and Chemical Engineering, Graduate School of Science and Engineering, Kagoshima University, 1-21-40 Korimoto, Kagoshima 890-0065, Japan
3 National Research Center for the Control and Prevention of Infectious Diseases (CCPID), Nagasaki University, 1-12-4 Sakamoto, Nagasaki 852-8523, Japan
* Correspondence: ishi@cb.kagoshima-u.ac.jp (T.I.); shuzourata@nagasaki-u.ac.jp (S.U.); Tel.: +81-99-285-8334 (T.I.); +81-95-819-7970 (S.U.)

Abstract: Arenaviruses and coronaviruses include several human pathogenic viruses, such as Lassa virus, Lymphocytic choriomeningitis virus (LCMV), SARS-CoV, MERS-CoV, and SARS-CoV-2. Although these viruses belong to different virus families, they possess a common motif, the DED/EDh motif, known as an exonuclease (ExoN) motif. In this study, proof-of-concept studies, in which the DED/EDh motif in these viral proteins, NP for arenaviruses, and nsp14 for coronaviruses, could be a drug target, were performed. Docking simulation studies between two structurally different chemical compounds, ATA and PV6R, and the DED/EDh motifs in these viral proteins indicated that these compounds target DED/EDh motifs. The concentration which exhibited modest cell toxicity was used with these compounds to treat LCMV and SARS-CoV-2 infections in two different cell lines, A549 and Vero 76 cells. Both ATA and PV6R inhibited the post-entry step of LCMV and SARS-CoV-2 infection. These studies strongly suggest that DED/EDh motifs in these viral proteins could be a drug target to combat two distinct viral families, arenaviruses and coronaviruses.

Keywords: Lymphocytic choriomeningitis virus (LCMV); SARS-CoV-2; antivirals; exonuclease (ExoN) motif; DED/EDh motif

Citation: Ngwe Tun, M.M.; Morita, K.; Ishikawa, T.; Urata, S. The Antiviral Effect of the Chemical Compounds Targeting DED/EDh Motifs of the Viral Proteins on Lymphocytic Choriomeningitis Virus and SARS-CoV-2. *Viruses* **2021**, *13*, 1220. https://doi.org/10.3390/v13071220

Academic Editors: Michael B.A. Oldstone and Juan C. De la Torre

Received: 18 May 2021
Accepted: 23 June 2021
Published: 24 June 2021

1. Introduction

Several arenaviruses cause hemorrhagic fever in humans. Lassa virus (LASV) is the causative agent of Lassa fever. In addition, evidence indicates that the globally distributed prototypic arenavirus Lymphocytic choriomeningitis virus (LCMV), is a neglected human pathogen of clinical significance in congenital viral infection and transplant recipients' infection. Current arenavirus antiviral drug therapy is restricted using the nucleoside analog ribavirin (Rib), which is only partially effective, and is associated with significant side effects. Arenavirus possesses a negative-strand RNA as a viral genome and encodes only four viral proteins: a glycoprotein precursor, GPC; an RNA-dependent RNA polymerase (RdRp), L; a matrix protein, Z; and the nucleoprotein, NP. NP is the most abundant and multifunctional protein during viral infection, with key roles in host immunosuppression, viral replication, and encapsidation of the viral genome [1].

Coronaviruses are also a major threat to public health. The most notable infections are severe acute respiratory syndrome (SARS), Middle East respiratory syndrome (MERS), and Coronavirus disease 2019 (COVID-19), caused by SARS-CoV, MERS-CoV, and SARS-CoV-2, respectively. Due to the large outbreak of COVID-19 worldwide, many efforts have been made to develop vaccines and antivirals against COVID-19. However, the appearance of the escape mutant virus against these developed vaccines and antivirals has been a significant concern. Therefore, the development of an effective treatment against COVID-19 is urgently needed. Coronaviruses are enveloped viruses with single-stranded positive-sense RNA genomes of approximately 30 kb. There are approximately 15 predicted open reading frames (ORFs), including ORF1a and ORF1b, spike (S), envelope (E), membrane (M), and nucleocapsid (N). Among these ORFs, nsp14 is known to have bifunctional roles, 3′ to 5′ exoribonuclease (ExoN) and guanine-N7-methyltransferase domains [2].

Arenaviruses NP and coronaviruses nsp14 possess a common motif known as a DED/EDh motif, which has ExoN activity as described above. The role of ExoN activity in the viral life cycle is known to avoid the innate immune response by degrading produced viral RNA, which can be recognized by the RNA sensors. Recent studies have shown the roles of these DED/EDh motifs in viral replication. In this study, proof-of-concept experiments were performed using chemical compounds, which could bind to the DED/EDh motifs in LCMV NP and SARS-CoV-2 nsp14 in silico, respectively, and inhibited the post-entry step of the LCMV and SARS-CoV-2. These results strongly suggest the potential of the DED/EDh motifs as a novel drug target to combat several highly pathogenic viruses belonging to different virus families, including *Arenaviridae* and *Coronaviridae*.

2. Materials and Methods

2.1. Cells and Compounds

Vero 76, Vero E6, and A549 cells were obtained as described previously [3] and maintained in Dulbecco's Modified Eagle's Medium (DMEM) supplemented with 10% fetal bovine serum (FBS) and 1% penicillin/streptomycin. Aurintricarboxylic acid (ATA) and pontacyl violet 6R (PV6R) were obtained from Santa Cruz (sc-3525A) and Tokyo Chemical Industry (P0593-1G), respectively.

2.2. Viruses

Tri-segmented recombinant LCMV (3rLCMV/GFP), which expresses green fluorescent protein (GFP) upon infection, was rescued using BHK-21 cells as described previously [4]. The SARS-CoV-2 (JPN/TY/WK-521), which was isolated and provided by the National Institute of Infectious Diseases, Japan [5], was propagated in Vero E6 cells.

2.3. Cell Viability Assay

Cell viability of A549 and Vero 76 cells after ATA and PV6R treatment was assessed using the CellTiter-Glo Luminescent Cell Viability Assay (Promega, Madison, WI, USA), which determines the number of viable cells in a culture based on ATP levels. A549 and Vero 76 cells (2×10^4 cells/well) were seeded in 96 well plate to form a monolayer. After cell seeding, the cells were treated with ATA, PV6R, and DMSO as a control. At 24 and 48 h post-treatment, the culture supernatant was removed, and CellTiter-Glo reagent was added. Thereafter, the assay was performed according to the manufacturer's recommendations using SpectraMAX iD5 (Molecular Device, San Jose, CA, USA). The viability of DMSO-treated control cells was set to 1.0.

2.4. Compounds' Treatments

A549 and Vero 76 cells (2×10^4 cells/well) were seeded in 96 well plates. The cells were infected with 3rLCMV/GFP or SARS-CoV-2. At 1 h post-infection (p.i.), media was replaced with fresh media containing DMSO, ATA, or PV6R. At 24 h p.i. for 3rLCMV/GFP and at 48 h p.i. for the SARS-CoV-2, culture supernatant was collected to measure the viral titer. To visualize the infected cells, the 3rLCMV/GFP infected cells were fixed with 4% paraformaldehyde (PFA) at 20 h p.i.

2.5. Titration of 3rLCMV/GFP

Vero 76 cells (2×10^4 cells/well) were seeded one day prior to virus infection in 96 well plates. Culture supernatant upon the compounds' treatment was collected. Ten times dilution of the virus samples were loaded on top of the Vero 76 cells after removal of the culture medium. At 20 h p.i., cells were fixed with 4% PFA, and the number of GFP-positive cells was counted manually and normalized as fluorescent focus unit (FFU)/mL.

2.6. Plaque Assay for SARS-CoV-2

Plaque assays were performed using Vero E6 cell monolayer in a 24-well plate. Briefly, all samples were diluted from 1:10 to 1:10,000 dilutions with a cell culture medium and inoculated on top of the cells in quadruplicate at 200 μL per well. After incubation at 37 °C for 1 h, infected cells were overlaid with 500 μL of 1.25% methylcellulose 4000 in 2% FCS MEM. The plates were then incubated at 37 °C for 4 days. The plates were washed with PBS (-) to remove the methylcellulose, fixed with 4% PFA (Wako, Osaka, Japan) for 30 min at room temperature, rinsed, and stained with crystal violet. Plaques were counted and expressed as plaque forming units (PFU/mL).

2.7. Measurement of SARS-CoV-2 RNA in the Cell

Isogen-II (Nippon Gene, Toyama, Japan) was used to isolate and prepare RNA from cells according to the manufacturer's instruction. A volume of 5 μL of RNA was used for quantitative real-time RT-PCR (qRT-PCR), and amplification of the nucleocapsid (N) gene was performed using a total of 20 μL of reaction mixture, consisting of 5 μL of Taqman master mix, 7 μL of nuclease-free water, 1 μL of 0.5 μM forward and reverse primers, and 1 μL of 0.25 μM probe with SARS-CoV-2 N primers of TaqMan Fast Virus 1-Step Master Mix (Life Technologies, Carlsbad, CA, USA). The primers and probes were described in a previous report [5].

2.8. Docking Simulation

The atomic coordinates of the X-ray structure of the LCMV NP were downloaded from the PDB (PDB-ID: 4O6H) [6]. On the other hand, homology modeling was performed to obtain the atomic coordinates of SARS-CoV-2 nsp14 using Modeller10.0 [7] because X-ray structures were not reported. In the homology modeling, the X-ray structure of SARS-CoV nsp14 (PDB-ID: 5C8U) was used as a template [8]. Binding structures of ATA and PV6R to the DED/EDh motif of LCMV and SARS-CoV-2 were predicted by docking simulation using AutoDock Vina [9]. A cubic region of 26.3 × 26.3 × 26.3 Å, whose center was located at the center of mass of the five amino acids of the DED/EDh motif, was selected as a search region. EXHAUSTIVENESS was set to 20. The initial atomic coordinates of ATA and PV6R were downloaded from PubChem (CIDs were 2259 and 112806, respectively).

2.9. Statistical Analysis

Excel and GraphPad Prism 5 (GraphPad Software, Inc., San Diego, CA, USA) software were used for all statistical analyses. Quantitative data were presented as the mean ± SD from at least three independent experiments (unless indicated otherwise). For all calculations, $p < 0.05$, was considered significant and was represented using an asterisk (*). Group comparisons were performed using one-way analysis of variance (ANOVA), followed by Dunnett's multiple comparison test. Welch's *t*-test was used to compare the two groups.

3. Results

3.1. Cell Viability against ATA and PV6R Treatment

To examine whether the DED/EDh motifs could be antiviral targets of arenavirus and coronavirus, two chemical compounds (aurintricarboxylic acid [ATA, Figure 1A] and pontacyl violet 6R [PV6R, Figure 1B]), which were previously shown to bind to the DED/EDh motifs and to inhibit ExoN activity [10], were selected. The chemical structures of ATA and PV6R are shown in Figure 1A and 1B, respectively. First, the cell viabilities of two cell lines, A549 (Figure 1C) and Vero 76 (Figure 1D), were examined after ATA and PV6R treatments. Both cell lines were treated with 500 μM (ATA) and 200 μM (PV6R) for 24 and 48 h, and cell viability was measured using CellTiter-Glo as described in the Materials and Methods. In A549 cell lines, cell viability after ATA treatment was 73.4% and 81.7% at 24 and 48 h post-treatment, respectively, compared to the DMSO control treatment. PV6R treatment in A549 cells reduced cell viability to 81.1% and 94.1%, respectively, compared to the DMSO control treatment. In Vero 76 cells, although ATA treatment slightly increased the cell viability (119.7%) at 24 h post-treatment, the same treatment reduced the cell viability to 92.4% at 48 h post-treatment. Cell viabilities in Vero 76 cells upon PV6R treatment were 80.5% and 95.4% at 24 and 48 h post-treatment, respectively.

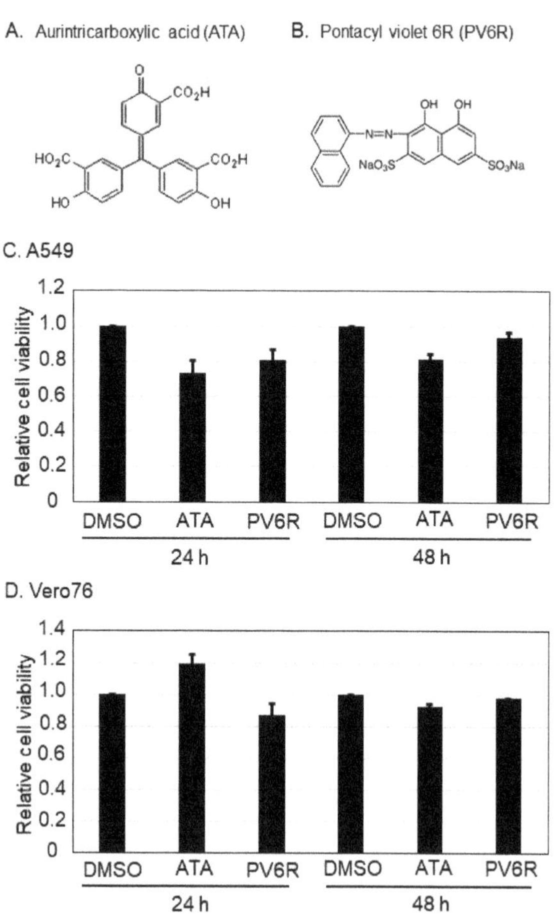

Figure 1. Chemical structure of aurintricarboxylic acid (ATA) and pontacyl violet 6R (PV6R), and their cell viabilities upon the treatments on A549 and Vero 76 cell lines. Chemical structure of ATA (**A**) and PV6R (**B**). Either A549 (**C**) or Vero 76 (**D**) cells were treated with 500 μM of ATA and 200 μM of PV6R for 24 h or 48 h, and their cell viabilities were measured using CellTiter-Glo. Relative cell viability which was normalized with the DMSO control treatment as 1.0 was shown. Data were collected from at least three independent experiments, and provided data correspond to the mean +/− SD.

3.2. Both ATA and PV6R Inhibited the Post-Entry Step of the LCMV Infection

The effect of the DEDDh motif on the structural stability of LCMV NP was examined. The structure of the DEDDh motif of LCMV NP is shown in Figure 2A, and the effects of structural stability upon compound binding were calculated in silico. The calculated binding free energies between the ATA and DEDDh motifs of LCMV NP, and the PV6R and DEDDh motifs of LCMV NP were −8.2 and −7.7 kcal/mol, respectively (Figure 2B,C). With the concentration that did not exhibit apparent cell toxicity, the effects of ATA and PV6R on LCMV infection were examined. To assess this point, tri-segment recombinant LCMV expressing GFP upon infection (3rLCMV/GFP) was used [4]. GFP expression was reflected in the replication and transcription of LCMV in the cell. A549 cells were infected with 3rLCMV/GFP at an MOI of 0.1, and Vero 76 was infected with 3rLCMV/GFP at an MOI of 0.01. At 1.5 h post-infection (p.i.), the culture media was replaced with fresh media containing DMSO, ATA, or PV6R. Cells were fixed with 4% paraformaldehyde (PFA) at 20 h p.i., captured using a fluorescent microscope (Figure 2D), and GFP-positive cells were counted (Figure 2E,F). Treatment with ATA reduced the number of GFP-positive cells to 20.1% and 25.2% compared to DMSO treatment in A549 and Vero 76 cells, respectively. Treatment with PV6R reduced the number of GFP-positive cells to 44.2% and 37.8% compared to DMSO treatment in A549 and Vero 76 cells, respectively.

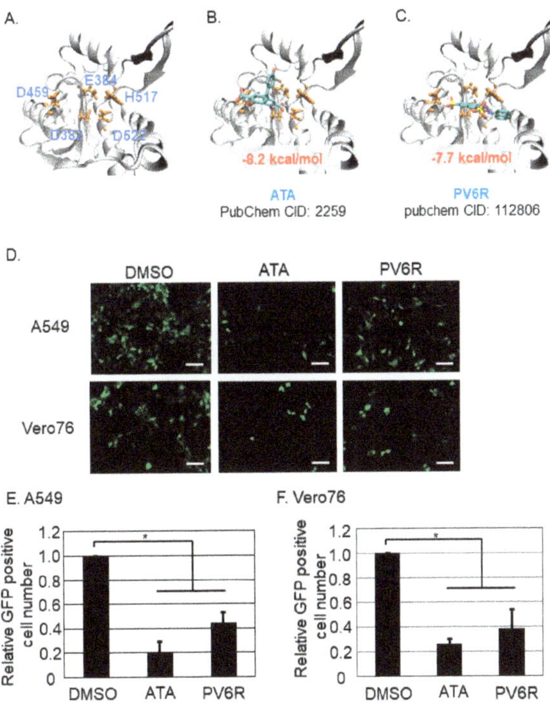

Figure 2. Both ATA and PV6R inhibited the post-entry step of the LCMV infection. (**A**) The structure of the DEDDh motif of LCMV NP. Amino acid number which corresponded to the DEDDh motif was described in BLUE font, together with the amino acid number. (**B**) Docking simulation of the ATA and the DEDDh motif of LCMV NP. (**C**) Docking simulation of the PV6R and the DEDDh motif of LCMV NP. (**D**) Either A549 cells or Vero 76 cells were infected with 3rLCMV/GFP and treated with DMSO, ATA, or PV6R. At 20 h post infection, the cells were fixed and the GFP positive cells were observed with the fluorescent microscope. Bar; 100 µm. GFP positive cell number in A549 (**E**) and in Vero 76 (**F**) were counted and normalized with the GFP positive cell number in DMSO control treatment as 1.0. Data were collected from at least three independent experiments, and shown data correspond to the mean +/− SD (*; $p < 0.05$).

3.3. ATA, but Not PV6R, Inhibited the Propagation of the LCMV

Because a significant reduction in the number of GFP-positive cells was observed in ATA and PV6R treatments in A549 and Vero 76 cells, the viral production from the ATA and PV6R treated Vero 76 cells was also evaluated. Vero 76 cells were infected with 3rLCMV/GFP for 1.5 h, and the culture medium was replaced with fresh media containing DMSO, ATA (500 μM), or PV6R (200 μM). At 24 h p.i., the culture supernatant was collected to measure their viral titer using Vero 76 cells. ATA treatment resulted in an approximately 100-fold reduction in the viral titer, compared to DMSO treatment. In contrast, PV6R treatment increased the viral titer by approximately 10 times, compared to the DMSO control treatment (Figure 3A). To observe if PV6R affects steps other than the replication step in the cells of the LCMV replication, PV6R was incubated with both 3rLCMV/GFP and Vero 76 cells, respectively, before the infection. After 1 h incubation, the virus with PV6R and Vero 76 cells with PV6R were mixed together, allowing the infection. At 1.5 h p.i., infected cells were washed and kept in incubation without any compounds. Infected cells were fixed at 20 h p.i. and observed under the fluorescent microscope (Figure 3B). PV6R treatment significantly increased the GFP positive cell number (6.7 times) compared to that of the DMSO treatment (Figure 3C).

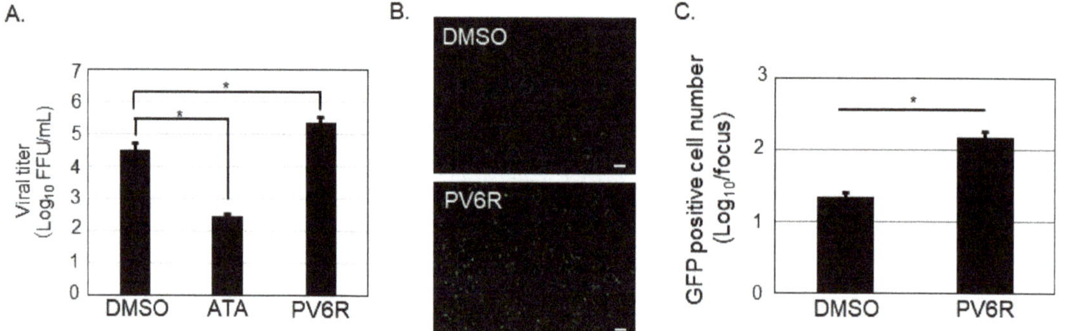

Figure 3. ATA, but not PV6R, inhibited LCMV production. (**A**) Culture supernatant of the Vero 76 cells infected with 3rLCMV/GFP and treated with DMSO, ATA (500 μM), or PV6R (200 μM), were used to measure the viral titer. (**B**,**C**) 3rLCMV/GFP and Vero 76 cells were pre-treated with PV6R (200 μM) and allowed for the infection. PV6R was not included in the culture media after the infection. At 20 h post infection, the cells were fixed and the GFP positive cells were observed with the fluorescent microscope. Bar; 200 μM (**B**). Counted GFP positive cells were plotted from both DMSO and PV6R treatments (**C**). Data were collected from at least three independent experiments, and shown data correspond to the mean +/− SD (* $p < 0.05$).

3.4. Both ATA and PV6R Inhibited SARS-CoV-2 Replication and Propagation

Coronaviruses' nsp14 possess the DEEDh motif (equivalent to the DEDDh motif), which contributes to anti-interferon (IFN) activity and virus replication [11]. The crystal structure of the nsp14 of SARS-CoV-2 has not been reported; thus, homology modeling was performed to predict SARS-CoV-2 nsp14 structure using SARS-CoV nsp14 (Figure 4A) [8]. The predicted homology model of SARS-CoV-2 nsp14 overlapped with the DEDDh motif of LCMV NP (Figure 4B). Furthermore, docking simulation between SARS-CoV-2 nsp14 and the compounds (ATA and PV6R) was also performed (Figure 4C,D). For both compounds, the binding structures that strongly interacted with the 5 amino acids of the DEEDh motif were obtained. The calculated binding free energies were −6.8 (ATA) and −7.7 (PV6R) kcal/mol, respectively. To examine if ATA and PV6R inhibited the SARS-CoV-2 in vitro, Vero76 cells were infected with SARS-CoV-2, and at 1.5 h p.i., the culture media was replaced with fresh media containing DMSO, ATA (500 μM), or PV6R (200 μM). At 24 h p.i.,

RNA was collected from the infected cells to quantify the viral RNA produced from the SARS-CoV-2 infection (Figure 4E). ATA and PV6R treatments reduced 66.6-fold and 25-fold of the viral genome copy number compared to the DMSO treatment, respectively (Figure 4E). Furthermore, to examine if these compounds inhibited the propagation of the SARS-CoV-2, virus infection was proceeded in the same way as the above experiment and the culture supernatant was collected to evaluate the produced infectious virion in the culture supernatant. In Vero 76 cells, ATA and PV6R treatments reduced the production of the infectious virion to 0.35% and 4.2%, respectively, compared to the DMSO control treatment.

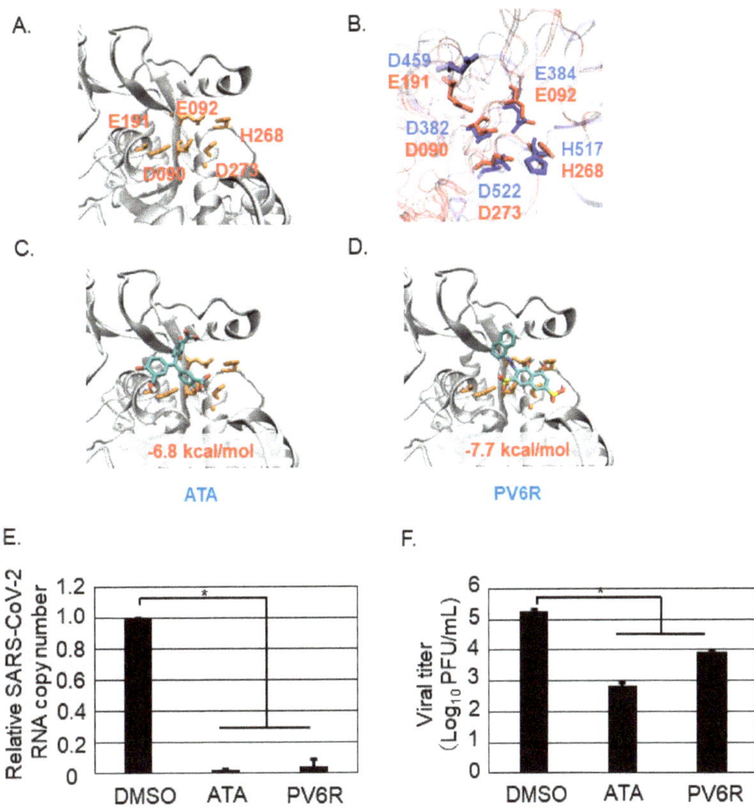

Figure 4. Both ATA and PV6R inhibited SARS-CoV-2 infection and propagation. (**A**) Homology modeling of the DEEDh motif of the SARS-CoV-2 nsp14. Amino acid number which corresponded to the DEEDh motif was described in RED font together with the amino acid number. (**B**) Overlap of the SARS-CoV-2 nsp14 and LCMV NP DED/EDh motifs was shown. Amino acids of the DED/EDh motifs in SARS-CoV-2 nsp14 and LCMV NP were colored in RED and BLUE, respectively. (**C**) Docking simulation of the ATA and the homology model of the SARS-CoV-2 nsp14. (**D**) Docking simulation of the PV6R and the homology model of the SARS-CoV-2 nsp14. Vero 76 cells were infected with SARS-CoV-2 and treated with DMSO, ATA, or PV6R, respectively. The viral copy number in the cells at 24 h p.i. was measured. Relative viral copy number normalized with the DMSO control treatment in Vero 76 cells was shown (**E**). Vero 76 cells were infected with SARS-CoV-2 and treated with DMSO, ATA, or PV6R, respectively, and the viral titer was measured at 48 h p.i. (**F**). Data were collected from at least three independent experiments, and shown data correspond to the mean +/− SD (* $p < 0.05$).

4. Discussion

Arenaviruses and coronaviruses are important human pathogens. LASV is the major pathogen of Lassa fever and belongs to the *Areanaviridae* family. LCMV also belongs to *Arenaviridae* and is harmful to pregnant women and transplant recipients [1]. SARS-CoV, MERS-CoV, and SARS-CoV-2 are other important human pathogens that belong to *Coronaviridae*. Although these two virus families (*Arenaviridae* and *Coronaviridae*) have different features, arenavirus possesses single-stranded negative-sense RNA and coronavirus possesses single-stranded positive-sense RNA as viral genomes, arenavirus NP and coronavirus nsp14 encode DED/EDh motifs, which have been shown to play important roles in antagonizing the innate immune response (also known as an ExoN motif) and in viral genome replication.

The first discovery of ExoN activity in human pathogenic viruses was from the coronavirus. The DEEDh motif of nsp14 as an ExoN was shown by biochemical experiments [12]. It is now widely accepted that the ExoN of nsp14 is conserved among coronaviruses [13]. In addition, the DEEDh motif of nsp14 is important for viral replication apart from the anti-IFN function [11]. Determining the structure of the SARS-CoV nsp14-nsp10 complex also provided evidence of its ExoN activity [8]. Accordingly, the ExoN activity of coronavirus nsp14 appears to be an important drug target to combat coronaviruses, including SARS-CoV-2 [14]. The structure of SARS-CoV-2 nsp14 has not yet been published in the protein data bank [15]; therefore, a predicted structure based on the SARS-CoV nsp14 was constructed (Figure 4A). Using this model, we calculated the stability of the DEEDh motif of SARS-CoV-2 nsp14 upon binding two chemical compounds, ATA and PV6R, in silico. During this study, a similar attempt was reported, and the interaction between PV6R and SARS-CoV-2 nsp14 was examined in silico as well [16]. Our study is the first to reveal that both ATA and PV6R inhibited SARS-CoV-2 replication and propagation in vitro (Figure 4E,F), as predicted from the in silico studies.

In the case of arenaviruses, a study was performed to identify specific amino acids (D382 and G385) in LCMV NP, which corresponded to antagonizing the IFN activity and was involved in genome replication using a reverse genetics system and the minigenome system [17]. It was demonstrated that the point mutations in D382 and G385 in LCMV NP did not affect genome replication; however, it affected the anti-IFN activity. In contrast, deletion of these amino acids (D382 and G385) abolished the genome replication and an anti-IFN activity. These results suggest that the overall structure of the DEDDh motif in NP could affect not only anti-IFN activity, but also genome replication. A similar approach was used with LASV to show the same motif on anti-IFN activity in NP (D389 and G392) [18]. The authors of this article described that LASV NP-G392A recombinant virus was not rescued, although NP-D389A/G392A and NP-D389T/G392A double mutants were rescued. These results suggest that disruption of the DEDDh motif in NP could affect genome replication. Upon determination of the LASV and LCMV NP structures, it became apparent that the DEDDh motif in arenavirus NP corresponds to coronavirus nsp14 as an ExoN motif [6,19–21].

Based on the increasing evidence that the DED/EDh motifs of arenavirus NP and coronavirus nsp14 could be a target to inhibit virus replication and propagation, this study attempted in vitro analysis to assess this point as a proof-of-concept study. Although we did not experimentally show that ATA and PV6R bind to the arenavirus NP and coronavirus nsp14 in this study, we reported the antiviral effects of ATA and PV6R, which were previously shown to inhibit ExoN activity in biochemical studies [10], on the post-entry step of LCMV and SARS-CoV-2. Due to our experimental design, we could not precisely point out the compounds' target of the post-entry step in the viral life cycle. However, our in silico study showed that both compounds exhibited a strong potential to bind to the DEDDh motif of LCMV NP and accordingly inhibited the post-entry step of LCMV (Figure 2E,F). Furthermore, ATA and PV6R exhibited a strong potential to bind to the DEEDh motif of SARS-CoV-2 nsp14 (Figure 4C,D) and affected protein stability. Accordingly, these compounds also inhibited the post-entry step (Figure 4E) and propagation (Figure 4F) of SARS-CoV-2. These results supported the idea that both ATA and PV6R inhibited the arenavirus and coronavirus replication through binding ExoN motif in each viral proteins.

It is worth noting that ATA affects cellular events other than ExoN activity. ATA has been reported to inhibit ZIKV [22], vaccinia virus [23], hepatitis C virus [24], and influenza A and B viruses [25], all of which do not possess the DED/EDh motif. The potential of the ATA to bind SARS-CoV proteins other than the nsp14 [26,27] has also been reported, suggesting that the antiviral effect we observed was not purely due to the disruption of the DEEDh motif of SARS-CoV-2 nsp14. However, the antiviral effect of PV6R against SARS-CoV-2 supports the idea that the DEEDh motif of SARS-CoV-2 nsp14 could be an antiviral target. Although the concentration of the compounds used in this study showed a modest effect on cell viability (Figure 1C,D), in which the newly synthesized ATP was measured, these chemical compounds could still affect the host factors. The 3′-to-5′ ExoN plays major roles in prokaryotic and eukaryotic cells, and among these exonucleases, the DEDDh superfamily is involved in 3′ maturation, nuclear mRNA surveillance, and mRNA decay [28]. Indeed, ATA is known to inhibit nucleases [29], nucleic acid processing enzymes [30], and phosphorylation of extracellular signal-regulated kinase 1/2 [23]. The results from the PV6R, whose chemical structure is different from that of ATA, supported the notion that DED/EDh motifs of arenavirus NP and coronavirus nsp14 could be antiviral targets. The chemical information of PV6R is limited; therefore, it is difficult to discuss how PV6R affects cellular events. The results obtained from our study indicated that PV6R could facilitate the entry process of LCMV infection (Figure 3B,C). Further experiments are required to explore the precise mechanism how PV6R facilitate the LCMV entry process.

In conclusion, this study proved that the DED/EDh motif of two different virus families, arenavirus and coronavirus, could be a drug target. Although ExoNs are also expressed in the host, targeting slight differences in these ExoNs between the host and viral ExoNs, such as the existence of the zinc finger motif, could lead to develop a specific drug [31]. In addition, treating these viral infections with dual and/or triple drugs could increase the antiviral effect and reduce the possible appearance of drug-resistant viruses.

Author Contributions: Conceptualization, S.U.; methodology, S.U. and T.I.; investigation, S.U., M.M.N.T. and T.I.; resources, S.U. and M.M.N.T.; data curation, S.U.; writing—original draft preparation, S.U.; writing—review and editing, M.M.N.T., K.M. and T.I.; visualization, S.U. and T.I.; supervision, S.U.; funding acquisition, S.U., M.M.N.T., K.M. and T.I. All authors have read and agreed to the published version of the manuscript.

Funding: This research was funded by Japan Agency for Medical Research and Development (AMED), U.S.-Japan Cooperative Medical Sciences Program (grant number 20JK0210001h0001).

Institutional Review Board Statement: Not applicable.

Informed Consent Statement: Not applicable.

Data Availability Statement: Not applicable.

Acknowledgments: Authors thank to the member of the Urata-unit, Department of Human Resource Development, National Research Center for the Control and Prevention of Infectious Diseases (CCPID), Nagasaki University. Authors also appreciate to Juan Carlos de la Torre (The Scripps Research Institute, La Jolla, CA, USA) for sharing plasmids to rescue the 3rLCMV/GFP.

Conflicts of Interest: The authors declare no conflict of interest.

References

1. Radoshitzky, S.R.; Buchmeier, M.J.; de la Torre, J.C. Arenaviridae: The Viruses and Their Replication. In *Fields Virology*, 7th ed.; Springer: Alphen aan den Rijn, The Netherlands, 2020; Volume 1, pp. 784–809.
2. Perlman, S.; Masters, P.S. Coronaviridae: The Viruses and Their Replication. In *Fields Virology*, 7th ed.; Springer: Alphen aan den Rijn, The Netherlands, 2020; Volume 1, pp. 410–448.
3. Watanabe, K.; Ishikawa, T.; Otaki, H.; Mizuta, S.; Hamada, T.; Nakagaki, T.; Ishibashi, D.; Urata, S.; Yasuda, J.; Tanaka, Y.; et al. Structure-based drug discovery for combating influenza virus by targeting the PA-PB1 interaction. *Sci. Rep.* **2017**, *7*, 9500. [CrossRef] [PubMed]

4. Emonet, S.F.; Garidou, L.; McGavern, D.B.; de la Torre, J.C. Generation of recombinant lymphocytic choriomeningitis viruses with trisegmented genomes stably expressing two additional genes of interest. *Proc. Natl. Acad. Sci. USA* **2009**, *106*, 3473–3478. [CrossRef]
5. Shirato, K.; Nao, N.; Katano, H.; Takayama, I.; Saito, S.; Kato, F.; Katoh, H.; Sakata, M.; Nakatsu, Y.; Mori, Y.; et al. Development of Genetic Diagnostic Methods for Detection for Novel Coronavirus 2019(nCoV-2019) in Japan. *Jpn. J. Infect. Dis.* **2020**, *73*, 304–307. [CrossRef] [PubMed]
6. West, B.R.; Hastie, K.M.; Saphire, E.O. Structure of the LCMV nucleoprotein provides a template for understanding arenavirus replication and immunosuppression. *Acta Cryst. D Biol. Cryst.* **2014**, *70*, 1764–1769. [CrossRef] [PubMed]
7. Sali, A.; Blundell, T.L. Comparative protein modelling by satisfaction of spatial restraints. *J. Mol. Biol.* **1993**, *234*, 779–815. [CrossRef]
8. Ma, Y.; Wu, L.; Shaw, N.; Gao, Y.; Wang, J.; Sun, Y.; Lou, Z.; Yan, L.; Zhang, R.; Rao, Z. Structural basis and functional analysis of the SARS coronavirus nsp14-nsp10 complex. *Proc. Natl. Acad. Sci. USA* **2015**, *112*, 9436–9441. [CrossRef]
9. Trott, O.; Olson, A.J. AutoDock Vina: Improving the speed and accuracy of docking with a new scoring function, efficient optimization, and multithreading. *J. Comput. Chem.* **2010**, *31*, 455–461. [CrossRef]
10. Huang, K.W.; Hsu, K.C.; Chu, L.Y.; Yang, J.M.; Yuan, H.S.; Hsiao, Y.Y. Identification of Inhibitors for the DEDDh Family of Exonucleases and a Unique Inhibition Mechanism by Crystal Structure Analysis of CRN-4 Bound with 2-Morpholin-4-ylethanesulfonate (MES). *J. Med. Chem.* **2016**, *59*, 8019–8029. [CrossRef]
11. Ogando, N.S.; Zevenhoven-Dobbe, J.C.; van der Meer, Y.; Bredenbeek, P.J.; Posthuma, C.C.; Snijder, E.J. The Enzymatic Activity of the nsp14 Exoribonuclease Is Critical for Replication of MERS-CoV and SARS-CoV-2. *J. Virol.* **2020**, *94*. [CrossRef] [PubMed]
12. Minskaia, E.; Hertzig, T.; Gorbalenya, A.E.; Campanacci, V.; Cambillau, C.; Canard, B.; Ziebuhr, J. Discovery of an RNA virus 3′->5′ exoribonuclease that is critically involved in coronavirus RNA synthesis. *Proc. Natl. Acad. Sci. USA* **2006**, *103*, 5108–5113. [CrossRef] [PubMed]
13. Case, J.B.; Li, Y.; Elliott, R.; Lu, X.; Graepel, K.W.; Sexton, N.R.; Smith, E.C.; Weiss, S.R.; Denison, M.R. Murine Hepatitis Virus nsp14 Exoribonuclease Activity Is Required for Resistance to Innate Immunity. *J. Virol.* **2018**, *92*. [CrossRef]
14. Pruijssers, A.J.; Denison, M.R. Nucleoside analogues for the treatment of coronavirus infections. *Curr. Opin. Virol.* **2019**, *35*, 57–62. [CrossRef]
15. Moeller, N.H.; Shi, K.; Demir, O.; Banerjee, S.; Yin, L.; Belica, C.; Durfee, C.; Amaro, R.E.; Aihara, H. Structure and dynamics of SARS-CoV-2 proofreading exoribonuclease ExoN. *bioRxiv* **2021**. [CrossRef]
16. Munaweera, R.; Hu, Y.S. Computational Characterizations of the Interactions Between the Pontacyl Violet 6R and Exoribonuclease as a Potential Drug Target Against SARS-CoV-2. *Front. Chem.* **2020**, *8*, 627340. [CrossRef]
17. Martinez-Sobrido, L.; Emonet, S.; Giannakas, P.; Cubitt, B.; Garcia-Sastre, A.; de la Torre, J.C. Identification of amino acid residues critical for the anti-interferon activity of the nucleoprotein of the prototypic arenavirus lymphocytic choriomeningitis virus. *J. Virol.* **2009**, *83*, 11330–11340. [CrossRef] [PubMed]
18. Carnec, X.; Baize, S.; Reynard, S.; Diancourt, L.; Caro, V.; Tordo, N.; Bouloy, M. Lassa virus nucleoprotein mutants generated by reverse genetics induce a robust type I interferon response in human dendritic cells and macrophages. *J. Virol.* **2011**, *85*, 12093–12097. [CrossRef]
19. Qi, X.; Lan, S.; Wang, W.; Schelde, L.M.; Dong, H.; Wallat, G.D.; Ly, H.; Liang, Y.; Dong, C. Cap binding and immune evasion revealed by Lassa nucleoprotein structure. *Nature* **2010**, *468*, 779–783. [CrossRef] [PubMed]
20. Hastie, K.M.; Kimberlin, C.R.; Zandonatti, M.A.; MacRae, I.J.; Saphire, E.O. Structure of the Lassa virus nucleoprotein reveals a dsRNA-specific 3′ to 5′ exonuclease activity essential for immune suppression. *Proc. Natl. Acad. Sci. USA* **2011**, *108*, 2396–2401. [CrossRef]
21. Jiang, X.; Huang, Q.; Wang, W.; Dong, H.; Ly, H.; Liang, Y.; Dong, C. Structures of arenaviral nucleoproteins with triphosphate dsRNA reveal a unique mechanism of immune suppression. *J. Biol. Chem.* **2013**, *288*, 16949–16959. [CrossRef] [PubMed]
22. Park, J.G.; Avila-Perez, G.; Madere, F.; Hilimire, T.A.; Nogales, A.; Almazan, F.; Martinez-Sobrido, L. Potent Inhibition of Zika Virus Replication by Aurintricarboxylic Acid. *Front. Microbiol.* **2019**, *10*, 718. [CrossRef]
23. Myskiw, C.; Deschambault, Y.; Jefferies, K.; He, R.; Cao, J. Aurintricarboxylic acid inhibits the early stage of vaccinia virus replication by targeting both cellular and viral factors. *J. Virol.* **2007**, *81*, 3027–3032. [CrossRef]
24. Chen, Y.; Bopda-Waffo, A.; Basu, A.; Krishnan, R.; Silberstein, E.; Taylor, D.R.; Talele, T.T.; Arora, P.; Kaushik-Basu, N. Characterization of aurintricarboxylic acid as a potent hepatitis C virus replicase inhibitor. *Antivir. Chem. Chemother.* **2009**, *20*, 19–36. [CrossRef]
25. Hashem, A.M.; Flaman, A.S.; Farnsworth, A.; Brown, E.G.; Van Domselaar, G.; He, R.; Li, X. Aurintricarboxylic acid is a potent inhibitor of influenza A and B virus neuraminidases. *PLoS ONE* **2009**, *4*, e8350. [CrossRef] [PubMed]
26. He, R.; Adonov, A.; Traykova-Adonova, M.; Cao, J.; Cutts, T.; Grudesky, E.; Deschambaul, Y.; Berry, J.; Drebot, M.; Li, X. Potent and selective inhibition of SARS coronavirus replication by aurintricarboxylic acid. *Biochem. Biophys. Res. Commun.* **2004**, *320*, 1199–1203. [CrossRef] [PubMed]
27. Yap, Y.; Zhang, X.; Andonov, A.; He, R. Structural analysis of inhibition mechanisms of aurintricarboxylic acid on SARS-CoV polymerase and other proteins. *Comput. Biol. Chem.* **2005**, *29*, 212–219. [CrossRef]
28. Ibrahim, H.; Wilusz, J.; Wilusz, C.J. RNA recognition by 3′-to-5′ exonucleases: The substrate perspective. *Biochim. Biophys. Acta* **2008**, *1779*, 256–265. [CrossRef]

29. Walther, W.; Stein, U.; Siegel, R.; Fichtner, I.; Schlag, P.M. Use of the nuclease inhibitor aurintricarboxylic acid (ATA) for improved non-viral intratumoral in vivo gene transfer by jet-injection. *J. Gene. Med.* **2005**, *7*, 477–485. [CrossRef]
30. Cushman, M.; Wang, P.L.; Chang, S.H.; Wild, C.; De Clercq, E.; Schols, D.; Goldman, M.E.; Bowen, J.A. Preparation and anti-HIV activities of aurintricarboxylic acid fractions and analogues: Direct correlation of antiviral potency with molecular weight. *J. Med. Chem.* **1991**, *34*, 329–337. [CrossRef] [PubMed]
31. Cruz-Gonzalez, A.; Munoz-Velasco, I.; Cottom-Salas, W.; Becerra, A.; Campillo-Balderas, J.A.; Hernandez-Morales, R.; Vazquez-Salazar, A.; Jacome, R.; Lazcano, A. Structural analysis of viral ExoN domains reveals polyphyletic hijacking events. *PLoS ONE* **2021**, *16*, e0246981. [CrossRef] [PubMed]

Review

Progress in Anti-Mammarenavirus Drug Development

Yu-Jin Kim [1], Victor Venturini [1,2] and Juan C. de la Torre [1,*]

1 Department of Immunology and Microbiology, The Scripps Research Institute, La Jolla, CA 92037, USA; yujin@scripps.edu (Y.-J.K.); v.venturini@outlook.es (V.V.)
2 Department of Biotechnology, Faculty of Experimental Sciences, Francisco de Vitoria University (UFV), Carretera Pozuelo-Majadahonda, Km 1,800, Pozuelo de Alarcón, 28223 Madrid, Spain
* Correspondence: juanct@scripps.edu

Abstract: Mammarenaviruses are prevalent pathogens distributed worldwide, and several strains cause severe cases of human infections with high morbidity and significant mortality. Currently, there is no FDA-approved antiviral drugs and vaccines against mammarenavirus and the potential treatment option is limited to an off-label use of ribavirin that shows only partial protective effect and associates with side effects. For the past few decades, extensive research has reported potential anti-mammarenaviral drugs and their mechanisms of action in host as well as vaccine candidates. This review describes current knowledge about mammarenavirus virology, progress of antiviral drug development, and technical strategies of drug screening.

Keywords: mammarenavirus; antiviral drug; drug repurposing; high-throughput screening

Citation: Kim, Y.-J.; Venturini, V.; de la Torre, J.C. Progress in Anti-Mammarenavirus Drug Development. *Viruses* **2021**, *13*, 1187. https://doi.org/10.3390/v13071187

Academic Editor: Luis Martinez-Sobrido

Received: 25 May 2021
Accepted: 19 June 2021
Published: 22 June 2021

1. Introduction

Members of the family *Arenaviridae* are classified into four genera based on phylogenetic analysis of their RNA-directed RNA polymerase (L protein) and nucleoprotein (NP) sequences: *Antennavirus*, *Hartmanivirus*, *Mammareanvirus*, and *Reptarenavirus* [1]. Antennaviruses (2 species) were discovered in actinopterygian fish by next-generation sequencing, and no biological isolate has been reported yet. Hartmaniviruses (4 species) and reptarenaviruses (5 species) infect captive snakes, and some of them have been associated with boid inclusion body disease (BIBD). Mammarenaviruses (39 species) infect mainly rodents, and the infection is generally asymptomatic. Current knowledge about the biology of snake and fish arenaviruses is very limited, and their zoonotic potential unknown. In contrast, some mammarenaviruses have been found to infect and cause disease in humans.

Mammarenaviruses are enveloped viruses with a bi-segmented single-stranded negative-sense RNA genome [1]. Mammarenaviruses cause chronic infections of their rodent natural reservoirs across the world, but some of them have zoonotic potential. Human infections occur through mucosal exposure to aerosols or by direct contact of abraded skin with infectious materials [2]. Based on their antigenic properties, mammarenaviruses have been classified into two distinct groups, Old world (OW) mammarenaviruses, aka Lassa-lymphocytic choriomeningitis virus serocomplex," including viruses present in Africa and the worldwide distributed lymphocytic choriomeningitis virus (LCMV), and the New World (NW) mammarenaviruses, aka "Tacaribe serocomplex", including viruses indigenous to the Americas [3]. Both OW and NW mammarenaviruses include several species members that can cause severe hemorrhagic fever (HF) diseases in humans that are associated with high morbidity and significant mortality; these viruses include Lassa (LASV), Junin (JUNV), Machupo (MACV), Guanarito (GTOV), Sabia (SABV), Chapare (CHPV), and Lujo (LUJV) [4]. Concerns posed by human pathogenic mammarenaviruses are exacerbated by the overall lack of FDA-licensed vaccines and current anti-mammareanavirus therapy being limited to off-label use of ribavirin that is only partially effective, has a narrow therapeutic window, and can be associated with side effects [5]. The only mammarenavirus vaccine tested in humans is the live-attenuated Candid#1 strain of

JUNV that has been shown to be safe and provide effective protection against Argentine HF (AHF) disease caused by JUNV [6,7]. Accordingly, Candid#1 is approved in Argentina for use in populations at high risk of JUNV infection [8].

This article presents a concise review of our current understanding of the mammarenavirus life cycle at the molecular and cellular level and of progress on antiviral drugs targeting specific steps of the mammarenavirus life cycle and their implications for potential therapeutic strategies against human pathogenic mammarenaviruses.

2. Mammarenavirus Molecular and Cell Biology

2.1. Virus Particle, Viral Proteins and Genomic Organization

Mammarenavirus virions are spherical lipid-enveloped particles with a wide diameter range (400 to 2000 Å) that enclose the bi-segmented negative-stranded (NS) RNA viral genome [9]. Each genome segment, L and S, uses an ambisense coding strategy to direct the synthesis of two polypeptides [10]. The small (S, ca 3.5 kb) genome segment encodes the viral nucleoprotein (NP) and the glycoprotein precursor GPC that, upon co-and post-translational processing, generates the GP complex that forms the spikes present at the virion surface and that mediate virus cell entry. The large (L, ca 7.5 kb) genome segment encodes the viral RNA-dependent RNA polymerase (RdRp; L protein) and the small RING finger protein Z that is a bona fide matrix protein (Figure 1A). NP is the most abundant viral protein in infected cells and virions. NP encapsidates the virus genome and antigenome RNA species to form a nucleocapsid complex that, in association with the viral L polymerase, forms the viral ribonucleoprotein complex (vRNP) that directs the biosynthetic processes of replication and transcription of the viral genome [11]. NP has also been shown to counteract the induction of type I interferon (IFN-I) response that plays a critical role in the cell innate immune response to infection [10,12–14]. This anti-IFN-I activity of NP has been linked to a functional DEDDh 3′–5′ exoribonuclease domain present within the NP C-terminal region [15,16]. GPC is co-translationally cleaved by cellular signal peptidases to generate a 58 amino acid-long stable signal peptide (SSP) and the GP1/2 precursor that is post-translationally processed by the cellular protease subtilisin kexin isozyme-1 (SKI-1)/site 1 protease (S1P) to generate the mature virion surface glycoproteins GP1 and GP2 [17]. The SSP, together with GP1 and GP2, form the spikes (GP complex) located at the surface of mature virions and that mediate virus cell entry. The GP1 subunit is located at the top of the spike away from the membrane and mediates virion interaction with host cell-surface receptors [18,19]. The GP2 subunit is involved in virion-cell membrane fusion by means of a conformational change triggered by the acidic environment of the endosome [20,21]. The SSP has been implicated in the trafficking and processing of the viral envelope glycoproteins and in the GP2-mediated pH-dependent fusion process [3]. N-glycosylation modifications of GPC are required for SKI-1/S1P cleavage [22], whereas SSP must be myristoylated for the GP2-mediated fusion activity [23]. The L protein mediates replication and transcription of the virus RNA genome by the vRNP. The L protein also has an endonuclease activity that mediates the cap-snatching process required for transcription initiation [11,24]. The RING finger protein Z serves as a bona fide matrix protein [25]. Z has been shown to interact with various host cell proteins, including the eukaryote translation initiation factor 4E (eIF4E) and the promyelocytic leukemia protein (PML), interactions that might contribute to facilitate virus multiplication [26–30]. As with other matrix proteins of NS RNA viruses, Z also negatively regulates the activity of the virus polymerase [31,32], mediates interactions between the vRNP and GP2 required for assembly of matured viral particles [33,34], and is the main driving force of viral budding from the plasma membrane, a process that includes Z's interaction with several members of the endosomal sorting complex required for transport (ESCRT) [25,35]. Z has also been shown to interact with the cytosolic pathogen recognition receptor (PRR) RIG-I and counteracts the induction of the IFN-I response.

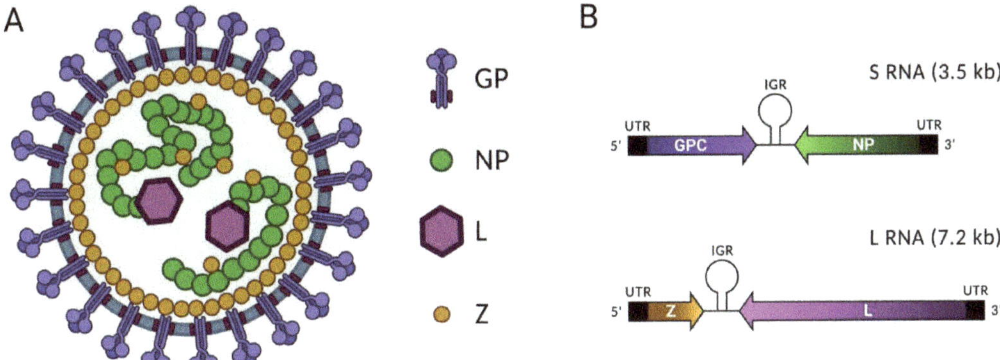

Figure 1. Mammarenavirus virion and genome organization. (**A**) Schematic diagram of mammarenavirus virion. The virion is enveloped and contains four types of viral proteins; glycoprotein (GP), nucleoprotein (NP), RdRp (L), the RING finger protein Z. (**B**) Genome organization of mammarenaviruses. The bi-segmented negative sense RNA genome consists of the large (L) and small (S) segments. This ambisense genome organization encodes two independent viral proteins in each segment; GPC and NP in the S segment, Z and L in the L segment.

Both genome segments contain non-coding cis-regulatory elements required for replication and transcription of the virus genome, including the 5′ and 3′ untranslated regions (UTR) present at the ends of both RNA segments and the non-coding intergenic regions (IGR) separating the two open reading frames present in each genome segment [36]. Mammarenaviruses have highly conserved sequences at the 3′-end of the L and S RNA genome segments that contain the genome promoter recognized by the virus polymerase to initiate RNA synthesis [37]. Genomes and antigenome RNAs are highly complementary between their 5′- and 3′-termini, with both L and S genome segments, predicted to form panhandle structures, which is consistent with EM images showing the existence of circular RNP complexes within mammarenavirions. Mammarenavirus IGRs are predicted to fold into stable hairpin structures that constitute bona fide transcription termination signals. There are significant differences in sequence and predicted folded structure between the S and L IGR, but among isolates and strains of the same arenavirus, the S and L IGR sequences are highly conserved. The IGR has been also implicated in virus assembly or budding, or both, being required for the generation of infectious particles [38,39].

2.2. Mammarenavirus Life Cycle

2.2.1. Cell Entry

Mammarenavirus enter susceptible cells mainly via receptor-mediated endocytosis. The acidic environment of the late endosome facilitates a pH-dependent conformational change in the GP complex that induces a GP2-mediated fusion step between viral and cell membranes. Following fusion, the viral RNP is released into the cytoplasm, where it directs both replication and transcription of the viral genome (Figure 2). Different cell surface receptors are used by Old World (OW) and New World (NW) mammarenaviruses. OW mammarenaviruses, including lymphocytic choriomeningitis virus (LCMV) and Lassa virus (LASV), as well as clade C NW mammarenaviruses, use α-dystroglycan as the major high-affinity receptor to which GP1 attaches [40,41]. In contrast, pathogenic clade B NW mammarenaviruses including Guanarito (GTOV), Sabia (SABV), Junin (JUNV), Machupo (MACV) viruses use human transferrin receptor 1 (hTfR1) as the main receptor for entry [42]. A cell entry receptor for clade A NW mammarenaviruses has not been identified yet [43]. The OW hemorrhagic fever mammarenavirus Lujo virus (LUJV) LUJV was shown to use neuropilin (NRP)-2 as its main cell entry receptor [44]. Interestingly, NRP-2 is highly expressed on microvascular endothelial cells and alveolar macrophages, which may explain the extent of coagulopathy observed in LUJV-induced clinical disease and aerosol transmission of LUJV [45–47]. Once GP1 binds to the cellular receptor, the virion is endocytosed following different pathways by NW and OW mammarenaviruses. NW mammarenaviruses are internalized via clathrin-dependent endocytosis [48,49], whereas OW arenaviruses proceed through a clathrin-independent pathway that involves PI3K-mediated formation of multivesicular bodies (MVB) during late endocytosis and the endosomal sorting complex required for transport (ESCRT) proteins [50]. Notably, completion of the cell entry process for LASV and LUJV requires a late endosomal receptor switch mechanism. For LASV, the lysosome-associated membrane protein 1 (LAMP1) mediates a cholesterol-dependent pH-receptor switch mechanism [51], which stimulates GP1 dissociation from the GP complex and a conformational change in GP2 that drives the fusion of viral and cell membranes [20,52]. In the case of LUJV, this process is mediated by the tetraspanin CD63 [44].

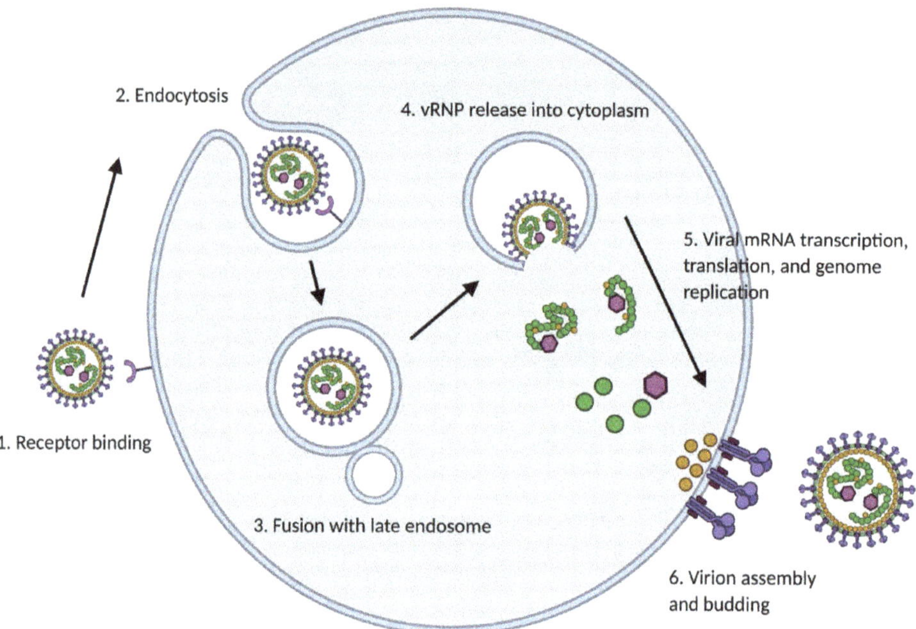

Figure 2. Mammarenavirus life cycle. Virus cell entry is initiated by the interaction between GP and different receptors present at the cell surface (1). Virus uptake into the cell is mediated by endocytosis (2). The acidic environment of the late endosome promotes fusion between viral and host cell membranes (3). Following the pH-dependent membrane fusion event, the vRNP is released into the cytoplasm, where it directs the biosynthetic processes of replication and transcription of the viral genome (4,5). Viral assembly takes place in the cell cytoplasm, and virions bud from the plasma membrane. Z plays critical roles in both these processes (6).

2.2.2. Genome Replication and Transcription

Following its release from the late endosome into the cytoplasm, the vRNP directs the biosynthetic processes of replication and transcription of the viral genome RNAs (Figure 2). L and NP are the minimal viral trans-acting factors required for these processes [53,54], whereas Z has been shown to exhibit a dose-dependent inhibitory effect on the vRNP activity, likely by interacting with the viral L polymerase and disabling its catalytic activity [31,53].

Mammarenaviruses generate three different RNA species during replication and transcription: genomic (gRNA) and antigenomic (agRNA) RNA species and sub-genomic viral mRNA species. Synthesis of viral mRNAs starts at the genome promoter and terminates within the distal region of the hairpin-structured IGRs [4]. Due to the ambisense coding strategy that characterizes mammarenavirus genome, NP and L mRNAs are transcribed from the genomic RNA, while GPC and Z mRNAs are transcribed from the corresponding antigenome RNA species. This ambisense coding strategy has been suggested to produce a hierarchal pattern of gene expression where NP and L proteins are produced at very early times of infection, and GPC and Z proteins at later times [4,55].

Mutation-function analysis using cell-based minireplicon assays for several mammarenaviruses indicated that the activity of the mammarenavirus genomic promoter requires both sequence specificity within the highly conserved 3′-terminal 19 nucleotides of arenavirus genomes and the integrity of the predicted panhandle structure formed via sequence complementarity between the 5′- and 3′-termini of viral genome RNAs [37,56]. IGRs serve as bona fide transcription termination signals, but the synthesis of translation-competent viral mRNAs does not strictly require the presence of the IGR [38,39,57].

Mammarenavirus transcription initiation uses a cap-snatching mechanism, where short 5′-capped primers derived from host-cell mRNAs are used by the virus polymerase complex to prime the synthesis of viral mRNAs [4,58]. The 5′-capped primers are generated via cleavage of cellular mRNAs by the endonuclease activity associated with the N-terminus of L [59]. The 5′-end of mammarenavirus genome and antigenome RNAs each contain a non-templated G residue that likely reflects a prime-and-realign mechanism for RNA replication mediated by L [60,61]. Mammarenavirus genome and antigenome, sequence terminal complementarity combined with the prime-and-realign mechanism for replication initiation generates double-stranded RNAs with overhanging 5′-ppp nucleotides. These structures can act as RIG-I decoys, thereby diminishing RIG-I-mediated interferon induction in mammarenavirus-infected cells [62]. NP mRNA of the NW mammarenavirus Tacaribe (TCRV) was detected at early times of infection and in the presence of inhibitors of protein synthesis, suggesting that, unlike the closely related bunyaviruses, but similarly to orthomyxoviruses, rhabdoviruses, and paramyxoviruses, primary transcription directed by the incoming mammarenavirus vRNP can proceed in the absence of translation [4,63].

2.2.3. Assembly and Budding

As with many other enveloped NS RNA viruses, mammarenavirus assembly and budding require the participation of the matrix protein Z [64]. Z interacts with GP2 [33] and NP [65], which facilitates vRNP-GP complex interactions required for the assembly of mature viral particles (Figure 2). As with other matrix proteins, Z has the capability to self-assemble and direct budding of VLPs [35]. Z budding activity is driven by canonical viral late (L) domains within the C-terminus of Z, which mediate Z interaction with components of the cellular ESCRT complexes [66] known to play essentials roles in the budding process of different viruses [67,68].

3. Drugs Targeting Different Steps of Mammarenavirus Life Cycle

3.1. Cell Entry

Completion of mammarenavirus cell entry requires a GP2-mediated fusion event between viral and cellular membranes, a process triggered by the acidic environment of the late endosome. Therefore, targeting the GP2-mediated fusion event is an attractive strategy to inhibit mammarenavirus infection. Accordingly, a high-throughput screen (HTS) of a small molecule library identified a series of molecules (ST-193, ST-294, and ST-336) that target GP2 and inhibit viral entry [69,70]. ST-193 has been shown to confer significant protection against LASV infection in a guinea pig model [71], and the optimized chemical analog of ST-193 compound, LHF-535, was shown to be a potent broad-spectrum inhibitor of HF mammarenaviruses via targeting the GP2-mediated fusion step of virus cell entry [72]. Another GP2-mediated fusion inhibitor, F3406, was shown to inhibit LCMV multiplication [73]. AVP-p, a peptide derived from the Pichinde virus (PICV) GP2 subunit, was found to bind the prefusion state of the GP complex, which arrests the GP2-mediated fusion event [74]. Arbidol (umifenovir) was developed as an inhibitor of influenza virus infection, but it has been shown to exhibit broad-spectrum antiviral activity against different viruses by interfering with different steps, including cell entry of the virus life cycle [75] Arbidol inhibits both LASV and Ebola virus (EBOV) GP-mediated fusion events required to complete the virus cell entry process [76,77]. Screening of a natural product library identified tangeretin as a cell entry inhibitor of seven different HF-causing viruses [78].

Several compounds have been shown to inhibit mammarenavirus cell entry via off-target activities distinct from their intended therapeutic effects. The clotrimazole-derivative TRAM-34 is an ion channel blocker that antagonizes the calcium-activated potassium channel KCa3.1. TRAM-34 specifically inhibited mammarenavirus GP2-mediated fusion, and this anti-mammarenaviral effect was independent of its channel blocking activity [79]. Losmapimod, an p38 MAPK inhibitor, was developed as a therapeutic drug for chronic obstructive pulmonary disease (COPD) and has been reported to inhibit LASV entry by blocking the pH-dependent GP2-mediated fusion without requiring inhibition of p38 MAPK [80]. NH125, a selective eEF-2 kinase inhibitor, inhibited cell entry of recombinant VSVs expressing envelope glycoproteins of avian influenza virus, EBOV, and LASV due to the compound lysosomotropic properties and independently of eEF-2 kinase inhibition [81]. ZCL278 was identified as an inhibitor of Cdc42, a small GTPase that regulates actin polymerization [82]. Subsequently, ZCL278 was shown to exhibit broad-spectrum anti-viral activity, including against mammarenaviruses, by redistributing viral particles from endosomal to lysosomal membranes, and this antiviral activity was not dependent on the downregulation of Cdc42 activity [83]. The antifungal drug isavuconazole, which was approved by the FDA as an orphan drug for aspergillosis, mucormycosis, and candidiasis, was shown to inhibit viral fusion, targeting the SSP-GP2 interface of LASV [84]. The adamantyl diphenyl piperazine 3.3 was developed to target the lysosome-associated membrane protein 1 (LAMP1) by competing with cholesterol, preventing the interaction between LAMP1 and LASV GP [85].

3.2. *Viral Genome Replication*

The nucleoside analog ribavirin has been shown to inhibit mammarenavirus replication in cell culture systems, as well as to have clinical benefits when used to treat LF cases at early times after the onset of clinical symptoms [86,87]. Several different mechanisms of action have been proposed to account for the antiviral activity of ribavirin, including up-regulation of interferon responses [88], GTP pool depletion by inhibiting IMP dehydrogenase (IMPDH), and direct inhibition of viral RNA-directed RNA polymerase (RdRp) activity [89]. Due to its broad-spectrum antiviral activity against multiple RNA and DNA viruses, most likely different mechanisms of action, and combination of them, contribute to the antiviral activity of ribavirin. Notably, a recent study has shown that the main mechanism of the antiviral activity of ribavirin against LASV, is by protecting LASV-infected cells from death [90].

Favipiravir was initially developed as an antiviral drug targeting the influenza virus polymerase and subsequently shown to exhibit a broad-spectrum activity against different RNA viruses, including mammarenaviruses, bunyaviruses, flaviviruses, alphaviruses, picornaviruses, and noroviruses [91,92]. BCX4430 is another broad-spectrum inhibitor of many types of RNA viruses, including mammarenaviruses, which interferes with the activity of the viral RdRp by acting as an RNA chain terminator [93].

Peptide-conjugated phosphorodiamidate morpholino oligomers (PPMO), designed to target conserved regions within mammarenavirus genome RNA, were effective against multiple mammarenaviruses, including JUNV, LCMV, TCRV, and PICV [94]. The inhibitors of DEDDh family of $3'$-$5'$ exonucleases aurintricarboxylic acid (ATA) and pontacyl violet 6R (PV6R) have been reported to inhibit the $3'$-$5'$ exonuclease activity of LASV NP [15], which in addition to its role counteracting the host cell IFN-I response, has been shown to play a critical role in viral fitness in IFN-I deficient cells [95]. As metal-chelating pharmacophores, diketo acids, polyphenols, and *N*-hydroxyisoquinoline-1,3-diones were able to inhibit the endonuclease activity of arenavirus L protein, which resulted in inhibition of the cap-snatching mechanism used by mammarenavirus polymerases to initiate transcription of viral mRNAs [96]. A carboxamide-derivatized disulfide, NSC4492, was reported to show antiviral activity against JUNV and TCRV [97] and this compound was shown to impair viral RNA synthesis of JUNV via targeting the replication complex [98]. KP-146 was shown to have dual roles in its antiviral activity against LCMV, not only interfering with vRNP activity responsible for directing LCMV genome replication and gene transcription but also inhibiting Z protein-mediated budding process [95].

3.3. Processing of GPC

The mammarenavirus GPC precursor is co-translationally cleaved by cellular signal peptidases to generate a 58 amino acid-long stable signal peptide (SSP) and the immature GP1/GP2 precursor [99]. Subsequent processing of GP1/GP2 by the cellular subtilisin kexin isozyme-1 (SKI-1)/site-1 protease (S1P) into GP1 and GP2 is required for the production of infectious progeny [17,19]. GP1, GP2, and the SSP form the mature trimeric GP spike complex [99]. Decanoyl-RRLL-chloromethylketone (dec-RRLL-CMK) was developed as a SKI-1/S1P inhibitor based on the cleavage recognition site present within LASV GPC [100] and was shown to exert a potent antiviral activity against LCMV as well as additive antiviral drug activity in combination with ribavirin [101]. PF-429242, a small molecule inhibitor of S1P, was shown to interfere with the proteolytic processing of GPC, which correlated with the compound's ability of inhibiting multiplication of LCMV and LASV in cultured cells [102]. In S1P-deficient cells, wild-type LCMV consistently underwent extinction without emergence of S1P-independent escape variants [101]. Moreover, PF-429242 efficiently and rapidly cleared persistent virus from infected cells, and interruption of drug treatment did not result in re-emergence of infection, indicating that PF-429242 treatment resulted in virus extinction [103]. These findings indicate a high genetic barrier for the emergence of viral variants capable of using an alternative host cellular protease for the processing of GPC, thus making S1P a very attractive target for the development of antiviral drugs against mammarenaviruses.

3.4. Virion Assembly and Cell Egress

Assembly and cell release of infectious mammarenavirus progeny involves Z-L, Z-NP, and Z-GP interactions to facilitate the co-localization of all viral proteins for the assembly of mature infectious particles [30]. Functional studies have shown that the matrix Z protein plays a key role in mammareanvirus budding, a process mediated by the interaction of Z late (L) domain motifs, PTAP and PPPY, with components of the cellular ESCRT complexes [104]. Z-mediated budding also requires myristoylation of the Z protein at a glycine (G) in position 2 to target Z to the plasma membrane, the location of arenavirus budding [105]. Accordingly, treatment with 2-hydroxymyristic acid, an inhibitor of the N-myristoyltransferase (NMT), impaired Z budding activity and production of mammarenavirus infectious progeny [105]. Valproic acid (VPA), a short-chain fatty acid, used in anti-epileptic therapy [106], was shown to inhibit Z-mediated budding of LCMV, likely due to VPA-mediated alteration of the lipid composition of cellular membranes, which is critical in virus budding [107]. Compound 0013, identified as a potent inhibitor of the interaction between the PTAP L domain and Tsg101, a member of the host cell ESCRT complex proteins, was shown to inhibit viral budding by blocking the Z-Tsg101 interaction [108]. The ubiquitin ligase Nedd4 E3 is also a component of the ESCRT complex [109]. A small molecule termed compound1 was identified as an inhibitor of Z-Nedd4 interaction, resulting in inhibition of viral budding [110]. BEZ-235, a phosphatidylinositol 3-kinase (PI3K) inhibitor, was shown to inhibit Z protein-mediated budding of LCMV and LASV by a mechanism of action yet to be determined [111]. In addition to driving virion assembly and release, Z proteins have been reported to modulate various host mechanisms such as repression of translation by binding and counteracting eIF4E [26] and suppression of host innate immune responses [112], suggesting potential host targets for developing antiviral drugs. Recent Z interactome study has identified human proteins that interact with arenavirus Z and validated potential host targets for antiviral therapeutics, including ADP ribosylation factor 1 (ARF1), ATP synthase, H transporting mitochondrial F1 complex beta polypeptide (ATP5B), ATPase H transporting lysosomal 38-kDa V0 subunit d1 (ATP6V0D1), inosine monophosphate dehydrogenase 2 (IMPDH2), peroxiredoxin 3 (PRDX3), and Ras-related protein Rab5c [113].

3.5. Monoclonal Antibodies

Monoclonal antibody-based therapies represent an attractive strategy to treat infections by highly pathogenic mammarenaviruses. A monoclonal antibody specific for human transferrin receptor 1 (hTfR1), the receptor used by pathogenic NW mammarenaviruses, inhibits viral entry of several NW mammarenaviruses, including JUNV, GTOV, CHAV, SABV, and MACV [114]. In addition, monoclonal antibodies targeting mammarenavirus GP have been reported to have potent neutralizing activity against MACV and LASV that correlated with inhibition of virus multiplication [115,116]. A recent study documented the isolation of human monoclonal antibodies from LF survivors and characterized their epitope and neutralization profiles, showing that 80% of the monoclonal antibodies with neutralizing activity targeted complex epitopes involving LASV GP1 and GP2 subunits [117]. Importantly, this study identified several human monoclonal antibodies with neutralizing activities against members of the main four lineages of LASV, and some of them showed cross-reactivity to LCMV, LUJO, and MACV. This finding has provided insights to develop therapeutic strategies based on the use of broadly reactive monoclonal antibodies.

3.6. Targeting Host Factors

As strict parasites, viruses rely on many host cell factors to complete their life cycles. Therefore, there is increasing interest in targeting host cell factors required for virus multiplication as an antiviral drug strategy. Direct-acting antivirals (DAAs) that target specific viral gene products and functions are likely to be well tolerated by the infected host cell. Still, they are limited by the common problem in antiviral therapy posed by the emergence of drug-resistant variants. In contrast, the emergence of viral variants resistant to host-targeting antivirals (HTAs) is usually significantly reduced or entirely absent, but HTAs can be associated with significant side effects. However, side effects associated with the use of HTAs might be manageable in the case of acute infections, such as HF disease caused by arenaviruses, where the duration of the treatment would be rather short.

Dihydroorotate dehydrogenase (DHODH) small molecule inhibitors A3 and A77172 interfere with de novo pyrimidine biosynthesis and exhibit potent antiviral activity against LCMV and JUNV [118,119]. Likewise, de novo purine biosynthesis is a potential cellular target for the development of HTAs. Inhibition of inosine monophosphate dehydrogenase (IMPDH), a key enzyme in the purine biosynthesis pathway, was shown to be associated with a broad-spectrum antiviral activity against RNA viruses including JUNV [120,121]. S-adenosylhomocysteine hydrolase (SAHH) is an important cellular enzyme for regulating viral mRNA capped structures, and inhibition of SAHH activity by 3-deazaneplanocin was associated with potent antiviral activity against TCRV and PICV [122]. ATPase Na+/K+ transporting subunit alpha 1 (ATP1A1) and prohibitin (PHB) were identified as mammarenavirus NP-interacting host cell proteins and act as pro-viral factors that promote mammarenavirus multiplication. Accordingly, the ATP1A1 inhibitors, ouabain and bufalin, as well as the PHB inhibitor, rocaglamide A, exhibit potent antiviral activity against LCMV and LASV infection [123].

Anti-mammarenaviral drugs discussed in this section have been summarized in Table 1.

Table 1. Drugs with anti-mammarenaviral activity.

Target	Drug Name	Mechanism
Viral entry	ST-193 ST-294 ST-336 LHF-535 F3406 AVP-p arbidol tangeretin	- GP2 targeting compounds - Inhibition of pH-dependent membrane fusion
	TRAM-34	- Calcium-activated potassium channel blocker - Inhibition of pH-dependent membrane fusion
	losmapimod	- p38 MAPK inhibitor - Inhibition of pH-dependent membrane fusion in p38 MAPK down regulation-independent manner
	NH125	- eEF-2 kinase inhibitor - GP-mediated fusion inhibition due to lysosomotropic properies
	ZCL278	- Cdc42 inhibitor - Fusion inhibition by redistributing viral particles from endosomal to lysosomal membranes
	isavuconazole	- Antifungal drug for aspergillosis - Inhibition of pH-dependent membrane fusion by targeting the SSP-GP2 interface
	dec-RRLL-CMK PF-429242	- Blocking the proteolytic processing of GPC by inhibition of host protease S1P
	adamantyl diphenyl piperazine 3.3	- Blocking the interaction between host LAMP1 and viral GP

Table 1. *Conts.*

Target	Drug Name	Mechanism
Viral genomereplication	ribavirin	- Upregulation of host IFN responses - Cellular GTP depletion - Viral RdRp inhibition
	favipiravir BCX4430	- Viral RdRp inhibition
	PPMO	- Interfering with viral RNA synthesis and translation
	ATA PV6R	- Inhibition of NP exonuclease activity
	diketo acids polyphenols N-hydroxyisoquinoline-1,3-diones	- Inhibition of L endonuclease activity
	NSC4492 KP-146	- Targeting the vRNP to impair viral RNA synthesis
	A3 A771726	- DHODH inhibitor - Inhibition of de novo pyrimidine biosynthesis
	acridone bredinin	- IMPDH inhibitor - Inhibition of de novo purine biosynthesis
	3-deazaneplanocin	- Inhibition of SAHH activity which is important in viral mRNA capped structure
	ouabain bufalin	- ATP1A1 inhibitor - Preventing interaction with viral NP
	rocaglamide	- PHB inhibitor - Preventing interaction with viral NP
Virion assembly and budding	2-hydroxymyristic acid	- Inhibition of NMT which mediates myristoylation of Z protein - Blocking Z-mediated budding
	valproic acid	- Altering lipid composition of cellular membranes which is critical in virus budding
	compound0013	- Tsg101 inhibitor - Blocking interaction between Tsg101 and viral Z protein that is required for virion egress
	compound1	- Inhibiting the interaction between Nedd4 and Z protein, blocking viral budding
	BEZ-235	- PI3K inhibitor - Inhibition of Z-mediated budding via unknown mechanism
	KP-146	- Dual roles in viral genome replication and budding

4. High-Throughput Screening for Discovery of Anti-Mammarenaviral Drugs

The development of mammarenavirus reverse genetics (RG) systems has provided investigators with a novel and powerful approach for the investigation of the cis-acting sequences and trans-acting factors that control arenavirus replication, gene expression, and budding, as well as the rescue of infectious mammarenaviruses from cloned cDNAs. These advances in mammarenavirus molecular genetics have facilitated the generation of recombinant mammarenaviruses expressing reporter genes of interest that have enabled the development of cell-based assays and HTS strategies to identify novel anti-mammarenaviral drugs and assess how they target each of the different steps of the virus life cycle [36]. Different strategies, including the development of tri-segmented mammarenaviruses [36] and the use of the self-cleaving P2A linker to facilitate expression of a reporter gene and NP from the same bi-cistronic NP mRNA [73], have been used to generate these recombinant mammarenaviruses.

However, the use of recombinant infectious HF mammarenaviruses expressing reporter genes for antiviral drug screening campaigns or the investigation of drug mechanism of action would face the complications imposed by the requirement of BSL4 biocontainment. Hence, the advantage of the implementation of RG approaches to develop non-infectious cell-based assays recreating each of the key steps of the virus life cycle and that is amenable to HTS, which can be used without the need of a high-level biocontainment facility. Thus, a number of different platforms have been developed to screen for compounds capable of inhibiting cell entry mediated by GPs of different HF mammarenaviruses, using pseudotype viruses [124–126]. Likewise, the biosynthetic processes of replication and transcription of HF mammarenavirus genomes can be mimicked using cell-based minireplicon, or minigenome (MG) systems [127]. These cell-based MG systems are based on the intracellular reconstitution of a functional vRNP directing expression of a reporter gene whose expression level serves as a surrogate of the vRNP activity. For this, cells are transfected with plasmids expressing the viral trans-acting factors L and NP, together with a plasmid that allows for intracellular synthesis of an RNA containing the open reading frame of a reporter gene under the control of the cis-acting regulatory sequences of the S or L genome RNA [54]. In addition to these transient transfection-based MG systems, cell lines have been generated to constitutively express a functional, non-infectious vRNP of LCMV or LASV [128]. This overcomes some technical complications related to transient transfection in the context of HTS. Thus, cell lines expressing LCMV and LASV functional vRNPs were successfully used to screen different compound libraries, resulting in the identification of a number of hits that were confirmed to exhibit antiviral activity against infectious LCMV and LASV [128].

Generation of infectious virus-like particle (iVLP) systems containing functional virus MGs allows for modeling of not only viral genome replication but also cell entry and budding [129]. A chimeric protein consisting of a secretion deficient form of *Gaussia* luciferase (GLuc) fused to the C-terminus of LASV Z protein was successfully used to develop a cell-based assay to quantify Z-mediated budding activity. This assay has features compatible with its use in HTS [130], as levels of GLuc in the tissue culture supernatant serve as an accurate surrogate of Z budding activity.

5. Drug Repurposing Strategy

The Discovery and development of novel drugs require significant investments and resources and an average processing time for market authorization of 10 to 17 years [131]. The rapid development of antiviral therapeutics is important to combat emerging viruses. Finding novel applications of clinically approved drugs can accelerate the drug development process and significantly reduce risks during clinical trials assessing the new drug application. Accordingly, repurposing existing drugs is considered an attractive strategy to combat emerging viral infections [132]. This has been illustrated by efforts to combat the current COVID-19 pandemic, where screening of libraries of already approved drugs resulted in the rapid identification of anti-SARS-CoV-2 drug candidates that were very rapidly advanced to clinical trials [133]. Among the listed compounds in Table 1, ribavirin, arbidol and favipiravir are currently being tested in COVID-19 patients in clinical trials [133].

Screening of a library of FDA-approved drugs using VSV pseudotyped with LASV GP identified a number of inhibitors of LASV GP-mediated cell entry [125,134]. Likewise, screening of the Repurposing, Focused Rescue, and Accelerated Medchem (ReFRAME) library identified several potent anti-mammarenaviral compounds [135]. Importantly, selected hits initially identified based on their anti-LCMV activity, which were confirmed to show potent antiviral activity against the HF causing mammarenaviruses LASV and JUNV. These compounds exerted their antiviral activity via targeting host cellular factors, including enzymes required for pyrimidine and purine biosynthesis, regulators of apoptosis, and the mitochondrial electron transport complex III [135]. Recently, this ReFRAME library was used to screen for antiviral drugs against SARS-CoV-2, and the existing pharmacological and safety data on the identified hits will facilitate their rapid testing in the clinic [136]. In addition to HTS formats to rapidly identify novel targets and antiviral drug candidates, function-focus based assays have also been successfully used to identify compounds that could be repurposed as antiviral drugs. For example, screening of a collection of kinase inhibitors identified several cellular kinases that were involved in LASV GP-mediated viral entry, including protein kinase C, phosphoinositide 3-kinase, and human hepatocyte growth factor receptor (HGFR), which is a receptor tyrosine kinase [137].

6. Conclusions and Future Perspectives

As documented in this review, significant efforts have been dedicated to finding effective antiviral drugs against human pathogenic mammarenaviruses. Different screening platforms have identified a number of antiviral drug candidates with potent activity in cell-based infection assays. However, for the majority of the identified hits, there is only very limited information regarding their in vivo efficacy. To advance the development of novel effective antiviral drugs, further validation should be conducted using appropriate in vivo models of mammarenavirus disease, including non-human primates.

Drug repurposing approaches have identified a number of host cell factors as attractive antiviral targets for which drugs with a good safety profile have been already documented, which should facilitate the assessment of their efficacy in vivo using appropriate animal models of mammarenavirus induced disease. Since different viruses may share some key host cell functions to complete their life cycle, a host-targeting strategy would be an attractive approach for the development of broad-spectrum antiviral therapeutics. Synergistic antiviral effects have been documented in combination therapies of approved drugs, as illustrated by the results of combination therapy of ribavirin and favipiravir against LASV infection in pre-clinical [87] and clinical [138] studies. Likewise, combination therapy of arbidol with aripiprazole or sertraline resulted in synergistic inhibition of pseudotyped viruses with GPs from LASV and JUNV [139]. Synergistic effects are likely to be facilitated by combination therapy with drugs targeting different steps of the virus life cycle. To identify combinations for antiviral therapeutics, modern computational approaches, including available data libraries and analytical resources [140], would be promising tools by which data and text mining could identify potential drug combinations for further experimental validations.

Author Contributions: Conceptualization, Y.-J.K., V.V. and J.C.d.l.T.; writing—original draft preparation, Y.-J.K. and V.V.; writing—review and editing, Y.-J.K., V.V. and J.C.d.l.T.; visualization, Y.-J.K.; supervision, J.C.d.l.T.; project administration, J.C.d.l.T.; funding acquisition, J.C.d.l.T. All authors have read and agreed to the published version of the manuscript.

Funding: This work was supported by the NIH/NIAID grant RO1AI125626.

Institutional Review Board Statement: Not applicable.

Informed Consent Statement: Not applicable.

Acknowledgments: All figures were created with BioRender.com (accessed on 1 May 2021). This is the manuscript #30091 from The Scripps Research Institute.

Conflicts of Interest: The authors declare no conflict of interest.

References

1. Radoshitzky, S.R.; Buchmeier, M.J.; Charrel, R.N.; Clegg, J.C.S.; Gonzalez, J.J.; Gunther, S.; Hepojoki, J.; Kuhn, J.H.; Lukashevich, I.S.; Romanowski, V.; et al. ICTV Virus Taxonomy Profile: Arenaviridae. *J. Gen. Virol.* **2019**, *100*, 1200–1201. [CrossRef]
2. Buchmeier, M.J.; Peters, C.J.; de la Torre, J.C. Arenaviridae: The viruses and their replication. In *Field's Virology*, 5th ed.; Knipe, D.M., Holey, P.M., Eds.; Lippincott Williams & Wilkins: Philadelphia, PA, USA, 2007; Volume 2, pp. 1791–1851.
3. Brisse, M.E.; Ly, H. Hemorrhagic Fever-Causing Arenaviruses: Lethal Pathogens and Potent Immune Suppressors. *Front. Immunol.* **2019**, *10*, 372. [CrossRef]
4. Shao, J.; Liang, Y.; Ly, H. Human hemorrhagic Fever causing arenaviruses: Molecular mechanisms contributing to virus virulence and disease pathogenesis. *Pathogens* **2015**, *4*, 283–306. [CrossRef]
5. Schieffelin, J. Treatment of Arenavirus Infections. *Curr. Treat. Options Infect. Dis.* **2015**, *7*, 261–270. [CrossRef]
6. McKee, K.T., Jr.; Oro, J.G.; Kuehne, A.I.; Spisso, J.A.; Mahlandt, B.G. Candid No. 1 Argentine hemorrhagic fever vaccine protects against lethal Junin virus challenge in rhesus macaques. *Intervirology* **1992**, *34*, 154–163. [CrossRef]
7. Maiztegui, J.I.; McKee, K.T., Jr.; Barrera Oro, J.G.; Harrison, L.H.; Gibbs, P.H.; Feuillade, M.R.; Enria, D.A.; Briggiler, A.M.; Levis, S.C.; Ambrosio, A.M.; et al. Protective efficacy of a live attenuated vaccine against Argentine hemorrhagic fever. AHF Study Group. *J. Infect. Dis.* **1998**, *177*, 277–283. [CrossRef] [PubMed]
8. Grant, A.; Seregin, A.; Huang, C.; Kolokoltsova, O.; Brasier, A.; Peters, C.; Paessler, S. Junin virus pathogenesis and virus replication. *Viruses* **2012**, *4*, 2317–2339. [CrossRef]
9. Neuman, B.W.; Adair, B.D.; Burns, J.W.; Milligan, R.A.; Buchmeier, M.J.; Yeager, M. Complementarity in the supramolecular design of arenaviruses and retroviruses revealed by electron cryomicroscopy and image analysis. *J. Virol.* **2005**, *79*, 3822–3830. [CrossRef] [PubMed]
10. Zhou, S.; Cerny, A.M.; Zacharia, A.; Fitzgerald, K.A.; Kurt-Jones, E.A.; Finberg, R.W. Induction and inhibition of type I interferon responses by distinct components of lymphocytic choriomeningitis virus. *J. Virol.* **2010**, *84*, 9452–9462. [CrossRef] [PubMed]
11. Ferron, F.; Weber, F.; de la Torre, J.C.; Reguera, J. Transcription and replication mechanisms of Bunyaviridae and Arenaviridae L proteins. *Virus Res.* **2017**, *234*, 118–134. [CrossRef] [PubMed]
12. Shao, J.; Liang, Y.; Ly, H. Roles of Arenavirus Z Protein in Mediating Virion Budding, Viral Transcription-Inhibition and Interferon-Beta Suppression. *Methods Mol. Biol.* **2018**, *1604*, 217–227. [CrossRef] [PubMed]
13. Li, S.; Sun, Z.; Pryce, R.; Parsy, M.L.; Fehling, S.K.; Schlie, K.; Siebert, C.A.; Garten, W.; Bowden, T.A.; Strecker, T.; et al. Acidic pH-Induced Conformations and LAMP1 Binding of the Lassa Virus Glycoprotein Spike. *PLoS Pathog.* **2016**, *12*, e1005418. [CrossRef]
14. Fehling, S.K.; Lennartz, F.; Strecker, T. Multifunctional nature of the arenavirus RING finger protein Z. *Viruses* **2012**, *4*, 2973–3011. [CrossRef] [PubMed]
15. Hastie, K.M.; Kimberlin, C.R.; Zandonatti, M.A.; MacRae, I.J.; Saphire, E.O. Structure of the Lassa virus nucleoprotein reveals a dsRNA-specific 3′ to 5′ exonuclease activity essential for immune suppression. *Proc. Natl. Acad. Sci. USA* **2011**, *108*, 2396–2401. [CrossRef]
16. Martinez-Sobrido, L.; Zuniga, E.I.; Rosario, D.; Garcia-Sastre, A.; de la Torre, J.C. Inhibition of the type I interferon response by the nucleoprotein of the prototypic arenavirus lymphocytic choriomeningitis virus. *J. Virol.* **2006**, *80*, 9192–9199. [CrossRef]
17. Lenz, O.; ter Meulen, J.; Klenk, H.D.; Seidah, N.G.; Garten, W. The Lassa virus glycoprotein precursor GP-C is proteolytically processed by subtilase SKI-1/S1P. *Proc. Natl. Acad. Sci. USA* **2001**, *98*, 12701–12705. [CrossRef] [PubMed]
18. Eschli, B.; Quirin, K.; Wepf, A.; Weber, J.; Zinkernagel, R.; Hengartner, H. Identification of an N-terminal trimeric coiled-coil core within arenavirus glycoprotein 2 permits assignment to class I viral fusion proteins. *J. Virol.* **2006**, *80*, 5897–5907. [CrossRef] [PubMed]
19. Kunz, S.; Edelmann, K.H.; de la Torre, J.C.; Gorney, R.; Oldstone, M.B. Mechanisms for lymphocytic choriomeningitis virus glycoprotein cleavage, transport, and incorporation into virions. *Virology* **2003**, *314*, 168–178. [CrossRef]
20. Di Simone, C.; Zandonatti, M.A.; Buchmeier, M.J. Acidic pH triggers LCMV membrane fusion activity and conformational change in the glycoprotein spike. *Virology* **1994**, *198*, 455–465. [CrossRef]
21. Igonet, S.; Vaney, M.C.; Vonrhein, C.; Bricogne, G.; Stura, E.A.; Hengartner, H.; Eschli, B.; Rey, F.A. X-ray structure of the arenavirus glycoprotein GP2 in its postfusion hairpin conformation. *Proc. Natl. Acad. Sci. USA* **2011**, *108*, 19967–19972. [CrossRef]
22. Eichler, R.; Lenz, O.; Garten, W.; Strecker, T. The role of single N-glycans in proteolytic processing and cell surface transport of the Lassa virus glycoprotein GP-C. *Virol. J.* **2006**, *3*, 41. [CrossRef] [PubMed]
23. York, J.; Romanowski, V.; Lu, M.; Nunberg, J.H. The signal peptide of the Junin arenavirus envelope glycoprotein is myristoylated and forms an essential subunit of the mature G1-G2 complex. *J. Virol.* **2004**, *78*, 10783–10792. [CrossRef] [PubMed]
24. Ortin, J.; Martin-Benito, J. The RNA synthesis machinery of negative-stranded RNA viruses. *Virology* **2015**, *479–480*, 532–544. [CrossRef] [PubMed]
25. Perez, M.; Craven, R.C.; de la Torre, J.C. The small RING finger protein Z drives arenavirus budding: Implications for antiviral strategies. *Proc. Natl. Acad. Sci. USA* **2003**, *100*, 12978–12983. [CrossRef]
26. Campbell Dwyer, E.J.; Lai, H.; MacDonald, R.C.; Salvato, M.S.; Borden, K.L. The lymphocytic choriomeningitis virus RING protein Z associates with eukaryotic initiation factor 4E and selectively represses translation in a RING-dependent manner. *J. Virol.* **2000**, *74*, 3293–3300. [CrossRef]
27. Volpon, L.; Osborne, M.J.; Capul, A.A.; de la Torre, J.C.; Borden, K.L. Structural characterization of the Z RING-eIF4E complex reveals a distinct mode of control for eIF4E. *Proc. Natl. Acad. Sci. USA* **2010**, *107*, 5441–5446. [CrossRef]

28. Borden, K.L.; Campbell Dwyer, E.J.; Salvato, M.S. An arenavirus RING (zinc-binding) protein binds the oncoprotein promyelocyte leukemia protein (PML) and relocates PML nuclear bodies to the cytoplasm. *J. Virol.* **1998**, *72*, 758–766. [CrossRef]
29. Djavani, M.; Rodas, J.; Lukashevich, I.S.; Horejsh, D.; Pandolfi, P.P.; Borden, K.L.; Salvato, M.S. Role of the promyelocytic leukemia protein PML in the interferon sensitivity of lymphocytic choriomeningitis virus. *J. Virol.* **2001**, *75*, 6204–6208. [CrossRef]
30. Urata, S.; Yasuda, J. Molecular mechanism of arenavirus assembly and budding. *Viruses* **2012**, *4*, 2049–2079. [CrossRef] [PubMed]
31. Kranzusch, P.J.; Whelan, S.P. Arenavirus Z protein controls viral RNA synthesis by locking a polymerase-promoter complex. *Proc. Natl. Acad. Sci. USA* **2011**, *108*, 19743–19748. [CrossRef]
32. Cornu, T.I.; de la Torre, J.C. RING finger Z protein of lymphocytic choriomeningitis virus (LCMV) inhibits transcription and RNA replication of an LCMV S-segment minigenome. *J. Virol.* **2001**, *75*, 9415–9426. [CrossRef] [PubMed]
33. Capul, A.A.; Perez, M.; Burke, E.; Kunz, S.; Buchmeier, M.J.; de la Torre, J.C. Arenavirus Z-glycoprotein association requires Z myristoylation but not functional RING or late domains. *J. Virol.* **2007**, *81*, 9451–9460. [CrossRef] [PubMed]
34. Ortiz-Riano, E.; Cheng, B.Y.; de la Torre, J.C.; Martinez-Sobrido, L. The C-terminal region of lymphocytic choriomeningitis virus nucleoprotein contains distinct and segregable functional domains involved in NP-Z interaction and counteraction of the type I interferon response. *J. Virol.* **2011**, *85*, 13038–13048. [CrossRef]
35. Strecker, T.; Eichler, R.; Meulen, J.; Weissenhorn, W.; Dieter Klenk, H.; Garten, W.; Lenz, O. Lassa virus Z protein is a matrix protein and sufficient for the release of virus-like particles [corrected]. *J. Virol.* **2003**, *77*, 10700–10705. [CrossRef]
36. Martinez-Sobrido, L.; de la Torre, J.C. Reporter-Expressing, Replicating-Competent Recombinant Arenaviruses. *Viruses* **2016**, *8*. [CrossRef]
37. Perez, M.; de la Torre, J.C. Characterization of the genomic promoter of the prototypic arenavirus lymphocytic choriomeningitis virus. *J. Virol.* **2003**, *77*, 1184–1194. [CrossRef]
38. Pinschewer, D.D.; Perez, M.; de la Torre, J.C. Dual role of the lymphocytic choriomeningitis virus intergenic region in transcription termination and virus propagation. *J. Virol.* **2005**, *79*, 4519–4526. [CrossRef]
39. Iwasaki, M.; Cubitt, B.; Sullivan, B.M.; de la Torre, J.C. The High Degree of Sequence Plasticity of the Arenavirus Noncoding Intergenic Region (IGR) Enables the Use of a Nonviral Universal Synthetic IGR To Attenuate Arenaviruses. *J. Virol.* **2016**, *90*, 3187–3197. [CrossRef] [PubMed]
40. Spiropoulou, C.F.; Kunz, S.; Rollin, P.E.; Campbell, K.P.; Oldstone, M.B. New World arenavirus clade C, but not clade A and B viruses, utilizes alpha-dystroglycan as its major receptor. *J. Virol.* **2002**, *76*, 5140–5146. [CrossRef]
41. Cao, W.; Henry, M.D.; Borrow, P.; Yamada, H.; Elder, J.H.; Ravkov, E.V.; Nichol, S.T.; Compans, R.W.; Campbell, K.P.; Oldstone, M.B. Identification of alpha-dystroglycan as a receptor for lymphocytic choriomeningitis virus and Lassa fever virus. *Science* **1998**, *282*, 2079–2081. [CrossRef]
42. Radoshitzky, S.R.; Abraham, J.; Spiropoulou, C.F.; Kuhn, J.H.; Nguyen, D.; Li, W.; Nagel, J.; Schmidt, P.J.; Nunberg, J.H.; Andrews, N.C.; et al. Transferrin receptor 1 is a cellular receptor for New World haemorrhagic fever arenaviruses. *Nature* **2007**, *446*, 92–96. [CrossRef]
43. Hallam, S.J.; Koma, T.; Maruyama, J.; Paessler, S. Review of Mammarenavirus Biology and Replication. *Front. Microbiol.* **2018**, *9*, 1751. [CrossRef]
44. Raaben, M.; Jae, L.T.; Herbert, A.S.; Kuehne, A.I.; Stubbs, S.H.; Chou, Y.Y.; Blomen, V.A.; Kirchhausen, T.; Dye, J.M.; Brummelkamp, T.R.; et al. NRP2 and CD63 Are Host Factors for Lujo Virus Cell Entry. *Cell Host Microbe* **2017**, *22*, 688–696 e685. [CrossRef]
45. Bielenberg, D.R.; Pettaway, C.A.; Takashima, S.; Klagsbrun, M. Neuropilins in neoplasms: Expression, regulation, and function. *Exp. Cell Res.* **2006**, *312*, 584–593. [CrossRef] [PubMed]
46. Aung, N.Y.; Ohe, R.; Meng, H.; Kabasawa, T.; Yang, S.; Kato, T.; Yamakawa, M. Specific Neuropilins Expression in Alveolar Macrophages among Tissue-Specific Macrophages. *PLoS ONE* **2016**, *11*, e0147358. [CrossRef] [PubMed]
47. Immormino, R.M.; Lauzier, D.C.; Nakano, H.; Hernandez, M.L.; Alexis, N.E.; Ghio, A.J.; Tilley, S.L.; Doerschuk, C.M.; Peden, D.B.; Cook, D.N.; et al. Neuropilin-2 regulates airway inflammatory responses to inhaled lipopolysaccharide. *Am. J. Physiol. Lung Cell Mol. Physiol.* **2018**, *315*, L202–L211. [CrossRef]
48. Rojek, J.M.; Sanchez, A.B.; Nguyen, N.T.; de la Torre, J.C.; Kunz, S. Different mechanisms of cell entry by human-pathogenic Old World and New World arenaviruses. *J. Virol.* **2008**, *82*, 7677–7687. [CrossRef] [PubMed]
49. Rojek, J.M.; Perez, M.; Kunz, S. Cellular entry of lymphocytic choriomeningitis virus. *J. Virol.* **2008**, *82*, 1505–1517. [CrossRef]
50. Pasqual, G.; Rojek, J.M.; Masin, M.; Chatton, J.Y.; Kunz, S. Old world arenaviruses enter the host cell via the multivesicular body and depend on the endosomal sorting complex required for transport. *PLoS Pathog.* **2011**, *7*, e1002232. [CrossRef]
51. Jae, L.T.; Raaben, M.; Herbert, A.S.; Kuehne, A.I.; Wirchnianski, A.S.; Soh, T.K.; Stubbs, S.H.; Janssen, H.; Damme, M.; Saftig, P.; et al. Virus entry. Lassa virus entry requires a trigger-induced receptor switch. *Science* **2014**, *344*, 1506–1510. [CrossRef]
52. Di Simone, C.; Buchmeier, M.J. Kinetics and pH dependence of acid-induced structural changes in the lymphocytic choriomeningitis virus glycoprotein complex. *Virology* **1995**, *209*, 3–9. [CrossRef]
53. Lopez, N.; Jacamo, R.; Franze-Fernandez, M.T. Transcription and RNA replication of tacaribe virus genome and antigenome analogs require N and L proteins: Z protein is an inhibitor of these processes. *J. Virol.* **2001**, *75*, 12241–12251. [CrossRef] [PubMed]
54. Lee, K.J.; Novella, I.S.; Teng, M.N.; Oldstone, M.B.; de La Torre, J.C. NP and L proteins of lymphocytic choriomeningitis virus (LCMV) are sufficient for efficient transcription and replication of LCMV genomic RNA analogs. *J. Virol.* **2000**, *74*, 3470–3477. [CrossRef]

55. King, B.R.; Samacoits, A.; Eisenhauer, P.L.; Ziegler, C.M.; Bruce, E.A.; Zenklusen, D.; Zimmer, C.; Mueller, F.; Botten, J. Visualization of Arenavirus RNA Species in Individual Cells by Single-Molecule Fluorescence In Situ Hybridization Suggests a Model of Cyclical Infection and Clearance during Persistence. *J. Virol.* **2018**, *92*. [CrossRef] [PubMed]
56. Hass, M.; Golnitz, U.; Muller, S.; Becker-Ziaja, B.; Gunther, S. Replicon system for Lassa virus. *J. Virol.* **2004**, *78*, 13793–13803. [CrossRef]
57. Lopez, N.; Franze-Fernandez, M.T. A single stem-loop structure in Tacaribe arenavirus intergenic region is essential for transcription termination but is not required for a correct initiation of transcription and replication. *Virus Res.* **2007**, *124*, 237–244. [CrossRef]
58. Raju, R.; Raju, L.; Hacker, D.; Garcin, D.; Compans, R.; Kolakofsky, D. Nontemplated bases at the 5′ ends of Tacaribe virus mRNAs. *Virology* **1990**, *174*, 53–59. [CrossRef]
59. Lelke, M.; Brunotte, L.; Busch, C.; Gunther, S. An N-terminal region of Lassa virus L protein plays a critical role in transcription but not replication of the virus genome. *J. Virol.* **2010**, *84*, 1934–1944. [CrossRef] [PubMed]
60. Vogel, D.; Rosenthal, M.; Gogrefe, N.; Reindl, S.; Gunther, S. Biochemical characterization of the Lassa virus L protein. *J. Biol. Chem.* **2019**, *294*, 8088–8100. [CrossRef]
61. Emonet, S.E.; Urata, S.; de la Torre, J.C. Arenavirus reverse genetics: New approaches for the investigation of arenavirus biology and development of antiviral strategies. *Virology* **2011**, *411*, 416–425. [CrossRef] [PubMed]
62. Marq, J.B.; Hausmann, S.; Veillard, N.; Kolakofsky, D.; Garcin, D. Short double-stranded RNAs with an overhanging 5′ ppp-nucleotide, as found in arenavirus genomes, act as RIG-I decoys. *J. Biol. Chem.* **2011**, *286*, 6108–6116. [CrossRef]
63. Linero, F.; Welnowska, E.; Carrasco, L.; Scolaro, L. Participation of eIF4F complex in Junin virus infection: Blockage of eIF4E does not impair virus replication. *Cell Microbiol.* **2013**, *15*, 1766–1782. [CrossRef]
64. Bieniasz, P.D. Late budding domains and host proteins in enveloped virus release. *Virology* **2006**, *344*, 55–63. [CrossRef] [PubMed]
65. Shtanko, O.; Imai, M.; Goto, H.; Lukashevich, I.S.; Neumann, G.; Watanabe, T.; Kawaoka, Y. A role for the C terminus of Mopeia virus nucleoprotein in its incorporation into Z protein-induced virus-like particles. *J. Virol.* **2010**, *84*, 5415–5422. [CrossRef] [PubMed]
66. Urata, S.; Noda, T.; Kawaoka, Y.; Yokosawa, H.; Yasuda, J. Cellular factors required for Lassa virus budding. *J. Virol.* **2006**, *80*, 4191–4195. [CrossRef] [PubMed]
67. Votteler, J.; Sundquist, W.I. Virus budding and the ESCRT pathway. *Cell Host Microbe.* **2013**, *14*, 232–241. [CrossRef]
68. Meng, B.; Lever, A.M.L. The Interplay between ESCRT and Viral Factors in the Enveloped Virus Life Cycle. *Viruses* **2021**, *13*, 324. [CrossRef] [PubMed]
69. Bolken, T.C.; Laquerre, S.; Zhang, Y.; Bailey, T.R.; Pevear, D.C.; Kickner, S.S.; Sperzel, L.E.; Jones, K.F.; Warren, T.K.; Amanda Lund, S.; et al. Identification and characterization of potent small molecule inhibitor of hemorrhagic fever New World arenaviruses. *Antivir. Res.* **2006**, *69*, 86–97. [CrossRef]
70. Larson, R.A.; Dai, D.; Hosack, V.T.; Tan, Y.; Bolken, T.C.; Hruby, D.E.; Amberg, S.M. Identification of a broad-spectrum arenavirus entry inhibitor. *J. Virol.* **2008**, *82*, 10768–10775. [CrossRef]
71. Cashman, K.A.; Smith, M.A.; Twenhafel, N.A.; Larson, R.A.; Jones, K.F.; Allen, R.D., III; Dai, D.; Chinsangaram, J.; Bolken, T.C.; Hruby, D.E.; et al. Evaluation of Lassa antiviral compound ST-193 in a guinea pig model. *Antivir. Res.* **2011**, *90*, 70–79. [CrossRef]
72. Madu, I.G.; Files, M.; Gharaibeh, D.N.; Moore, A.L.; Jung, K.H.; Gowen, B.B.; Dai, D.; Jones, K.F.; Tyavanagimatt, S.R.; Burgeson, J.R.; et al. A potent Lassa virus antiviral targets an arenavirus virulence determinant. *PLoS Pathog.* **2018**, *14*, e1007439. [CrossRef]
73. Ngo, N.; Henthorn, K.S.; Cisneros, M.I.; Cubitt, B.; Iwasaki, M.; de la Torre, J.C.; Lama, J. Identification and Mechanism of Action of a Novel Small-Molecule Inhibitor of Arenavirus Multiplication. *J. Virol.* **2015**, *89*, 10924–10933. [CrossRef]
74. Spence, J.S.; Melnik, L.I.; Badani, H.; Wimley, W.C.; Garry, R.F. Inhibition of arenavirus infection by a glycoprotein-derived peptide with a novel mechanism. *J. Virol.* **2014**, *88*, 8556–8564. [CrossRef] [PubMed]
75. Blaising, J.; Polyak, S.J.; Pecheur, E.I. Arbidol as a broad-spectrum antiviral: An update. *Antivir. Res.* **2014**, *107*, 84–94. [CrossRef] [PubMed]
76. Pecheur, E.I.; Borisevich, V.; Halfmann, P.; Morrey, J.D.; Smee, D.F.; Prichard, M.; Mire, C.E.; Kawaoka, Y.; Geisbert, T.W.; Polyak, S.J. The Synthetic Antiviral Drug Arbidol Inhibits Globally Prevalent Pathogenic Viruses. *J. Virol.* **2016**, *90*, 3086–3092. [CrossRef] [PubMed]
77. Hulseberg, C.E.; Feneant, L.; Szymanska-de Wijs, K.M.; Kessler, N.P.; Nelson, E.A.; Shoemaker, C.J.; Schmaljohn, C.S.; Polyak, S.J.; White, J.M. Arbidol and Other Low-Molecular-Weight Drugs That Inhibit Lassa and Ebola Viruses. *J. Virol.* **2019**, *93*. [CrossRef]
78. Tang, K.; He, S.; Zhang, X.; Guo, J.; Chen, Q.; Yan, F.; Banadyga, L.; Zhu, W.; Qiu, X.; Guo, Y. Tangeretin, an extract from Citrus peels, blocks cellular entry of arenaviruses that cause viral hemorrhagic fever. *Antivir. Res.* **2018**, *160*, 87–93. [CrossRef] [PubMed]
79. Torriani, G.; Trofimenko, E.; Mayor, J.; Fedeli, C.; Moreno, H.; Michel, S.; Heulot, M.; Chevalier, N.; Zimmer, G.; Shrestha, N.; et al. Identification of Clotrimazole Derivatives as Specific Inhibitors of Arenavirus Fusion. *J. Virol.* **2019**, *93*. [CrossRef]
80. Zhang, X.; Yan, F.; Tang, K.; Chen, Q.; Guo, J.; Zhu, W.; He, S.; Banadyga, L.; Qiu, X.; Guo, Y. Identification of a clinical compound losmapimod that blocks Lassa virus entry. *Antivir. Res.* **2019**, *167*, 68–77. [CrossRef] [PubMed]
81. Moeschler, S.; Locher, S.; Zimmer, G. 1-Benzyl-3-cetyl-2-methylimidazolium Iodide (NH125) Is a Broad-Spectrum Inhibitor of Virus Entry with Lysosomotropic Features. *Viruses* **2018**, *10*. [CrossRef]
82. Friesland, A.; Zhao, Y.; Chen, Y.H.; Wang, L.; Zhou, H.; Lu, Q. Small molecule targeting Cdc42-intersectin interaction disrupts Golgi organization and suppresses cell motility. *Proc. Natl. Acad. Sci. USA* **2013**, *110*, 1261–1266. [CrossRef]
83. Chou, Y.Y.; Cuevas, C.; Carocci, M.; Stubbs, S.H.; Ma, M.; Cureton, D.K.; Chao, L.; Evesson, F.; He, K.; Yang, P.L.; et al. Identification and Characterization of a Novel Broad-Spectrum Virus Entry Inhibitor. *J. Virol.* **2016**, *90*, 4494–4510. [CrossRef]

84. Zhang, X.; Tang, K.; Guo, Y. The antifungal isavuconazole inhibits the entry of lassa virus by targeting the stable signal peptide-GP2 subunit interface of lassa virus glycoprotein. *Antivir. Res.* **2020**, *174*, 104701. [CrossRef]
85. Wang, M.K.; Ren, T.; Liu, H.; Lim, S.Y.; Lee, K.; Honko, A.; Zhou, H.; Dyall, J.; Hensley, L.; Gartin, A.K.; et al. Critical role for cholesterol in Lassa fever virus entry identified by a novel small molecule inhibitor targeting the viral receptor LAMP1. *PLoS Pathog.* **2018**, *14*, e1007322. [CrossRef] [PubMed]
86. McCormick, J.B.; King, I.J.; Webb, P.A.; Scribner, C.L.; Craven, R.B.; Johnson, K.M.; Elliott, L.H.; Belmont-Williams, R. Lassa fever. Effective therapy with ribavirin. *N. Engl. J. Med.* **1986**, *314*, 20–26. [CrossRef] [PubMed]
87. Oestereich, L.; Rieger, T.; Ludtke, A.; Ruibal, P.; Wurr, S.; Pallasch, E.; Bockholt, S.; Krasemann, S.; Munoz-Fontela, C.; Gunther, S. Efficacy of Favipiravir Alone and in Combination With Ribavirin in a Lethal, Immunocompetent Mouse Model of Lassa Fever. *J. Infect. Dis.* **2016**, *213*, 934–938. [CrossRef] [PubMed]
88. Feld, J.J.; Hoofnagle, J.H. Mechanism of action of interferon and ribavirin in treatment of hepatitis C. *Nature* **2005**, *436*, 967–972. [CrossRef]
89. Lau, J.Y.; Tam, R.C.; Liang, T.J.; Hong, Z. Mechanism of action of ribavirin in the combination treatment of chronic HCV infection. *Hepatology* **2002**, *35*, 1002–1009. [CrossRef]
90. Carrillo-Bustamante, P.; Nguyen, T.H.T.; Oestereich, L.; Gunther, S.; Guedj, J.; Graw, F. Determining Ribavirin's mechanism of action against Lassa virus infection. *Sci. Rep.* **2017**, *7*, 1–12. [CrossRef] [PubMed]
91. Furuta, Y.; Komeno, T.; Nakamura, T. Favipiravir (T-705), a broad spectrum inhibitor of viral RNA polymerase. *Proc. Jpn. Acad. Ser. B Phys. Biol. Sci.* **2017**, *93*, 449–463. [CrossRef]
92. Furuta, Y.; Gowen, B.B.; Takahashi, K.; Shiraki, K.; Smee, D.F.; Barnard, D.L. Favipiravir (T-705), a novel viral RNA polymerase inhibitor. *Antivir. Res.* **2013**, *100*, 446–454. [CrossRef] [PubMed]
93. Warren, T.K.; Wells, J.; Panchal, R.G.; Stuthman, K.S.; Garza, N.L.; Van Tongeren, S.A.; Dong, L.; Retterer, C.J.; Eaton, B.P.; Pegoraro, G.; et al. Protection against filovirus diseases by a novel broad-spectrum nucleoside analogue BCX4430. *Nature* **2014**, *508*, 402–405. [CrossRef] [PubMed]
94. Neuman, B.W.; Bederka, L.H.; Stein, D.A.; Ting, J.P.; Moulton, H.M.; Buchmeier, M.J. Development of peptide-conjugated morpholino oligomers as pan-arenavirus inhibitors. *Antimicrob. Agents Chemother.* **2011**, *55*, 4631–4638. [CrossRef] [PubMed]
95. Huang, K.W.; Hsu, K.C.; Chu, L.Y.; Yang, J.M.; Yuan, H.S.; Hsiao, Y.Y. Identification of Inhibitors for the DEDDh Family of Exonucleases and a Unique Inhibition Mechanism by Crystal Structure Analysis of CRN-4 Bound with 2-Morpholin-4-ylethanesulfonate (MES). *J. Med. Chem.* **2016**, *59*, 8019–8029. [CrossRef]
96. Saez-Ayala, M.; Laban Yekwa, E.; Mondielli, C.; Roux, L.; Hernandez, S.; Bailly, F.; Cotelle, P.; Rogolino, D.; Canard, B.; Ferron, F.; et al. Metal chelators for the inhibition of the lymphocytic choriomeningitis virus endonuclease domain. *Antivir. Res.* **2019**, *162*, 79–89. [CrossRef] [PubMed]
97. Sepulveda, C.S.; Garcia, C.C.; Damonte, E.B. Inhibition of arenavirus infection by thiuram and aromatic disulfides. *Antivir. Res.* **2010**, *87*, 329–337. [CrossRef]
98. Sepulveda, C.S.; Garcia, C.C.; Levingston Macleod, J.M.; Lopez, N.; Damonte, E.B. Targeting of arenavirus RNA synthesis by a carboxamide-derivatized aromatic disulfide with virucidal activity. *PLoS ONE* **2013**, *8*, e81251. [CrossRef] [PubMed]
99. Pasquato, A.; Cendron, L.; Kunz, S. Cleavage of the Glycoprotein of Arenaviruses. In *Activation of Viruses by Host Proteases*; Böttcher-Friebertshäuser, E., Garten, W., Klenk, H.D., Eds.; Springer International Publishing: Cham, Germany, 2018; pp. 47–70.
100. Pasquato, A.; Pullikotil, P.; Asselin, M.C.; Vacatello, M.; Paolillo, L.; Ghezzo, F.; Basso, F.; Di Bello, C.; Dettin, M.; Seidah, N.G. The proprotein convertase SKI-1/S1P. In vitro analysis of Lassa virus glycoprotein-derived substrates and ex vivo validation of irreversible peptide inhibitors. *J. Biol. Chem.* **2006**, *281*, 23471–23481. [CrossRef]
101. Rojek, J.M.; Pasqual, G.; Sanchez, A.B.; Nguyen, N.T.; de la Torre, J.C.; Kunz, S. Targeting the proteolytic processing of the viral glycoprotein precursor is a promising novel antiviral strategy against arenaviruses. *J. Virol.* **2010**, *84*, 573–584. [CrossRef]
102. Urata, S.; Yun, N.; Pasquato, A.; Paessler, S.; Kunz, S.; de la Torre, J.C. Antiviral activity of a small-molecule inhibitor of arenavirus glycoprotein processing by the cellular site 1 protease. *J. Virol.* **2011**, *85*, 795–803. [CrossRef]
103. Pasquato, A.; Rochat, C.; Burri, D.J.; Pasqual, G.; de la Torre, J.C.; Kunz, S. Evaluation of the anti-arenaviral activity of the subtilisin kexin isozyme-1/site-1 protease inhibitor PF-429242. *Virology* **2012**, *423*, 14–22. [CrossRef] [PubMed]
104. Urata, S.; de la Torre, J.C. Arenavirus budding. *Adv. Virol.* **2011**, *2011*, 180326. [CrossRef]
105. Perez, M.; Greenwald, D.L.; de la Torre, J.C. Myristoylation of the RING finger Z protein is essential for arenavirus budding. *J. Virol.* **2004**, *78*, 11443–11448. [CrossRef]
106. Russo, R.; Kemp, M.; Bhatti, U.F.; Pai, M.; Wakam, G.; Biesterveld, B.; Alam, H.B. Life on the battlefield: Valproic acid for combat applications. *J. Trauma Acute Care Surg.* **2020**, *89*, S69–S76. [CrossRef]
107. Vazquez-Calvo, A.; Martin-Acebes, M.A.; Saiz, J.C.; Ngo, N.; Sobrino, F.; de la Torre, J.C. Inhibition of multiplication of the prototypic arenavirus LCMV by valproic acid. *Antivir. Res.* **2013**, *99*, 172–179. [CrossRef] [PubMed]
108. Lu, J.; Han, Z.; Liu, Y.; Liu, W.; Lee, M.S.; Olson, M.A.; Ruthel, G.; Freedman, B.D.; Harty, R.N. A host-oriented inhibitor of Junin Argentine hemorrhagic fever virus egress. *J. Virol.* **2014**, *88*, 4736–4743. [CrossRef] [PubMed]
109. Yasuda, J.; Hunter, E.; Nakao, M.; Shida, H. Functional involvement of a novel Nedd4-like ubiquitin ligase on retrovirus budding. *EMBO Rep.* **2002**, *3*, 636–640. [CrossRef]

110. Han, Z.; Lu, J.; Liu, Y.; Davis, B.; Lee, M.S.; Olson, M.A.; Ruthel, G.; Freedman, B.D.; Schnell, M.J.; Wrobel, J.E.; et al. Small-molecule probes targeting the viral PPxY-host Nedd4 interface block egress of a broad range of RNA viruses. *J. Virol.* **2014**, *88*, 7294–7306. [CrossRef]
111. Urata, S.; Ngo, N.; de la Torre, J.C. The PI3K/Akt pathway contributes to arenavirus budding. *J. Virol.* **2012**, *86*, 4578–4585. [CrossRef] [PubMed]
112. Xing, J.; Ly, H.; Liang, Y. The Z proteins of pathogenic but not nonpathogenic arenaviruses inhibit RIG-I-like receptor-dependent interferon production. *J. Virol.* **2015**, *89*, 2944–2955. [CrossRef]
113. Ziegler, C.M.; Eisenhauer, P.; Kelly, J.A.; Dang, L.N.; Beganovic, V.; Bruce, E.A.; King, B.R.; Shirley, D.J.; Weir, M.E.; Ballif, B.A.; et al. A Proteomics Survey of Junin Virus Interactions with Human Proteins Reveals Host Factors Required for Arenavirus Replication. *J. Virol.* **2018**, *92*. [CrossRef]
114. Helguera, G.; Jemielity, S.; Abraham, J.; Cordo, S.M.; Martinez, M.G.; Rodriguez, J.A.; Bregni, C.; Wang, J.J.; Farzan, M.; Penichet, M.L.; et al. An antibody recognizing the apical domain of human transferrin receptor 1 efficiently inhibits the entry of all new world hemorrhagic Fever arenaviruses. *J. Virol.* **2012**, *86*, 4024–4028. [CrossRef]
115. Cross, R.W.; Mire, C.E.; Branco, L.M.; Geisbert, J.B.; Rowland, M.M.; Heinrich, M.L.; Goba, A.; Momoh, M.; Grant, D.S.; Fullah, M.; et al. Treatment of Lassa virus infection in outbred guinea pigs with first-in-class human monoclonal antibodies. *Antivir. Res.* **2016**, *133*, 218–222. [CrossRef] [PubMed]
116. Amanat, F.; Duehr, J.; Huang, C.; Paessler, S.; Tan, G.S.; Krammer, F. Monoclonal Antibodies with Neutralizing Activity and Fc-Effector Functions against the Machupo Virus Glycoprotein. *J. Virol.* **2020**, *94*. [CrossRef] [PubMed]
117. Robinson, J.E.; Hastie, K.M.; Cross, R.W.; Yenni, R.E.; Elliott, D.H.; Rouelle, J.A.; Kannadka, C.B.; Smira, A.A.; Garry, C.E.; Bradley, B.T.; et al. Most neutralizing human monoclonal antibodies target novel epitopes requiring both Lassa virus glycoprotein subunits. *Nat. Commun.* **2016**, *7*, 1–14. [CrossRef]
118. Ortiz-Riano, E.; Ngo, N.; Devito, S.; Eggink, D.; Munger, J.; Shaw, M.L.; de la Torre, J.C.; Martinez-Sobrido, L. Inhibition of arenavirus by A3, a pyrimidine biosynthesis inhibitor. *J. Virol.* **2014**, *88*, 878–889. [CrossRef] [PubMed]
119. Sepulveda, C.S.; Garcia, C.C.; Damonte, E.B. Antiviral activity of A771726, the active metabolite of leflunomide, against Junin virus. *J. Med. Virol.* **2018**, *90*, 819–827. [CrossRef] [PubMed]
120. Sepulveda, C.S.; Garcia, C.C.; Fascio, M.L.; D'Accorso, N.B.; Docampo Palacios, M.L.; Pellon, R.F.; Damonte, E.B. Inhibition of Junin virus RNA synthesis by an antiviral acridone derivative. *Antivir. Res.* **2012**, *93*, 16–22. [CrossRef] [PubMed]
121. Nair, V.; Chi, G.; Shu, Q.; Julander, J.; Smee, D.F. A heterocyclic molecule with significant activity against dengue virus. *Bioorg. Med. Chem. Lett.* **2009**, *19*, 1425–1427. [CrossRef] [PubMed]
122. Chen, Q.; Smith, A. l-like 3-deazaneplanocin analogues: Synthesis and antiviral properties. *Bioorg. Med. Chem. Lett.* **2019**, *29*, 126613. [CrossRef]
123. Iwasaki, M.; Minder, P.; Cai, Y.; Kuhn, J.H.; Yates, J.R., III; Torbett, B.E.; de la Torre, J.C. Interactome analysis of the lymphocytic choriomeningitis virus nucleoprotein in infected cells reveals ATPase Na^+/K^+ transporting subunit Alpha 1 and prohibitin as host-cell factors involved in the life cycle of mammarenaviruses. *PLoS Pathog.* **2018**, *14*, e1006892. [CrossRef] [PubMed]
124. Lee, A.M.; Rojek, J.M.; Spiropoulou, C.F.; Gundersen, A.T.; Jin, W.; Shaginian, A.; York, J.; Nunberg, J.H.; Boger, D.L.; Oldstone, M.B.; et al. Unique small molecule entry inhibitors of hemorrhagic fever arenaviruses. *J. Biol. Chem.* **2008**, *283*, 18734–18742. [CrossRef]
125. Wang, P.; Liu, Y.; Zhang, G.; Wang, S.; Guo, J.; Cao, J.; Jia, X.; Zhang, L.; Xiao, G.; Wang, W. Screening and Identification of Lassa Virus Entry Inhibitors from an FDA-Approved Drug Library. *J. Virol.* **2018**, *92*. [CrossRef] [PubMed]
126. Lay Mendoza, M.F.; Acciani, M.D.; Levit, C.N.; Santa Maria, C.; Brindley, M.A. Monitoring Viral Entry in Real-Time Using a Luciferase Recombinant Vesicular Stomatitis Virus Producing SARS-CoV-2, EBOV, LASV, CHIKV, and VSV Glycoproteins. *Viruses* **2020**, *12*. [CrossRef] [PubMed]
127. Martinez-Sobrido, L.; Cheng, B.Y.; de la Torre, J.C. Reverse Genetics Approaches to Control Arenavirus. *Methods Mol. Biol.* **2016**, *1403*, 313–351. [CrossRef] [PubMed]
128. Cubitt, B.; Ortiz-Riano, E.; Cheng, B.Y.; Kim, Y.J.; Yeh, C.D.; Chen, C.Z.; Southall, N.O.E.; Zheng, W.; Martinez-Sobrido, L.; de la Torre, J.C. A cell-based, infectious-free, platform to identify inhibitors of lassa virus ribonucleoprotein (vRNP) activity. *Antivir. Res.* **2020**, *173*, 104667. [CrossRef] [PubMed]
129. Wendt, L.; Bostedt, L.; Hoenen, T.; Groseth, A. High-throughput screening for negative-stranded hemorrhagic fever viruses using reverse genetics. *Antivir. Res.* **2019**, *170*, 104569. [CrossRef] [PubMed]
130. Capul, A.A.; de la Torre, J.C. A cell-based luciferase assay amenable to high-throughput screening of inhibitors of arenavirus budding. *Virology* **2008**, *382*, 107–114. [CrossRef] [PubMed]
131. Ashburn, T.T.; Thor, K.B. Drug repositioning: Identifying and developing new uses for existing drugs. *Nat. Rev. Drug Discov.* **2004**, *3*, 673–683. [CrossRef]
132. Mercorelli, B.; Palu, G.; Loregian, A. Drug Repurposing for Viral Infectious Diseases: How Far Are We? *Trends Microbiol.* **2018**, *26*, 865–876. [CrossRef]
133. Pandey, A.; Nikam, A.N.; Shreya, A.B.; Mutalik, S.P.; Gopalan, D.; Kulkarni, S.; Padya, B.S.; Fernandes, G.; Mutalik, S.; Prassl, R. Potential therapeutic targets for combating SARS-CoV-2: Drug repurposing, clinical trials and recent advancements. *Life Sci.* **2020**, *256*, 117883. [CrossRef] [PubMed]
134. Madrid, P.B.; Chopra, S.; Manger, I.D.; Gilfillan, L.; Keepers, T.R.; Shurtleff, A.C.; Green, C.E.; Iyer, L.V.; Dilks, H.H.; Davey, R.A.; et al. A systematic screen of FDA-approved drugs for inhibitors of biological threat agents. *PLoS ONE* **2013**, *8*, e60579. [CrossRef]

135. Kim, Y.J.; Cubitt, B.; Chen, E.; Hull, M.V.; Chatterjee, A.K.; Cai, Y.; Kuhn, J.H.; de la Torre, J.C. The ReFRAME library as a comprehensive drug repurposing library to identify mammarenavirus inhibitors. *Antivir. Res.* **2019**, *169*, 104558. [CrossRef]
136. Riva, L.; Yuan, S.; Yin, X.; Martin-Sancho, L.; Matsunaga, N.; Pache, L.; Burgstaller-Muehlbacher, S.; De Jesus, P.D.; Teriete, P.; Hull, M.V.; et al. Discovery of SARS-CoV-2 antiviral drugs through large-scale compound repurposing. *Nature* **2020**, *586*, 113–119. [CrossRef] [PubMed]
137. Oppliger, J.; Torriani, G.; Herrador, A.; Kunz, S. Lassa Virus Cell Entry via Dystroglycan Involves an Unusual Pathway of Macropinocytosis. *J. Virol.* **2016**, *90*, 6412–6429. [CrossRef] [PubMed]
138. Raabe, V.N.; Kann, G.; Ribner, B.S.; Morales, A.; Varkey, J.B.; Mehta, A.K.; Lyon, G.M.; Vanairsdale, S.; Faber, K.; Becker, S.; et al. Favipiravir and Ribavirin Treatment of Epidemiologically Linked Cases of Lassa Fever. *Clin. Infect. Dis.* **2017**, *65*, 855–859. [CrossRef]
139. Herring, S.; Oda, J.M.; Wagoner, J.; Kirchmeier, D.; O'Connor, A.; Nelson, E.A.; Huang, Q.; Liang, Y.; DeWald, L.E.; Johansen, L.M.; et al. Inhibition of Arenaviruses by Combinations of Orally Available Approved Drugs. *Antimicrob. Agents Chemother.* **2021**, *65*. [CrossRef]
140. Sun, X.; Vilar, S.; Tatonetti, N.P. High-throughput methods for combinatorial drug discovery. *Sci. Transl. Med.* **2013**, *5*, 205rv1. [CrossRef]

Article

The Origins and Future of Sentinel: An Early-Warning System for Pandemic Preemption and Response

Yolanda Botti-Lodovico [1,†], Parvathy Nair [2,†], Dolo Nosamiefan [3,†], Matthew Stremlau [4,†], Stephen Schaffner [1,5,6,†], Sebastian V. Agignoae [1,7], John Oke Aiyepada [8], Fehintola V. Ajogbasile [3,9], George O. Akpede [8,10], Foday Alhasan [11], Kristian G. Andersen [12,13], Danny A. Asogun [14], Oladele Oluwafemi Ayodeji [15], Aida S. Badiane [16], Kayla Barnes [1,6], Matthew R. Bauer [17], Antoinette Bell-Kareem [18], Muoebonam Ekene Benard [8], Ebo Ohomoime Benevolence [8], Osiemi Blessing [8], Chloe K. Boehm [1,19], Matthew L. Boisen [20], Nell G. Bond [18], Luis M. Branco [20], Michael J. Butts [1,5], Amber Carter [1], Andres Colubri [1,21], Awa B. Deme [16], Katherine C. DeRuff [1], Younousse Diédhiou [16], Akhilomen Patience Edamhande [8], Siham Elhamoumi [1], Emily J. Engel [18], Philomena Eromon [3], Mosoka Fallah [22], Onikepe A. Folarin [3,9], Ben Fry [23], Robert Garry [18], Amy Gaye [16], Michael Gbakie [11], Sahr M. Gevao [24], Gabrielle Gionet [1], Adrianne Gladden-Young [1], Augustine Goba [25], Jules Francois Gomis [16], Anise N. Happi [3], Mary Houghton [1], Chikwe Ihekwuazu [8], Christopher Ojemiega Iruolagbe [26], Jonathan Jackson [27], Simbirie Jalloh [25], Jeremy Johnson [1], Lansana Kanneh [11], Adeyemi Kayode [3,9], Molly Kemball [1,5], Ojide Chiedozie Kingsley [28], Veronica Koroma [11], Dylan Kotliar [1,5], Samar Mehta [29], Hayden C. Metsky [1], Airende Michael [8], Marzieh Ezzaty Mirhashemi [1], Kayvon Modjarrad [30], Mambu Momoh [25,31], Cameron A. Myhrvold [1,19], Okonofua Grace Naregose [8], Tolla Ndiaye [16], Mouhamadou Ndiaye [16], Aliou Ndiaye [16], Erica Normandin [1,32], Ikponmwosa Odia [8], Judith Uche Oguzie [3,9], Sylvanus A. Okogbenin [8], Peter O. Okokhere [8,10,33], Johnson Okolie [3], Idowu B. Olawoye [3,9], Testimony J. Olumade [3,9], Paul E. Oluniyi [3,9], Omigie Omoregie [8,34], Daniel J. Park [1], Mariétou Faye Paye [1], Brittany Petros [1,35,36], Anthony A. Philippakis [1], Abechi Priscilla [3], Alan Ricks [37,38], Anne Rimoin [39], John Demby Sandi [25], John S. Schieffelin [40], Monica Schreiber [5], Mame Cheikh Seck [16], Sameed Siddiqui [1,41], Katherine Siddle [1], Allison R. Smither [42], Mouhamad Sy [16], Ngayo Sy [16], Christopher H. Tomkins-Tinch [1,5], Oyewale Tomori [8], Chinedu Ugwu [3,9], Jessica N. Uwanibe [3,9], Eghosasere Anthonia Uyigue [8], Dada Ireti Victoria [3,43], Anika Vinzé [1], Megan E. Vodzak [1,2], Nicole Welch [1,17], Haja Isatta Wurie [25], Daba Zoumarou [16], Donald S. Grant [11,25,‡], Daouda Ndiaye [9,16,‡], Bronwyn MacInnis [1,‡], Pardis C. Sabeti [1,2,5,6,44,45,*,‡] and Christian Happi [3,6,9,*,‡]

Citation: Botti-Lodovico, Y.; Nair, P.; Nosamiefan, D.; Stremlau, M.; Schaffner, S.; Agignoae, S.V.; Aiyepada, J.O.; Ajogbasile, F.V.; Akpede, G.O.; Alhasan, F.; et al. The Origins and Future of Sentinel: An Early-Warning System for Pandemic Preemption and Response. *Viruses* **2021**, *13*, 1605. https://doi.org/10.3390/v13081605

Academic Editors: Michael B. A. Oldstone and Juan C. De la Torre

Received: 7 June 2021
Accepted: 21 July 2021
Published: 13 August 2021

1 Broad Institute of Massachusetts Institute of Technology (MIT) and Harvard, Cambridge, MA 02142, USA; ybottilo@broadinstitute.org (Y.B.-L.); sfs@broadinstitute.org (S.S.); sagignoa@broadinstitute.org (S.V.A.); kbarnes@broadinstitute.org (K.B.); cboehm@broadinstitute.org (C.K.B.); mjbutts@broadinstitute.org (M.J.B.); acarter@broadinstitute.org (A.C.); andres@broadinstitute.org (A.C.); kderuff@broadinstitute.org (K.C.D.); selhamou@broadinstitute.org (S.E.); ggionet@broadinstitute.org (G.G.); agladden@broadinstitute.org (A.G.-Y.); mhoughto@broadinstitute.org (M.H.); jjohnson@broadinstitute.org (J.J.); mkemball@broadinstitute.org (M.K.); dkotliar@broadinstitute.org (D.K.); hmetsky@broadinstitute.org (H.C.M.); marzieh@broadinstitute.org (M.E.M.); cmyhrvol@princeton.edu (C.A.M.); enormand@broadinstitute.org (E.N.); dpark@broadinstitute.org (D.J.P.); mpaye@broadinstitute.org (M.F.P.); bpetros@broadinstitute.org (B.P.); aphilipp@broadinstitute.org (A.A.P.); ssiddiqu@broadinstitute.org (S.S.); kjsiddle@broadinstitute.org (K.S.); tomkinsc@broadinstitute.org (C.H.T.-T.); avinze@broadinstitute.org (A.V.); mvodzak@broadinstitute.org (M.E.V.); nwelch@broadinstitute.org (N.W.); bronwyn@broadinstitute.org (B.M.)
2 Howard Hughes Medical Institute, Chevy Chase, MD 20815, USA; pnair@broadinstitute.org
3 African Center of Excellence for Genomics of Infectious Diseases (ACEGID), Redeemer's University, Ede, Osun State, Nigeria; dolonosa@broadinstitute.org (D.N.); ajogbasilef@run.edu.ng (F.V.A.); 2mimijoe@gmail.com (P.E.); folarino@run.edu.ng (O.A.F.); anisehappi@yahoo.com (A.N.H.); kayodet@run.edu.ng (A.K.); oguziej@run.edu.ng (J.U.O.); okoliec@run.edu.ng (J.O.); olawoyei0303@run.edu.ng (I.B.O.); olumadet@run.edu.ng (T.J.O.); oluniyip@run.edu.ng (P.E.O.); abechip@run.edu.ng (A.P.); ugwuc@run.edu.ng (C.U.); uwanibej@run.edu.ng (J.N.U.); dadaireti@gmail.com (D.I.V.)
4 Equator Labs Incorporated, Washington, DC 20011, USA; stremlau@gmail.com
5 Department of Organismic and Evolutionary Biology, Harvard University, Cambridge, MA 02138, USA; mschreib@broadinstitute.org
6 Department of Immunology and Infectious Diseases, Harvard T.H. Chan School of Public Health, Harvard University, Boston, MA 02115, USA
7 Harvard Kennedy School, Cambridge, MA 02138, USA

[8] Institute of Lassa Fever, Research and Control, Irrua Specialist Teaching Hospital, Irrua, Edo State, Nigeria; aiyepadaoke82@yahoo.com (J.O.A.); georgeakpede@yahoo.co.uk (G.O.A.); nwata2007@gmail.com (M.E.B.); ebobene77@yahoo.com (E.O.B.); blessingosiemi@rocketmail.com (O.B.); patienceedamhande@yahoo.com (A.P.E.); chikwe.ihekweazu@gmail.com (C.I.); airemikelis@gmail.com (A.M.); grace4ambi84@gmail.com (O.G.N.); iodia905@gmail.com (I.O.); okogbenins@yahoo.com (S.A.O.); pitaokokhere@yahoo.com (P.O.O.); oomigie@ymail.com (O.O.); oyewaletomori@gmail.com (O.T.); diamondeghe@gmail.com (E.A.U.)

[9] Department of Biological Sciences, Redeemer's University, Ede, Osun State, Nigeria; daouda.ndiaye@ucad.edu.sn

[10] Department of Medicine, Faculty of Clinical Sciences, College of Medicine, Ambrose Alli University, Ekpoma, Edo State, Nigeria

[11] Viral Hemorrhagic Fever Program, Kenema Government Hospital, Ministry of Health and Sanitation, Kenema, Sierra Leone; fodayalhasan37@gmail.com (F.A.); gbakiemichael@gmail.com (M.G.); lansanakanneh@gmail.com (L.K.); vjkoroma@gmail.com (V.K.); donkumfel@yahoo.co.uk (D.S.G.)

[12] Department of Immunology and Microbiology, The Scripps Research Institute, La Jolla, CA 92037, USA; andersen@scripps.edu

[13] Scripps Research Translational Institute, La Jolla, CA 92037, USA

[14] Ambrose Alli University, Ekpoma, Edo State, Nigeria; asogun2001@yahoo.com

[15] Federal Medical Centre, Owo, Ondo State, Nigeria; femiayodeji@yahoo.com

[16] Université Cheikh Anta Diop, BP 5005, Dakar, Senegal; asbadiane@gmail.com (A.S.B.); deme.awa@gmail.com (A.B.D.); ydjedju@yahoo.fr (Y.D.); myanaa08@gmail.com (A.G.); jules.gomis@gmail.com (J.F.G.); ndiayetola@gmail.com (T.N.); mouhamadou.ndiaye@ucad.edu.sn (M.N.); aliou.ndiaye10@yahoo.fr (A.N.); mcseck203@yahoo.fr (M.C.S.); symouhamad92@gmail.com (M.S.); ngayosy50@hotmail.com (N.S.); rwallisz@yahoo.fr (D.Z.)

[17] Division of Medical Sciences, Harvard Medical School, Boston, MA 02115, USA; mbauer@broadinstitute.org

[18] Tulane University Medical Center, New Orleans, LA 70112, USA; abell13@tulane.edu (A.B.-K.); nbond@tulane.edu (N.G.B.); eengel@tulane.edu (E.J.E.); rfgarry@tulane.edu (R.G.)

[19] Department of Molecular Biology, Princeton University, Princeton, NJ 08540, USA

[20] Zalgen Labs, Germantown, MD 20876, USA; mboisen24@gmail.com (M.L.B.); lbranco@zalgenlabs.com (L.M.B.)

[21] University of Massachusetts Medical School, Worcester, MA 01605, USA

[22] Refuge Place International, Bassa Town, Lower Johnsonville, Liberia; mfallah1969@gmail.com

[23] Fathom Information Design, Boston, MA 02114, USA; ben@fathom.info

[24] University of Sierra Leone, Freetown, Sierra Leone; gevaosm@yahoo.com

[25] College of Medicine and Allied Health Sciences, University of Sierra Leone, Freetown, Sierra Leone; augstgoba@yahoo.com (A.G.); simbiriej5@gmail.com (S.J.); mambumomoh@gmail.com (M.M.); johnatsandi@gmail.com (J.D.S.); imwurie@icloud.com (H.I.W.)

[26] Department of Internal Medicine, Irrua Specialist Teaching Hospital, Irrua, Edo State, Nigeria; chriojem@gmail.com

[27] Dimagi, Inc., Cambridge, MA 02139, USA; jjackson@dimagi.com

[28] Department of Medical Microbiology, Alex-Ekwueme Federal University Teaching Hospital Abakaliki, Abakaliki, Ebonyi, Nigeria; edomann2001@yahoo.com

[29] Department of Critical Care Medicine, University of Maryland Medical Center, Baltimore, MA 21201, USA; sbmehta@gmail.com

[30] Walter Reed Army Institute of Research, Silver Spring, MD 20910, USA; kmodjarrad@eidresearch.org

[31] Eastern Polytechnic College, Kenema, Sierra Leone

[32] Department of Systems Biology, Harvard Medical School, Boston, MA 02115, USA

[33] Department of Medicine, Irrua Specialist Teaching Hospital, Irrua, Edo State, Nigeria

[34] West African Examinations Council, Yaba, Lagos State, Nigeria

[35] Harvard-MIT Health Sciences and Technology, Cambridge, MA 02139, USA

[36] Harvard/MIT MD-PhD Program, Boston, MA 02115, USA

[37] MASS Design Group, Boston, MA 02116, USA; alan@mass-group.org

[38] Yale School of Architecture, New Haven, CT 06511, USA

[39] Department of Epidemiology, Jonathan and Karing Fielding School of Public Health, University of California, Los Angeles, CA 90095, USA; arimoin@g.ucla.edu

[40] Section of Infectious Disease, Department of Pediatrics, Tulane University School of Medicine, New Orleans, LA 70112, USA; jschieff@tulane.edu

[41] Computational and Systems Biology Program, Massachusetts Institute of Technology, Cambridge, MA 02139, USA

[42] Department of Microbiology and Immunology, Tulane University School of Medicine, New Orleans, LA 70112, USA; asmither@tulane.edu

[43] Ikorodu General Hospital, Ikorodu, Lagos State, Nigeria

[44] Department of Medicine, Division of Infectious Diseases, Massachusetts General Hospital, Boston, MA 02114, USA
[45] Massachusetts Consortium on Pathogen Readiness, Boston, MA 02115, USA
* Correspondence: pardis@broadinstitute.org (P.C.S.); happic@run.edu.ng (C.H.); Tel.: +1-617-384-5335 (P.C.S.)
† These authors contributed equally to the work.
‡ These authors jointly supervised the work.

Abstract: While investigating a signal of adaptive evolution in humans at the gene LARGE, we encountered an intriguing finding by Dr. Stefan Kunz that the gene plays a critical role in Lassa virus binding and entry. This led us to pursue field work to test our hypothesis that natural selection acting on LARGE—detected in the Yoruba population of Nigeria—conferred resistance to Lassa Fever in some West African populations. As we delved further, we conjectured that the "emerging" nature of recently discovered diseases like Lassa fever is related to a newfound capacity for detection, rather than a novel viral presence, and that humans have in fact been exposed to the viruses that cause such diseases for much longer than previously suspected. Dr. Stefan Kunz's critical efforts not only laid the groundwork for this discovery, but also inspired and catalyzed a series of events that birthed Sentinel, an ambitious and large-scale pandemic prevention effort in West Africa. Sentinel aims to detect and characterize deadly pathogens before they spread across the globe, through implementation of its three fundamental pillars: Detect, Connect, and Empower. More specifically, Sentinel is designed to detect known and novel infections rapidly, connect and share information in real time to identify emerging threats, and empower the public health community to improve pandemic preparedness and response anywhere in the world. We are proud to dedicate this work to Stefan Kunz, and eagerly invite new collaborators, experts, and others to join us in our efforts.

Keywords: pandemic preemption; pandemic response; diagnostic tools; bioinformatics; genomic surveillance; infectious disease; Lassa virus; Lassa fever; Ebola; LARGE

1. Background

In 2007, our team set out to investigate an intriguing hypothesis, catalyzed by the research of Dr. Stefan Kunz, about the potentially ancient origins of Lassa virus. Lassa virus is the cause of a viral hemorrhagic fever known as Lassa fever and a category A bioterror threat. Despite its deadly nature and designation as an "emerging" infectious disease, we had reason to believe, based on our investigations of the human genome, that Lassa virus had circulated for millennia in Nigeria, conferring a selective pressure on human populations. Our work pursuing this hypothesis has led to a multidisciplinary, multinational viral genomics research program, and eventually our creation of *Sentinel*: an ambitious and innovative pandemic prevention effort now being piloted in West and Central Africa. Here we describe the history and principle behind Sentinel.

It began as a signal of human adaptation, connected to a gene called LARGE. As part of a genome survey of human variation in Europe, Asia, and Africa with the International Haplotype Map Consortium, Dr. Pardis Sabeti's team had developed methods for investigating signals of recent (within the last ~10 k years) natural selection in the human genome [1]. The team identified the locus surrounding LARGE as the strongest signal of recent selection in the survey; the signal was only visible in the African portion of the data, which came from the Yoruba people in Nigeria. From the literature, we learned that LARGE is a glycosylase that post-translationally modifies α-dystroglycan. This led us to a paper by Kunz, Michael Oldstone, and colleagues showing that LARGE's modification of α-dystroglycan, the cellular receptor for Lassa virus, was critical for efficient virus binding and entry to mammalian cells [2,3]. Knocking out the gene encoding LARGE greatly tempered Lassa virus's ability to infect cells. Taken together, these observations raised the possibility that polymorphisms in LARGE were under positive selection because they disrupted the protein's function, thereby preventing Lassa virus entry into host cells and conferring resistance to Lassa fever. This realization further led us to hypothesize that

Lassa virus may in fact have ancient origins, but circulated unnoticed in part due to genetic resistance to the virus in populations where Lassa fever is endemic.

To test this hypothesis, we had to collect fundamental information about the virus such as its age, geographic location, as well as the range and length of time that it had been circulating in affected zones. In particular, since the common variant in LARGE appeared to arise within the last 10,000 years, dating the age of the virus was essential to identifying the virus as a potential driver of natural selection for the gene.

We obtained the first pieces of evidence to support our hypothesis when we learned that the first geographic location in which Lassa fever was initially discovered was Nigeria in 1969 [4], where the signal of selection was also detected. We also learned that Lassa fever is endemic to West Africa [5]. In subsequent investigations of other global populations, we found that the signal of natural selection at LARGE was only present among populations in which Lassa fever was known to be present [6]. Today, Lassa fever is estimated to afflict 100,000–300,000 people each year, hospitalize 100,000, and kill 5000, although diagnostic testing remains limited and estimates vary widely [7]. Intriguingly, however, natural resistance to Lassa fever in parts of West Africa is common, with 50–90% of individuals showing mild or no symptoms when infected, while others suffer a complicated course that can lead to fever, encephalitis, deafness, and death [5]. It was unknown, however, why Lassa fever infection in some individuals was asymptomatic, while others developed severe disease.

To examine differential susceptibility to Lassa fever in individuals, the age and diversity of the virus, and whether Lassa virus was driving natural selection for thousands of years, we would need to study the genomes of both the virus and patients affected by it. We believed that with new tools to assay genetic variation across the human genome, we could examine and characterize any genetic evidence of resistance to Lassa fever, a study which we then set out to perform.

2. Studying Lassa Fever in Nigeria

Investigating a poorly understood, category A virus in rural Nigeria would require us to overcome many challenges. Study of Lassa virus and other high containment (biosafety level (BSL) 4) pathogens is complicated due to their lethality, lack of effective treatment, potential for aerosol transmission, and bioterror use, requiring rigorous precautions by highly trained teams. Moreover, at the time of our research, these diseases often appeared as sporadic outbreaks in remote settings, where infrastructure and transportation were lacking. However, with increased globalization and an ever-expanding human population, the need for large-scale research initiatives on BSL-4 pathogens remains acute [8,9]. Further, as only one BSL-4 lab exists in the entire region of West Africa [10] even today, transnational partnerships are critical to allow ongoing investigation of BSL-4 pathogen samples.

In order to study Lassa virus, Dr. Sabeti teamed up with Dr. Christian Happi, a colleague from previous studies on malaria, who had just begun doing viral research in Edo State, Nigeria, in the heart of the Lassa fever endemic zone in Nigeria. The nearby Irrua Specialist Teaching Hospital (ISTH) in Irrua, Nigeria was also the site of an upcoming government-funded center of excellence for Lassa fever, which at the time did not have diagnostics or treatment available onsite for the virus. Together, Happi and Sabeti's team established a partnership with ISTH in 2008, to develop diagnostics capacity onsite and facilitate acquisition of Ribavirin, the WHO essential therapeutic for Lassa fever.

ISTH treats ~16,000 patients each year, 80% of whom suffer from febrile illness [11,12]. The region suffers annual outbreaks of Lassa fever, but due to the lack of diagnostic tools or local capacity for surveillance, as well as the inaccessibility of clinics in remote regions, most local Lassa fever cases had gone undetected or were misdiagnosed as similarly presenting acute febrile illnesses. Drs. Happi and Sabeti worked alongside the Bernard Nocht Institute to establish on-site PCR diagnostics able to detect Lassa virus in patient blood, and to help source Ribavirin for the hospital. Rapid and accurate diagnostics led to more informed treatment plans for patients with Lassa fever, which facilitated an overall improvement in community engagement and health outcomes. The surrounding communities began hearing many stories of successful Lassa fever detection and recovery, and the site quickly became a referral center for patients with undiagnosed febrile illness within hundreds of kilometres.

Early in our work, we also gleaned an important insight after a man brought his young son to the hospital, where doctors diagnosed him with Lassa fever and successfully treated him. During a follow up investigation into the prevalence of Lassa virus in the patient's village, Dr. Happi learned that the child's mother, cousin, and neighbor had all succumbed to what appeared to be the same mysterious illness. In total, more than 20 members of that community who had died exhibited classical symptoms of clinical Lassa fever, without inciting any alarm or broader action by the public health community. One explanation for this may have been the fact that none of the deceased sought help from the nearby hospital, which previously lacked both diagnostic tests for Lassa fever, and sufficient capacity to treat the disease. Overall, the experience led us to contemplate the countless individuals who succumb to Lassa fever and other underdiagnosed diseases [13].

Based on findings from genetic and epidemiologic studies, we wrote a perspective entitled 'Emerging Disease or Diagnosis?' suggesting that we wrongly label some viruses as novel or rare, not because they truly are, but because we lack the necessary tools in the relevant places to detect them [9]. Our analysis of Lassa virus genomes suggested that the virus had been circulating for at least half a millenia in West Africa [14]. Contrary to popular belief, viral hemorrhagic fevers often exhibit nonspecific symptoms, like fever, nausea, headache, and malaise, which can complicate diagnosis. Without access to accurate diagnostics, healthcare workers are stymied in determining the etiology of fever, and many wrongly presume that malaria, typhoid, or shigella is responsible [9]. In the worst case scenario, lives are lost as a result of misdiagnosis, as in a 1989 Lassa fever outbreak in two Nigerian hospitals, which took the lives of 22 people [15]. This work led us to the realization that in many parts of the world, we are largely blind both to the prevalence of known infectious diseases and to the appearance of new threats. As we develop and equip local healthcare workers with proper diagnostic tools, however, we can detect areas of higher prevalence and subsequently develop more effective treatments, vaccines, surveillance tools, and research capacity where they are most needed, instead of awaiting the next outbreak.

3. ACEGID: Laying the Groundwork for Local Surveillance and Outbreak Response

Like Kunz, we began our work on Lassa virus with the goal of conducting molecular biology research on the mechanisms behind viral transmission and human resistance to Lassa fever disease. But to enable this research, we quickly recognized that responding to the basic diagnostic needs on the front line would be critical. More broadly, we also recognized that a sustained partnership between communities and their local healthcare clinics establishes trust, while the improved capacity of healthcare workers to detect and treat common diseases encourages more individuals to seek tests when needed, thereby increasing the number of samples available for important research [9].

As we continued to work with our partners at ISTH in Nigeria—and later extending to the Kenema Government Hospital (KGH) in Sierra Leone—to set up affordable diagnostics for routine use, we were able to establish this positive feedback loop, in which the higher number of diagnostic samples we collected at the point of care enabled us to conduct genomic sequencing efforts to identify and characterize other circulating viruses in the region. As new viral threats emerged, we were then able to quickly establish accurate diagnostics onsite to better serve communities. Since then, ISTH and KGH have become reference centers for management of viral hemorrhagic fevers in their respective regions. They have successfully detected more cases and saved more lives, thereby raising national awareness around these diseases. For example, in 2012 the Nigerian Federal Ministry of Health declared an increase in the number of suspected Lassa fever cases, an example of a possible "emergence of diagnosis"; with nearly 1000 cases reported by 41 local government agencies in 23 States [9]. These reports enabled rapid response measures by the government and public health community.

By standardizing and implementing this virtuous cycle as a broader system, we can ultimately build a much needed global surveillance capacity, strengthened by ongoing collaboration and trust between communities, local healthcare workers, hospitals like ISTH and KGH, and public health decision makers. These efforts have the potential to immediately benefit affected communities where certain diseases are endemic, while also helping to identify, monitor, and characterize emerging pathogens before they cause a regional outbreak or global pandemic.

As a first step in realizing this kind of collaborative surveillance model in West Africa, we merged these ideas into one blueprint, which laid the groundwork for our launch of the African Center of Excellence for Genomics of Infectious Disease (ACEGID). Based at Redeemer's University in Ede, Nigeria, ACEGID has built a network of partners in West Africa, including ISTH, KGH, and the Université Cheikh Anta Diop de Dakar in Senegal, supported by partners at Tulane University, and funded by the NIH Human Heredity and Health Africa Initiative and the World Bank. Its mission is organized into five specific mandates, including: (1) building research capacity among African scientists in genomics and molecular biology; (2) empowering researchers to apply genomic tools and data toward containing and defeating infectious diseases; (3) generating a curriculum to promote and support high-quality genomics research for the purpose of improving global health; (4) fostering a vibrant, collaborative, international research community, dedicated to high-quality, relevant, ethical, and responsive genomic research; and (5) engaging individuals and communities in outbreak prevention, education, and local public health practice, while also supporting clinical care and building a surveillance network for global health threats.

Just as we launched ACEGID in March 2014, however, Ebola began spreading through West Africa [8,16]. The outbreak went undiagnosed for months, reaching multiple countries, and mutating in ways that likely increased its infectivity [17–19]. By May, our KGH team, under the leadership of Dr. Humarr Khan, found itself on the frontlines of the local response efforts in Sierra Leone. We witnessed firsthand how a lack of proper diagnostics, particularly in remote areas, prevented healthcare workers from accurately determining the source of an infection, allowing it to spread far beyond local public health capacity. The outbreak eventually took 11,000 lives [16], including Dr. Khan and other members of our team who contracted Ebola while caring for patients. This served as a painful reminder of the deep need to detect and contain the broad range of deadly viruses that threaten the region every year.

ACEGID went on to make critical contributions in the response to the Ebola outbreak, the Zero Malaria campaign, regional efforts to reduce the burden of Lassa fever and other infectious diseases, and more [20]. When the first case of Ebola arrived at the Lagos International Airport in 2014, our team at Redeemer's diagnosed it, and helped stop a potentially devastating outbreak in its tracks [21]. Together, we have equipped over one thousand West African scientists with the education, training, and resources needed to monitor, study, and treat dangerous pathogens in their own region.

4. A New Approach to Pandemic Preparedness

The success of ACEGID laid the foundation for Sentinel, a pandemic preemption system designed to detect and characterize deadly pathogens before they spread across the globe. This initiative, supported by the TED Audacious Prize, was funded one month before the declaration of the COVID-19 outbreak as a pandemic.

Sentinel is organized around three core pillars, Detect, Connect, and Empower (Figure 1), that are as fundamental to preempting a pandemic as they are to responding to one that has already begun. More broadly, these pillars are interoperable and essential to supporting a public health system that can prevent a wide range of infectious diseases. The needs the pillars address remain as pressing today as they did when we conceived Sentinel, despite a year of vigorous public health effort to combat COVID-19. We describe the needs, as well as Sentinel's response to them, in depth below.

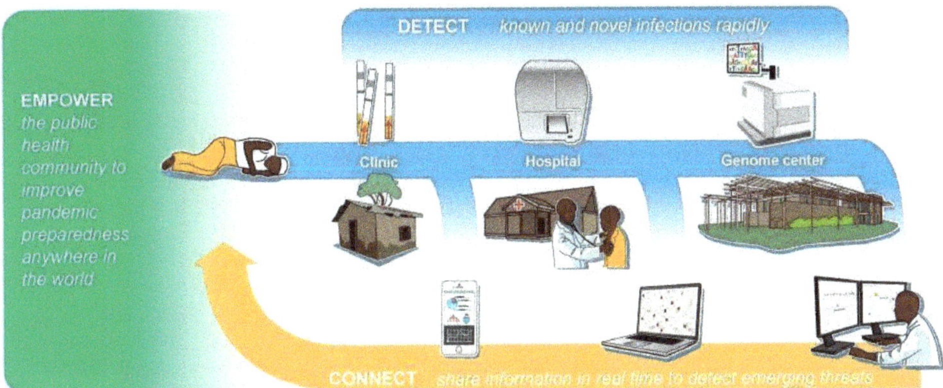

Figure 1. Sentinel will detect pathogens in any setting, from remote rural clinics to hospitals, and unknown pathogens will be identified and characterized at regional genome centers. Through a cloud-based system, Sentinel will share this information and connect healthcare workers, researchers, and disease control officials to track and predict threats. Finally, Sentinel will empower all stakeholders through a codified pandemic preemption system that can be scaled around the world.

5. Sentinel Pillar #1: Detect

Sentinel envisions a three-tiered approach to improving diagnosis and surveillance of viral infectious diseases. The first tier aims to detect the most likely pathogens at the point of care, using rapid, affordable diagnostic tests for the highest priority pathogens in a region. The second tier is designed to diagnose illnesses not covered by the point-of-care tests, and requires developing the capacity in local hospitals to identify a wide range of viruses in-house. The final tier relies on regional genomic sequencing centers to identify novel viruses and monitor known viruses; the resulting sequences can be useful for tracking changes to viruses, for monitoring spread of viral variants, and for developing new diagnostics and vaccines [22].

This approach places high demands on diagnostic technology. In the case of viral sequencing, the technology is already well-developed; the need is for creating sequencing capacity in places where it is currently unavailable, which is no easy task in itself. The technology can still benefit from enhancement, however, to increase scale, reduce cost, and improve sensitivity. The other tiers require new diagnostic technologies. The point-of-care tests, in particular, require the development of tests for many more viruses, tests that are rapid, affordable, and easy to use in settings with limited resources. Our team has already invested in a number of detection technologies to achieve these goals, focusing on CRISPR-based diagnostics that are sensitive, field-deployable, and rapidly programmable for a range of different pathogens. These CRISPR-based technologies can also be multiplexed to test for numerous viruses simultaneously in a single sample [23,24]. We are well aware, however, that the field of diagnostics is currently experiencing rapid changes: recent breakthroughs include novel point-of-care antigen capture technologies, nanotechnologies, and advances in isothermal amplification methods. Given the current pace of innovation, our goal is not to settle on a single technology in advance, but to have Sentinel serve as an innovation hub that can allow multiple technologies, including our own, to be piloted and optimized, retaining the flexibility to deploy the best technologies as they emerge.

6. Sentinel Pillar #2: Connect

In addition to improving our diagnostic arsenal, Sentinel aims to address another key challenge in pandemic prevention: the ability to connect various sources of diagnostic and surveillance data, and to share that information in real time with those who need it to guide critical public health strategy. Through Sentinel we are building tools to address these needs at multiple levels, from symptom surveillance of individuals, to diagnostic and clinical data integration by healthcare workers, to genomic surveillance data. Ultimately, we envision a future in which all of these data streams are themselves connected in real time, providing a holistic picture of known and emerging threats.

For years, our team has been developing tools and collaborating with others to make this vision a reality. Many of our solutions to data connectivity are cloud-based, mobile, and able to function in low connectivity environments. For example, the CommCare mobile app from Dimagi provides a means for frontline workers or individuals to share symptoms and diagnostic data and convey critical information between geographically separated clinics and labs in a way that is robust to intermittent and unreliable internet connectivity. Cloud genomic analysis platforms like Terra and DNAnexus bring strong compute resources and community-curated analysis pipelines to labs anywhere in the world, via a web-accessible interface. Powerful visualization tools of patient, epidemiologic, and genomic data sets, developed by Fathom, bring real time insights from field-generated data to those coordinating responses at the national level.

Our strategy operates on a number of principles, including incentivizing the sharing of quality data, creating easy-to-use, participatory, and flexible systems for collecting and sharing data, and building community trust in the tools that inform outbreak response. By adapting these principles to the specific needs of each level, we aim to ensure a continuous ability to access and analyze each input of data. Given the resource limitations of the regions where Sentinel will operate, many of these tools are designed for use by scientists, laboratory staff, and public health workers with limited computational experience or on-site hardware resources.

Ultimately, Sentinel's aim is to integrate each of these tools to enable data movement across an interoperable and secure data ecosystem that empowers all of the stakeholders in the surveillance system. While our team is presently working to further develop and integrate our own existing suite of information tools, future efforts for integration of the broader global catalog of information tools will require new funding initiatives, as well as combined private and public sector efforts to expand Internet connectivity and mobile data service to remote regions across the globe.

7. Sentinel Pillar #3: Empower

Pandemic preemption requires diagnostic and informatics technologies, but the technology is futile unless it is used and the results integrated into clinical care and public health decision-making. A successful system must include every stakeholder, from government and public health officials, to frontline healthcare workers and scientists, to individual patients and communities. Distributed local capacity and individual agency are critical to stopping a pandemic. But limited funding, community awareness, and support for capacity building often prevent the broad and equitable deployment of existing tools and the timely development of new ones.

Recognizing the importance of sustainability through local ownership, we regard training as paramount for building capacity in genomic sequencing, diagnostics, bioinformatics, and advanced genomic surveillance among local healthcare workers and laboratory scientists. Sentinel employs a train-the-trainers approach, organized through ACEGID, to generate sustainable and long-term capacity for use of Sentinel tools and implementation of each pillar. Leveraging our 10+ years of educational experience in West Africa, we aim to train 800 additional local healthcare workers and laboratory scientists to use our diagnostic and surveillance tools in pursuit of the Detect and Connect goals.

Sentinel also acknowledges the need for deep and longstanding partnerships to form integrated regional and global networks of experts and communities fully versed in pandemic preparedness and preemption. We know firsthand that this work is complex and challenging, and that our approach is only one of a myriad of efforts to support improved infectious disease diagnostics and surveillance. Having piloted this system in Nigeria and developed extensive partnerships through this work, we plan to grow our existing network and presence in Africa through cooperation with stakeholders like the Nigeria CDC, Africa CDC, and regional WHO offices in order to fulfill our role within this multi-stakeholder effort. As we anticipate the additional challenges that might arise over time, we will continually work to integrate into existing healthcare systems and empower local stakeholders, to understand and overcome reasons for resistance to our technologies, and to eventually build community trust, and encourage broad buy-in over time. Our proposed national innovation hubs in Nigeria, Senegal, and Sierra Leone serve as national genomics reference labs. The labs are closely connected to their respective national public health agencies and broader communities, which will ultimately drive rapid and strong local responses to outbreaks.

8. Remembering Dr. Kunz in Our Work Today

Over one year into the COVID-19 pandemic, Sentinel's focus on pandemic *preemption* continues to remain critically relevant. Dr. Stefan Kunz, whose groundbreaking work ignited our interest in Lassa fever, and led us onto a path towards new possibilities in genomic surveillance and our collective understanding of infectious disease, opened the potential for collaborative pandemic preemption efforts in Nigeria and beyond. There is still however much to do, and our path has not been a straightforward one. Our starting hypothesis, that variants in or around LARGE contribute to resistance to Lassa fever, remains an open question. We are still working to answer it, and to uncover other potential resistance variants. But the work has been complicated by multiple outbreaks, political instability, and violence at our hospitals, as well as the ongoing logistical challenges. Even as we have continued to pursue these efforts, we have also been responding to the more urgent priorities around infectious disease response on the ground. Nevertheless, as Sentinel matures, we are continuing on with this important research.

In the spirit of Dr. Kunz and his commitment to driving cutting-edge science and impactful partnerships for the improvement of global health, we remember the countless individuals who have worked alongside us, many of whom are still with us today, and some of whom we have tragically lost over this past decade. To honor their extraordinary contributions to public health and infectious disease research, we are proud to dedicate this work to Stefan Kunz, as well as our late colleagues who served on the frontlines of the Ebola response in Sierra Leone, including Alex Moigboi, Mohammed Fullah, Mbalu Fonnie, Vandi Sinnah, Alice Kovoma, S. Humarr Khan, Jacob Adu George Buanie, and Mohammed "Pa" Sow. Their tireless sacrifice, generosity of spirit, and unwavering dedication to their communities laid the groundwork for ongoing efforts to ensure that every individual has the right to health and quality care. As we continue to fight against infectious disease in West Africa and beyond, we remain grateful to them and the many partners who have made Sentinel possible, and we eagerly invite new collaborators, experts, and others to join us.

Author Contributions: Conceptualization, Y.B.-L., P.N., D.N. (Dolo Nosamiefan), M.S. (Matthew Stremlau), S.S. (Stephen Schaffner), A.C. (Andres Colubri), S.E., M.F., C.U., O.A.F., D.S.G., M.K., C.A.M., D.N. (Daouda Ndiaye), D.J.P., A.R. (Anne Rimoin), M.S. (Monica Schreiber), M.E.V., K.G.A., R.G., B.M., P.C.S., C.H.; software, A.C. (Andres Colubri), B.F., J.J. (Jonathan Jackson), C.A.M., D.J.P.; validation, A.C. (Andres Colubri); resources, S.S. (Stephen Schaffner), A.C. (Andres Colubri), S.E., P.E., M.F., C.U., A.N.H., O.A.F., B.F., D.S.G., M.K., C.A.M., D.N. (Daouda Ndiaye), S.A.O., P.O.O., D.J.P., A.A.P., A.R. (Alan Ricks), A.R. (Anne Rimoin), M.S. (Monica Schreiber), M.E.V., B.M., C.H., J.O., J.U.O., P.E.O., F.V.A., J.N.U., I.B.O., K.S., H.C.M., G.O.A., I.O., E.A.U., J.O.A., C.O.I., A.P.E., O.G.N., O.O., O.B., M.E.B., E.O.B., C.I., D.A.A., O.T., K.M., T.J.O., A.S.B., A.B.D., M.S. (Mouhamad Sy), N.S.,

D.Z., Y.D., A.G. (Amy Gaye), T.N., M.N., A.N., J.F.G., M.C.S., S.M., M.F.P., M.H., A.V., J.J. (Jeremy Johnson), C.K.B., B.P., M.R.B., E.N., M.E.M., S.S. (Sameed Siddiqui), K.C.D., N.W., A.C. (Amber Carter), G.G., M.J.B., S.V.A., C.H.T.-T., D.K., K.B., A.G.-Y., J.S.S., M.G., F.A., L.K., V.K., M.M., J.D.S., A.G. (Augustine Goba), S.J., N.G.B., E.J.E., A.R.S., A.B.-K., L.M.B., M.L.B., H.I.W., S.M.G., A.K., A.P., A.M.; data curation, O.O.A., A.C. (Andres Colubri), D.I.V., P.E., A.N.H., O.A.F., B.F., J.J. (Jonathan Jackson), O.C.K., S.A.O., P.O.O., D.J.P., A.A.P., P.C.S., C.H.; writing—original draft preparation, Y.B.-L., P.N., M.S. (Matthew Stremlau), S.S. (Stephen Schaffner), D.J.P., B.M., P.C.S., C.H.; writing—review and editing, Y.B.-L., P.N., S.S. (Stephen Schaffner), S.E., D.S.G., M.K., C.A.M., D.J.P., B.M., P.C.S., C.H.; visualization, A.C. (Andres Colubri), B.F., P.C.S.; supervision, S.S. (Stephen Schaffner), P.C.S., C.H.; project administration, D.N. (Dolo Nosamiefan), M.K., M.S. (Monica Schreiber), M.E.V.; funding acquisition, Y.B.-L., D.N. (Dolo Nosamiefan), M.S. (Matthew Stremlau), S.S. (Stephen Schaffner), S.E., P.E., M.F., D.S.G., M.K., C.A.M., D.N. (Daouda Ndiaye), D.J.P., A.R. (Anne Rimoin), M.S. (Monica Schreiber), M.E.V., K.G.A., R.G., B.M., P.C.S., C.H. All authors have read and agreed to the published version of the manuscript.

Funding: This work is made possible by support from Flu Lab and a cohort of generous donors through TED's Audacious Project, including the ELMA Foundation, MacKenzie Scott, the Skoll Foundation, and Open Philanthropy.

Institutional Review Board Statement: Not applicable.

Informed Consent Statement: Not Applicable.

Data Availability Statement: Data sharing is not applicable to this article.

Acknowledgments: We are also grateful to Theresa Ulrich, Lisa E. Hensley, Mary Carmichael, and Angela Page, in addition to many others, for their guidance and support.

Conflicts of Interest: Pardis Sabeti is a founder and shareholder of Sherlock Biosciences, and is both on the Board and serves as shareholder of the Danaher Corporation. Anthony Philippakis is a Venture Partner at GV. Jonathan Jackson is CEO of Dimagi. Robert Garry is co-founder, Matthew L. Boisen is director, and Luis M. Branco is co-founder and managing director of Zalgen Labs, a biotechnology company developing countermeasures to emerging viruses. Kristian G. Andersen has received consulting fees and compensated expert testimony on SARS-CoV-2 and the COVID-19 pandemic.

References

1. Sabeti, P.C.; Varilly, P.; Fry, B.; Lohmueller, J.; Hostetter, E.; Cotsapas, C.; Xie, X.; Byrne, E.H.; McCarroll, S.A.; Gaudet, R.; et al. Genome-Wide Detection and Characterization of Positive Selection in Human Populations. *Nature* **2007**, *449*, 913–918. [CrossRef] [PubMed]
2. Kunz, S.; Rojek, J.M.; Perez, M.; Spiropoulou, C.F.; Oldstone, M.B.A. Characterization of the Interaction of Lassa Fever Virus with Its Cellular Receptor α-Dystroglycan. *J. Virol.* **2005**, *79*, 5979–5987. [CrossRef] [PubMed]
3. Kunz, S.; Rojek, J.M.; Kanagawa, M.; Spiropoulou, C.F.; Barresi, R.; Campbell, K.P.; Oldstone, M.B.A. Posttranslational Modification of α-Dystroglycan, the Cellular Receptor for Arenaviruses, by the Glycosyltransferase LARGE Is Critical for Virus Binding. *J. Virol.* **2005**, *79*, 14282–14296. [CrossRef] [PubMed]
4. Troup, J.M.; White, H.A.; Fom, A.L.M.D.; Carey, D.E. An Outbreak of Lassa Fever on the Jos Plateau, Nigeria, in January–February 1970: A Preliminary Report. *Am. J. Trop. Med. Hyg.* **1970**, *19*, 695–696. [CrossRef] [PubMed]
5. McCormick, J.B.; Fisher-Hoch, S.P. Lassa Fever. In *Arenaviruses I: The Epidemiology, Molecular and Cell Biology of Arenaviruses*; Oldstone, M.B.A., Ed.; Current Topics in Microbiology and Immunology; Springer: Berlin/Heidelberg, Germany, 2002; pp. 75–109. ISBN 978-3-642-56029-3.
6. The 1000 Genomes Project Consortium. A global reference for human genetic variation. *Nature* **2015**, *526*, 68–74. [CrossRef] [PubMed]
7. Lassa Fever. CDC. Available online: https://www.cdc.gov/vhf/lassa/index.html (accessed on 31 May 2021).
8. Baize, S.; Pannetier, D.; Oestereich, L.; Rieger, T.; Koivogui, L.; Magassouba, N.; Soropogui, B.; Sow, M.S.; Keïta, S.; De Clerck, H.; et al. Emergence of Zaire Ebola Virus Disease in Guinea. *N. Engl. J. Med.* **2014**, *371*, 1418–1425. [CrossRef] [PubMed]
9. Gire, S.K.; Stremlau, M.; Andersen, K.G.; Schaffner, S.F.; Bjornson, Z.; Rubins, K.; Hensley, L.; McCormick, J.B.; Lander, E.S.; Garry, R.F.; et al. Emerging Disease or Diagnosis? *Science* **2012**, *338*, 750–752. [CrossRef] [PubMed]
10. Ultra-Secure Lab in Gabon Equipped for Ebola Studies | Voice of America—English. Available online: https://www.voanews.com/africa/ultra-secure-lab-gabon-equipped-ebola-studies (accessed on 8 July 2021).
11. Omilabu, S.A.; Badaru, S.O.; Okokhere, P.; Asogun, D.; Drosten, C.; Emmerich, P.; Becker-Ziaja, B.; Schmitz, H.; Günther, S. Lassa Fever, Nigeria, 2003 and 2004. *Emerg. Infect. Dis.* **2005**, *11*, 1642–1644. [CrossRef] [PubMed]

12. Asogun, D.A.; Adomeh, D.I.; Ehimuan, J.; Odia, I.; Hass, M.; Gabriel, M.; Ölschläger, S.; Becker-Ziaja, B.; Folarin, O.; Phelan, E.; et al. Molecular Diagnostics for Lassa Fever at Irrua Specialist Teaching Hospital, Nigeria: Lessons Learnt from Two Years of Laboratory Operation. *PLoS Negl. Trop. Dis.* **2012**, *6*, e1839. [CrossRef] [PubMed]
13. Sabeti, P.C. How Africa Is Fighting Back against Ebola. Available online: https://www.weforum.org/agenda/2014/05/fighting-ebola-forum-africa-2014/ (accessed on 31 May 2021).
14. Andersen, K.G.; Shapiro, B.J.; Matranga, C.B.; Sealfon, R.; Lin, A.E.; Moses, L.M.; Folarin, O.A.; Goba, A.; Odia, I.; Ehiane, P.E.; et al. Clinical Sequencing Uncovers Origins and Evolution of Lassa Virus. *Cell* **2015**, *162*, 738–750. [CrossRef]
15. Fisher-Hoch, S.P.; Khan, J.A.; Rehman, S.; Mirza, S.; Khurshid, M.; McCormick, J.B. Crimean Congo-Haemorrhagic Fever Treated with Oral Ribavirin. *Lancet* **1995**, *346*, 472–475. [CrossRef]
16. 2014–2016 Ebola Outbreak in West Africa. CDC. Available online: https://www.cdc.gov/vhf/ebola/history/2014-2016-outbreak/index.html (accessed on 31 May 2021).
17. Diehl, W.E.; Lin, A.E.; Grubaugh, N.D.; Carvalho, L.M.; Kim, K.; Kyawe, P.P.; McCauley, S.M.; Donnard, E.; Kucukural, A.; McDonel, P.; et al. Ebola Virus Glycoprotein with Increased Infectivity Dominated the 2013–2016 Epidemic. *Cell* **2016**, *167*, 1088–1098.e6. [CrossRef]
18. Wang, M.K.; Lim, S.-Y.; Lee, S.M.; Cunningham, J.M. Biochemical Basis for Increased Activity of Ebola Glycoprotein in the 2013–16 Epidemic. *Cell Host Microbe* **2017**, *21*, 367–375. [CrossRef]
19. Marzi, A.; Chadinah, S.; Haddock, E.; Feldmann, F.; Arndt, N.; Martellaro, C.; Scott, D.P.; Hanley, P.W.; Nyenswah, T.G.; Sow, S.; et al. Recently Identified Mutations in the Ebola Virus-Makona Genome Do Not Alter Pathogenicity in Animal Models. *Cell Rep.* **2018**, *23*, 1806–1816. [CrossRef]
20. About Us. ACEGID 2020. Available online: https://acegid.org/about-us/ (accessed on 31 May 2021).
21. Munshi, N. Christian Happi: 'With Pathogens, We Need to Play Offence'. *Financial Times*. 2021. Available online: https://www.ft.com/content/16c8d40f-39eb-496a-8f0f-e6761a10bbeb (accessed on 31 May 2021).
22. Siliezar, J. Responding to This Pandemic, Preparing for the Next. *Harvard Gazette*. 2020. Available online: https://news.harvard.edu/gazette/story/2020/05/pardis-sabetis-work-on-infectious-disease-coronavirus/ (accessed on 31 May 2021).
23. Ackerman, C.M.; Myhrvold, C.; Thakku, S.G.; Freije, C.A.; Metsky, H.C.; Yang, D.K.; Ye, S.H.; Boehm, C.K.; Kosoko-Thoroddsen, T.-S.F.; Kehe, J.; et al. Massively Multiplexed Nucleic Acid Detection with Cas13. *Nature* **2020**, *582*, 277–282. [CrossRef] [PubMed]
24. Myhrvold, C.; Freije, C.A.; Gootenberg, J.S.; Abudayyeh, O.O.; Metsky, H.C.; Durbin, A.F.; Kellner, M.J.; Tan, A.L.; Paul, L.M.; Parham, L.A.; et al. Field-Deployable Viral Diagnostics Using CRISPR-Cas13. *Science* **2018**, *360*, 444–448. [CrossRef] [PubMed]

Article

Lymphocytic Choriomeningitis Virus Alters the Expression of Male Mouse Scent Proteins

Michael B. A. Oldstone [1,*], Brian C. Ware [1], Amanda Davidson [2], Mark C. Prescott [3], Robert J. Beynon [3] and Jane L. Hurst [2,*]

1 Viral-Immunobiology Laboratory, Department of Immunology & Microbiology, The Scripps Research Institute, La Jolla, CA 92037, USA; brian.ware@cuanschutz.edu
2 Mammalian Behaviour & Evolution Group, Leahurst Campus, Institute of Infection, Veterinary and Ecological Sciences, University of Liverpool, Neston CH64 7TE, UK; adave@liverpool.ac.uk
3 Centre for Proteome Research, Institute of Systems, Molecular and Integrative Biology, University of Liverpool, Crown Street, Liverpool L69 7ZB, UK; markcprescott@icloud.com (M.C.P.); r.beynon@liverpool.ac.uk (R.J.B.)
* Correspondence: mbaobo@scripps.edu (M.B.A.O.); jane.hurst@liverpool.ac.uk (J.L.H.)

Abstract: Mature male mice produce a particularly high concentration of major urinary proteins (MUPs) in their scent marks that provide identity and status information to conspecifics. Darcin (MUP20) is inherently attractive to females and, by inducing rapid associative learning, leads to specific attraction to the individual male's odour and location. Other polymorphic central MUPs, produced at much higher abundance, bind volatile ligands that are slowly released from a male's scent marks, forming the male's individual odour that females learn. Here, we show that infection of C57BL/6 males with LCMV WE variants (v2.2 or v54) alters MUP expression according to a male's infection status and ability to clear the virus. MUP output is substantially reduced during acute adult infection with LCMV WE v2.2 and when males are persistently infected with LCMV WE v2.2 or v54. Infection differentially alters expression of darcin and, particularly, suppresses expression of a male's central MUP signature. However, following clearance of acute v2.2 infection through a robust virus-specific CD8 cytotoxic T cell response that leads to immunity to the virus, males regain their normal mature male MUP pattern and exhibit enhanced MUP output by 30 days post-infection relative to uninfected controls. We discuss the likely impact of these changes in male MUP signals on female attraction and mate selection. As LCMV infection during pregnancy can substantially reduce embryo survival and lead to lifelong infection in surviving offspring, we speculate that females use LCMV-induced changes in MUP expression both to avoid direct infection from a male and to select mates able to develop immunity to local variants that will be inherited by their offspring.

Keywords: pheromones; MUPs; darcin; sex; virus; CTL; selection

Citation: Oldstone, M.B.A.; Ware, B.C.; Davidson, A.; Prescott, M.C.; Beynon, R.J.; Hurst, J.L. Lymphocytic Choriomeningitis Virus Alters the Expression of Male Mouse Scent Proteins. *Viruses* **2021**, *13*, 1180. https://doi.org/10.3390/v13061180

Academic Editor: Luis Martinez-Sobrido

Received: 10 May 2021
Accepted: 18 June 2021
Published: 21 June 2021

1. Introduction

Mouse urine exhibits an obligate proteinuria, in the form of major urinary proteins (MUPs) [1–4]. These proteins, which account for over 99% of the urinary protein in a healthy male mouse, are synthesized in the liver and enter the urine via the glomerulus. MUPs are encoded by a multi gene complex on mouse chromosome 4, with at least 21 *Mup* genes that are transcriptionally active in the C57BL/6 laboratory mouse. Mouse MUPs form two subclasses: a cluster of "central" MUPs that exhibit a very high degree of sequence similarity and a smaller group of "peripheral" MUPs that each exhibit greater sequence divergence and are encoded by genes that flank the central MUP region. The major interest in MUPs stems from their biological activity in scent communication, as they provide both identity and status information about the donor animal [5–14].

Mice can detect MUPs directly on nasal contact with urine via specialized vomeronasal receptors [9]. MUPs also bind small volatile organic compounds (VOCs) in a central cavity with different affinities; the profile of expressed MUPs shapes an airborne odor profile as volatile ligands are slowly released and are detected through the main olfactory system [2–4,12]. One of the best defined MUPs is darcin (Uniprot: Q5FW60, MUP20_MOUSE), an 18,893 Da molecular weight protein that is expressed in adult male urine among wild house mice. Darcin acts as a sex pheromone that is responsible for female attraction to the male's scent [15–18] and was named after Jane Austen's romantic hero Fitzwilliam Darcy in the novel *Pride and Prejudice* [19]. Darcin also stimulates remembered attraction to the pheromone location and to the airborne odor signature that females associate with the pheromone, such that contact with darcin in a male's scent results in strong female attraction to that specific male [12,18,20]. Most of the other MUPs in male urine are the products of the central region of the *Mup* cluster. These provide distinctive MUP signatures in genetically heterogeneous wild mice that are used for individual and kin recognition [5,7,12,21]. While mice of both sexes produce urinary MUPs, investment in these scent components is around 2–4 times higher in males than in females [6], although there is a slightly greater sex bias in the *Mus musculus musculus* subspecies compared to *M.m. domesticus* [8,13,14]. Some MUPs, such as darcin and central MUP7 (Uniprot: Q58EV3_MOUSE), are produced only by males, other central MUPs can show strongly male-biased expression, while some are expressed at similar levels by both sexes [22,23].

There is some plasticity in MUP expression [6,10,11,14,24–26], but, thus far, we know relatively little about how social and environmental events influence MUP production and the biological consequences of this. Disease status can be a major factor influencing an animal's expenditure in sexual signals used in mate choice [27,28]. For example, darcin production by male house mice declines rapidly in response to immune challenge [10], although effects on the production of other MUPs have not been assessed.

Oldstone and Ware recently examined MUP production among male C57BL/6 laboratory mice infected with lymphocytic choriomeningitis virus (LCMV). They compared the effects of two variants cloned from the parental WE strain [29–31]. Variant 54 (identical to the parental WE strain) fails to generate virus-specific CD8 CTLs when inoculated into immunologically mature adult mice and leads to both a persistent infection and low MUP production in adult males [31]. By contrast, variant 2.2 (exhibiting a single amino acid change in glycoprotein residue 153) generates a robust CTL response that purges the virus and terminates the acute infection. MUP production is low during the acute infection phase, but greatly enhanced after viral clearance. There is no evidence of glomerular injury induced by infection of C57BL/6 mice with either variant, with no albumin present in urine samples during or following either infection. Instead, changes in urinary protein output specifically involve the production of MUPs [31]. As male mice invest heavily in MUPs in the scent signals they use to attract and compete for mates [6], changes in MUP expression with LCMV infection status and the ability of males to purge the virus may be signaled to females, altering a male's attractiveness as a potential mate. For female mice, assessing the LCMV status of potential mates may have immediate importance as infection during gestation can substantially reduce offspring survival in utero and lead to lifelong infection in their surviving offspring [32,33]. As the ability to develop immunity to LCMV variants depends on genotype, mate selection based on male immunocompetence would allow females to pass on good genes to their offspring. Thus, understanding the specific effects of LCMV infection on MUP signaling could provide important insight into the evolution of sexual signals and whether male MUP signals provide reliable information that could allow females to avoid a virus that could seriously impact their reproductive success. However, while the Oldstone and Ware study detected broad changes in the total amount of MUP in urine samples, estimated from densitometry on Western blots [31], this was not able to provide a more detailed understanding of how infection and recovery from infection influence the expression of the different MUP isoforms that play key roles in sexual and competitive signaling [34].

Here we report the changes in urinary MUPs that occur among immunocompetent adult males in response to acute v2.2 infection. We also assess MUP changes when adult males were challenged with v54 leading to persistent infection, or after males that had been inoculated with v2.2 at birth were given virus-specific CTL by adoptive transfer in adulthood [35,36]. We show that LCMV WE v2.2 infection has differential effects on darcin, male-specific central MUPs and MUP10, which is expressed similarly in both sexes. We find that expression of darcin is strongly influenced by infection status, but suppression of a male's central MUP identity signature is the strongest effect. However, clearance of acute infection by virus-specific CD8 CTLs, which results in lifelong immunity [31,33,35,36], leads to strong expression of darcin and the male's signature MUPs. We discuss the potential impact of these changes for female sexual attraction and present a hypothetical model showing how females might use these signals to avoid LCMV infection in themselves and their offspring.

2. Materials and Methods

Mice and Viruses: C57BL/6 male mice (6–8 weeks old) and pregnant females were obtained from the rodent breeding colony at The Scripps Research Institute. Additional urine samples were also obtained from uninfected colony males for comparison (7–60 weeks old). All mice were maintained in pathogen-free conditions and handling conformed to requirements of the NIH, The Scripps Research Institute Institutional Animal Care and Use Committee (IACUC) and the Association for Assessment and Accreditation of Laboratory Animal Care (AAALAC). MUP and darcin studies utilized only males. Newborn mice were inoculated with 1×10^3 PFU virus intracranially within 18 h following birth. Sexually mature males in late adolescence (7–8 weeks old) mice were injected with 2×10^6 PFU virus intravenously [31]. Viruses used were WE strain and its variants 2.2 and 54. The origin of these viruses, their growth and quantitation have been reported [29–31]. Virus carried in blood and tissues were quantified in a plaque forming assay [29–31]. Assayed samples were diluted ten-fold and tested in triplicate.

CTL Assay: virus-specific CTL assay was performed as previously reported [31]. Days 7−8 post-infection were the timing used to measure acute CTL response. Deletion of CD8 CTL utilized monoclonal antibody YTS.1694 [31]. Deletion was 98%. Generation of memory CD8 CTL, their harvest and i.p. transfer of 2×10^7 T cells to persistently infected mice have been reported [31,36].

Biochemical Assays: The virus overlay protein blot assay, Western blots and urine collections have been reported [31,37]. Protein assays, creatinine assays and electrospray ionization mass spectrometry (ESI-MS) to resolve different MUP phenotypes were performed as described [4,22,23,38–40]. For each urine sample, we calculated the total area of mass peaks in the ESI-MS spectra that corresponded to known MUP masses (range of 18,000–20,000 Da) and expressed each mass peak as a proportion of the total to examine changes in the relative proportion of different MUPs independent of total MUP output. Several MUPs that share very similar masses of 18,692–18,694 Da could not be distinguished in intact mass spectra (MUPs 1, 12, 2 and 15 and identical proteins 9, 11, 16, 18 and 19) and, thus, were analysed as a single group ($MUP9^{\#}$). To estimate the amount of each MUP mass expressed, the proportion of each mass peak was multiplied by the total urinary protein output [22].

Statistical Analysis: Data were analyzed using SPSS version 25 (IBM software). Groups were compared using ANOVAs, checking that residuals from each model approximated a normal distribution and log transforming variables where necessary to meet assumptions of parametric analyses. Repeated measures ANOVAs were used where samples were collected serially from the same individual males after CTL adoptive transfer. Bonferroni post-hoc comparisons were used to compare differences between multiple time points and a sequential Bonferroni adjustment was applied to significance levels across multiple MUP mass comparisons to correct appropriately for multiple comparisons.

3. Results

Earlier kinetic studies of response to LCMV WE v2.2 in sexually mature male mice [31] indicated that MUPs were present at low levels at days 5 or 10 post-infection but rose by day 20 and peaked at day 30. This strong MUP expression at day 30 post-v2.2 infection was maintained over the next 30 days of analysis. However, SDS-PAGE separation provides minimal resolution of different MUPs. Apart from a conformationally accelerated mobility of one MUP (darcin) and a slow migrating diffuse band of glycosylated MUPs, the majority of the abundant MUPs in urine migrate as a single, strongly staining band. The sequence similarity of the central MUPs makes individual identification by proteomics particularly challenging [4,22]. However, many of the MUPs have unique masses that are revealed by high resolution electrospray ionization mass spectrometry (ESI-MS) of intact MUPs in desalted urine samples. The ability to resolve several MUPs by mass offered an opportunity to explore the differential changes in expression of individual MUPs in response to viral infection and clearance. Here, we use ESI-MS to profile several MUP groups in urine during viral infection and after clearance. To allow assessment of changes in MUP output at a finer scale, we also determine total protein output corrected for variation in the dilution of urine samples by measurement of urinary creatinine [3,6].

3.1. *Normal Male MUP Pattern and Output*

To establish a reference pattern, we first characterised the typical MUP output and profile of MUP isoforms expressed by uninfected males in the C57BL/6 local stock colony. Urinary protein levels of stock uninfected adult males were very typical of males of this strain [3,41], at around 10 mg/mg creatinine (Figure 1A). MUPs account for almost all of this high urinary protein output among healthy male mice (Figure 1B). Adult mice express a relatively fixed profile of different central MUP isoforms according to MUP genotype that varies between the sexes [23,42]. Uninfected colony males showed a MUP profile that is typical of C57BL/6 strain males (Figure 1C). In addition to a characteristic pattern of central MUPs, adult breeding and non-breeding male C57BL/6 also expressed darcin (MUP20; 10.51 $\pm$ 0.74% of total MUP, Figure 1D). However, young non-breeding males less than 12 weeks old had significantly lower MUP output (Figure 1A,B); these males express an immature MUP profile that consisted largely of MUP10 (Figure 1C,D). MUP10 is a central MUP that is produced at similar levels by both male and female C57BL/6 mice [22]. These immature non-breeding males did not express appreciable levels of MUP7, MUP9$^{\#}$ (a group of MUPs sharing similar mass that cannot be separated by ESI-MS), MUP14, or darcin, all of which are expressed at much higher levels in mature adult males under androgen control. Notably, though, 8–10-week-old colony males that were housed with females for breeding had very similar MUP output to older colony males (Figure 1A,B) and expressed a pattern of central MUPs and darcin that is typical of mature adult males (Figure 1C,D).

3.2. *Clearance of LCMV Infection Leads to Enhanced MUP Output*

Infection of 8-week-old C57BL/6 males with LCMV WE v2.2 initially leads to low MUP output over the first 10 days, but this increases significantly after males clear infection and peaks at day 30 post-infection (see Figure 3). We quantified the changes in MUP output among 18 matched pairs of singly housed littermate males where one male of each pair was recovering from v2.2 infection while its littermate control sib was singly housed at the same time without infection. Urine was sampled at day 20 and day 30 post-infection.

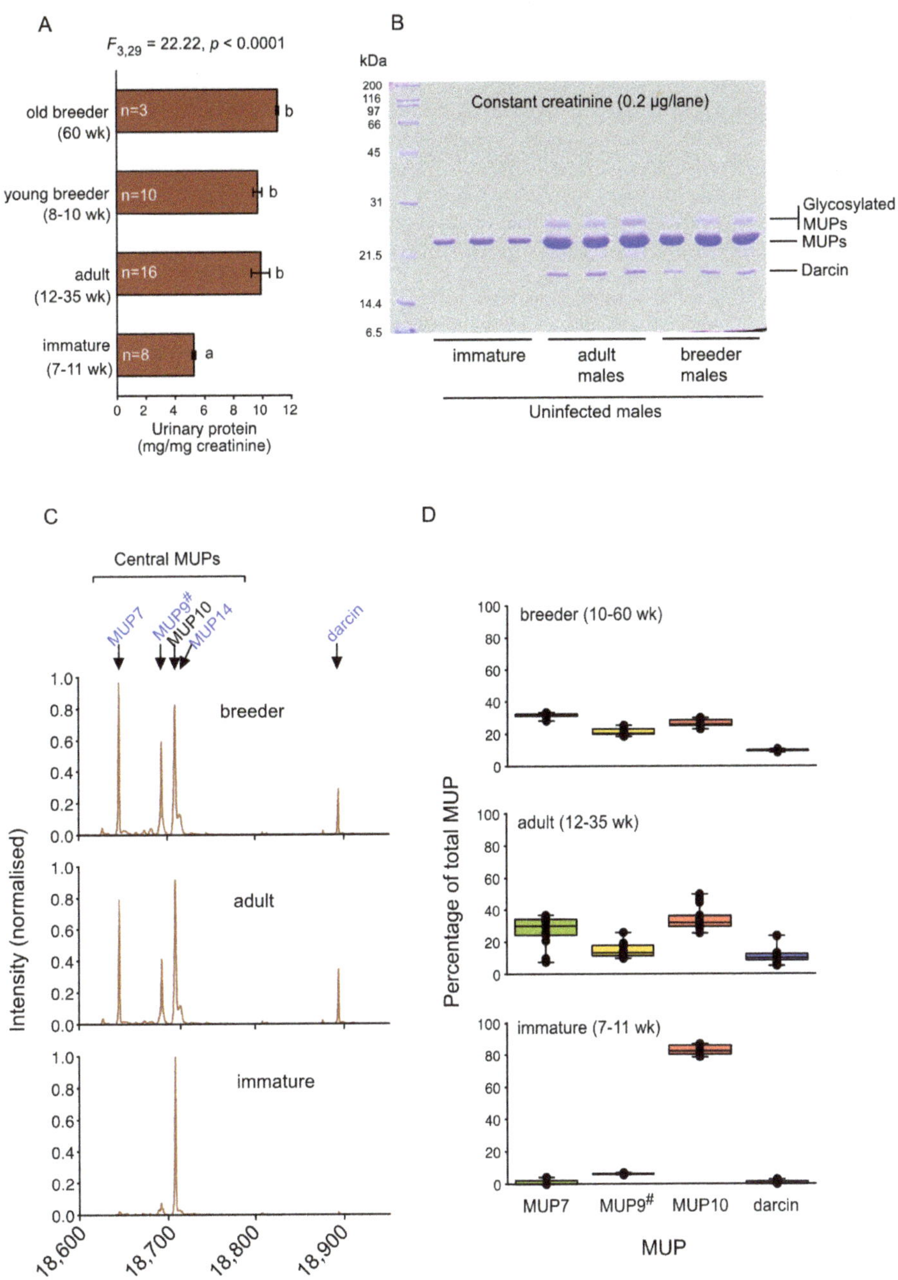

Figure 1. Urinary MUP output in urine of uninfected C57BL/6 male colony mice. (**A**) Total urinary protein output for a sample of immature and mature adult males housed in single-sex groups or with females for breeding. Protein is normalised to creatinine in urine to correct for variance in urine dilution. Different letters (a,b) indicate a significant difference between groups according to Bonferroni post-hoc comparisons ($p < 0.0005$). (**B**) Visualisation of urinary protein on reducing SDS PAGE, revealing the major MUP band, a high mobility darcin band, a slower mobility band of glycosylated MUPs and the

absence of any other major protein bands in the urine samples (immature males aged 7 weeks; adult males aged 12, 19 or 26 weeks; breeder males aged 8, 10 or 60 weeks). (**C**) Resolution of MUPs by ESI-MS. The plots are deconvoluted spectra of the high abundance MUPs averaged over all urine samples shown in panel A for each type of male. Each peak represents the mature mass of a known MUP or is a coalescence of multiple MUPs with identical or very similar masses ($MUP9^{\#}$). (**D**) Expression of each of the major MUP peaks as a percentage of total MUP in each sample for immature, adult and breeding males as shown in panel A (median, 25th, 75th percentiles, full range and individual data points). MUPs 13, 14 and 17 are expressed at low level in this strain and are not shown. $MUP9^{\#}$ refers to the group of central MUPs with masses of 18692–18694 Da that could not be resolved (MUPs 1, 12, 2 and 15 and identical proteins 9, 11, 16, 18 and 19). MUP3 is glycosylated and requires alternative approaches for quantification that were not conducted in this study.

Males recovering from v2.2 infection showed a significant increase in total MUP output between day 20 and day 30, while uninfected control littermates had high MUP output at day 20 (males aged 11 weeks) which reduced by day 30 (interaction between infection and time point, $F_{1,34} = 29.7$, $p < 0.0001$; Figure 2A). MUP output was very variable among infected males at day 20 post-infection, ranging from 2.1 to 17.4 mg protein/mg creatinine; approximately half the males had regained a level of MUP output similar to matched controls, but others still had low output (Figure 2A). However, by day 30, previously infected males consistently expressed a high level of MUP (range of 9.8–15.5 mg/mg creatinine), which was very similar to the level of output among control males at day 20 (Figure 2A). Surprisingly, though, the output of many of the uninfected sib control males declined to very low levels by day 30 (8/18 males), while others still had a fairly typical level of output for males of this strain.

To examine changes in the output of different MUPs, we used the area of each mass peak in the ESI-MS intact mass spectrum of each sample to estimate the proportional contribution of different MUPs to the total protein output. At day 20 post-infection, only MUP14 had significantly lower output in infected males compared to controls (Figure 2B), a male-specific MUP that has relatively low-level expression in C57BL/6 males [22]. Other MUP masses showed highly variable levels of expression between individuals, reflecting the variability in total protein output among both control and infected males at day 20. However, with the exception of MUP10, output of each MUP mass by day 30 was very significantly higher among males that had recovered from v2.2 infection compared to their uninfected control sibs ($p < 0.0001$; Figure 2C). The output of MUP10 did not change significantly with infection or between sample time points.

3.3. *Enhancement of Adult Male MUP Pattern and Darcin*

We used principal components analysis to examine the main patterns of variation in the total urinary protein output and the proportion of each MUP mass expressed. This derived two components with eigenvalues >1 that explained 79.9% of variance in the dataset. PC1 (62.1% of variance) reflected a strong contrast between the proportion of MUP10 expressed versus the total MUP output and the proportion of each of the other mass peaks in the male's profile (Figure 2D). This reflected a consistent change in MUP profile with the total amount of MUP expressed: the higher the total MUP output, the lower the proportional contribution of MUP10 and the greater the proportion of MUPs that are expressed specifically by mature adult males (see Figure 1). Scores for PC1 showed a strong interaction between infection and time post-infection (Figure 2E). At day 20, males recovering from v2.2 infection still had lower MUP output due to much lower expression of male-biased central MUPs compared to control males. However, by day 30 males that had recovered from v2.2 infection had much stronger MUP output and expression of male-biased central MUPs than control males. Notably, though, the loading of darcin on PC1 was relatively low.

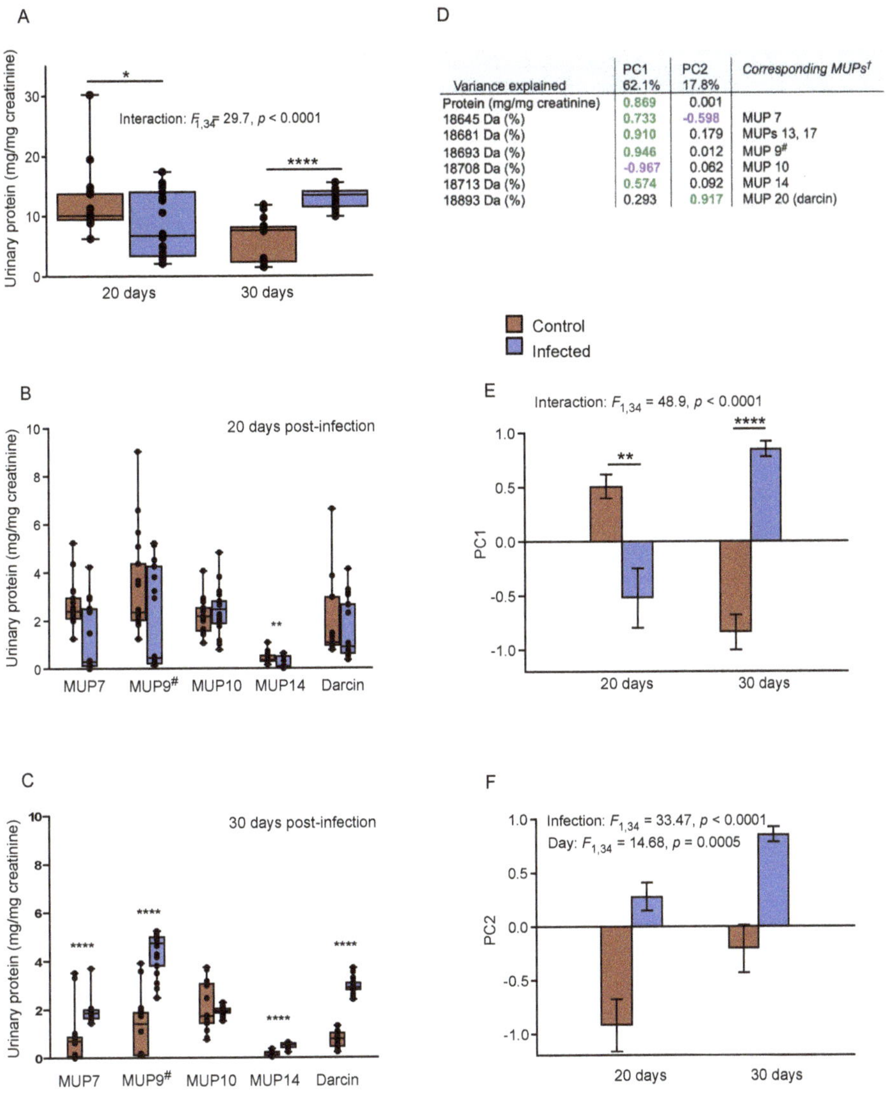

Figure 2. Changes in MUP output among adult males after acute LCMV v2.2 infection. Pairs of sib C57BL/6 males were singly housed at 8 weeks old and one male of each pair inoculated with v2.2 as previously described [31]. MUP output and profiles were analysed from urine samples taken at day 20 and day 30 post infection (N = 18 pairs per sample time point). Panel (**A**): total urinary protein output, corrected for urine dilution. Panels (**B**) and (**C**): amount of each major MUP peak expressed at day 20 or day 30 post infection, estimated from the total output and proportional area of each mass peak in the intact mass spectrum for each sample. Panels (**A–C**): plot medians, 25th, 75th percentiles, full range and individual data points. Panel (**D**): principal components analysis based on total protein output and proportion of each MUP mass, indicating loadings for each of the principal components derived (PC1 and PC2). Panels (**E**) and (**F**): PC1 and PC2 scores for infected and control males at day 20 and day 30 post-infection (mean ± s.e.m.). MUP9# refers to the group of central MUPs of very similar or identical mass (see Figure 1). * $p < 0.05$, ** $p < 0.01$, **** $p < 0.0001$. Sequential Bonferroni correction was applied to analyses in panels (**B**) and (**C**) to correct for comparison of multiple MUP peaks.

PC2 (17.8% of variance) was not influenced by total MUP output but instead reflected a strong contrast between the proportion of darcin versus the proportion of central MUP7 in a male's MUP profile. Both darcin and MUP7 are expressed almost exclusively by mature adult males among both laboratory and wild house mice [11,14,23,24,43,44]. However, these two MUPs have distinct functions, with MUP7 contributing to the pattern of very similar central MUPs in mouse urine that provides individually distinctive odour signatures in wild mice, while darcin is the sex pheromone that induces inherent female attraction to these odour signatures [12,18]. PC2 scores were significantly higher in males recovering from v2.2 infection compared to their uninfected control sibs at both day 20 and day 30 (Figure 2F). Thus, males expressed a higher proportion of darcin compared to central MUP7 following v2.2 infection. There was also a general increase in the proportion of darcin to MUP7 between day 20 and day 30 in both infected and control males (Figure 2F), consistent with an age-related increase in the proportion of darcin relative to MUP7. As both the total MUP output and the proportional expression of darcin in the profile had greatly increased in males by day 30 post-infection, there was a particularly strong difference in the total amount of darcin expressed by males after recovery from v2.2 infection compared to matched control sibs (Figure 2C).

Variance in the extent to which males infected with v2.2 had recovered a normal male MUP output by day 20 was high (Figure 2A,B). We examined whether this was related to differences in the viral titer that males experienced during infection, or to how quickly males cleared the infection. Serum viral titers were sampled at days 3, 7, 10, 20 and 30 after v2.2 inoculation. Viral titers were highest at day 7, ranging from 8500 to 24,900 PFU/mL serum. All males had cleared infection by day 20, but 10/18 males still expressed a very low serum viral titer at day 10 (200–2000 PFU/mL). There was no significant relationship between serum viral titer at day 7 and the total amount of MUP expressed at either day 20 or day 30 post-infection. However, males with the highest viral loads at day 7 tended to express the lowest proportion of darcin at day 20 ($r_{17} = -0.46$, $p = 0.053$), although there was no difference by day 30 ($r_{17} = 0.01$, $p = 0.97$). Males that had cleared infection by day 10 also tended to have slightly higher MUP output by day 30, but this was not statistically significant ($F_{1,16} = 3.46$, $p = 0.08$) and early clearance of infection did not lead to more darcin expressed at either day 20 ($F_{1,16} = 0.41$, $p = 0.53$) or day 30 ($F_{1,16} = 1.05$, $p = 0.32$). Thus, we found only limited evidence that viral titer and speed of clearance influenced individual variation in how quickly males regained a normal mature male MUP expression, but all males had a fully developed mature male output by day 30.

3.4. Adoptive Transfer of CTL during Persistent Infection

Virus-specific CD8 CTLs play an essential role in the clearance of infectious virus and in the subsequent enhancement of MUP output [31]. Males with persistent infection, either from inoculation with v54 in adulthood or with v2.2 within 18 h of birth, lack a functional CD8 CTL response to the virus and have low adult MUP output. Importantly, providing virus-specific CTL by adoptive transfer to mice with persistent infection leads to viral clearance and a subsequent increase in MUP output [31]. Here we look at the changes in MUP output with time after adoptive CTL transfer in v2.2 persistently infected mice (inoculated within 18 h of birth) and compare this with output after adult infection with v2.2 (which stimulates a natural CD8 CTL response and recovery) or with v54 (no virus-specific CD8 CTL response leading to persistent infection) among males of matched age (Figure 3).

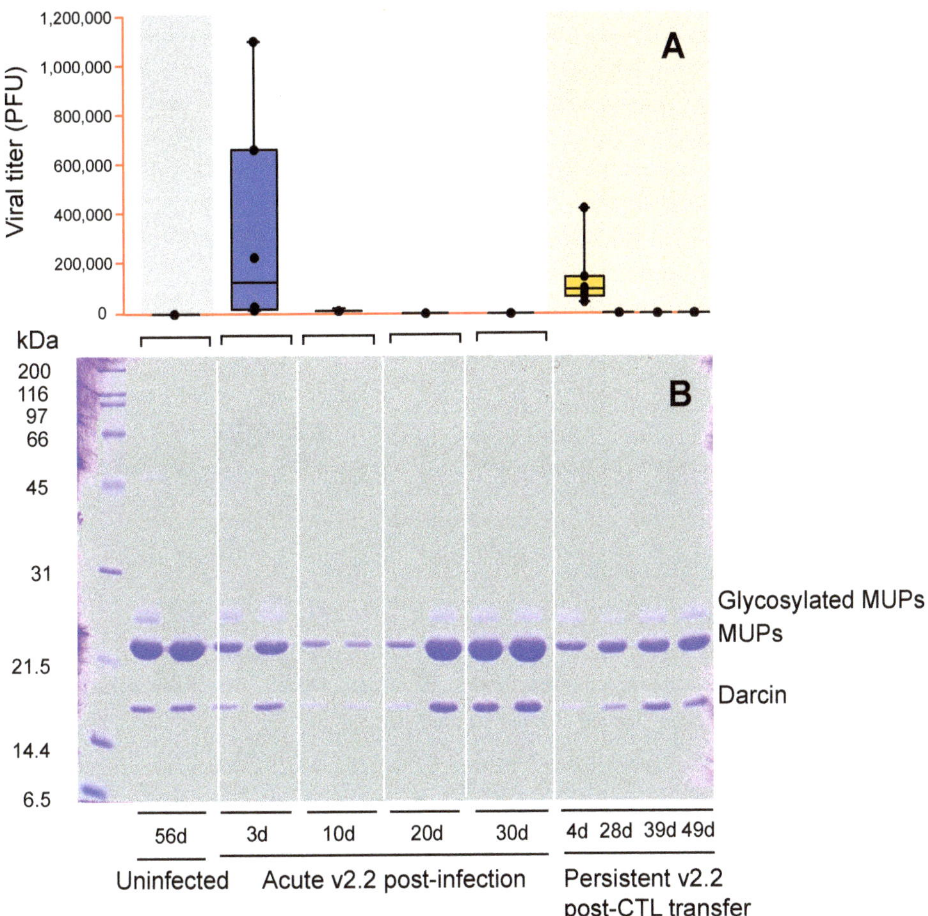

Figure 3. Typical urinary protein output during and after clearance of acute or persistent LCMV v2.2 infection. At 8 weeks of age, uninfected males were inoculated with v2.2 and blood and urine samples were analysed for viral titer (**A**) and urinary proteins ((**B**), MUPs resolved by SDS-PAGE) at days 3 or 7, 10, 20 or 30 (N = 6 males per time point). Another group of males (N = 7), inoculated with v2.2 within 18 h of birth to generate persistent infection, were treated with adoptive virus-specific CTL transfer at 8 weeks of age and sampled during infection and recovery.

Adoptive virus-specific CTL transfer into males with persistent v2.2 infection results in a substantial drop in viral titer by day 11 post-transfer, with no virus evident in serum at days 28, 39 and 49 (Figure 3A). Urinary protein output differed significantly between these time points ($F_{4,20} = 18.43$, $p < 0.0001$), due to a rise in MUP output that occurred between day 28 and day 39 post-transfer (Figures 3B and 4A). The rise in MUP output following clearance in adulthood of a persistent v2.2 infection that was initiated at birth was slower than that observed after natural clearance of acute v2.2 infection that was initiated in adult mice. The output did not quite reach the same high level of MUP investment that is typical of healthy adult males of this strain (Figure 4A).

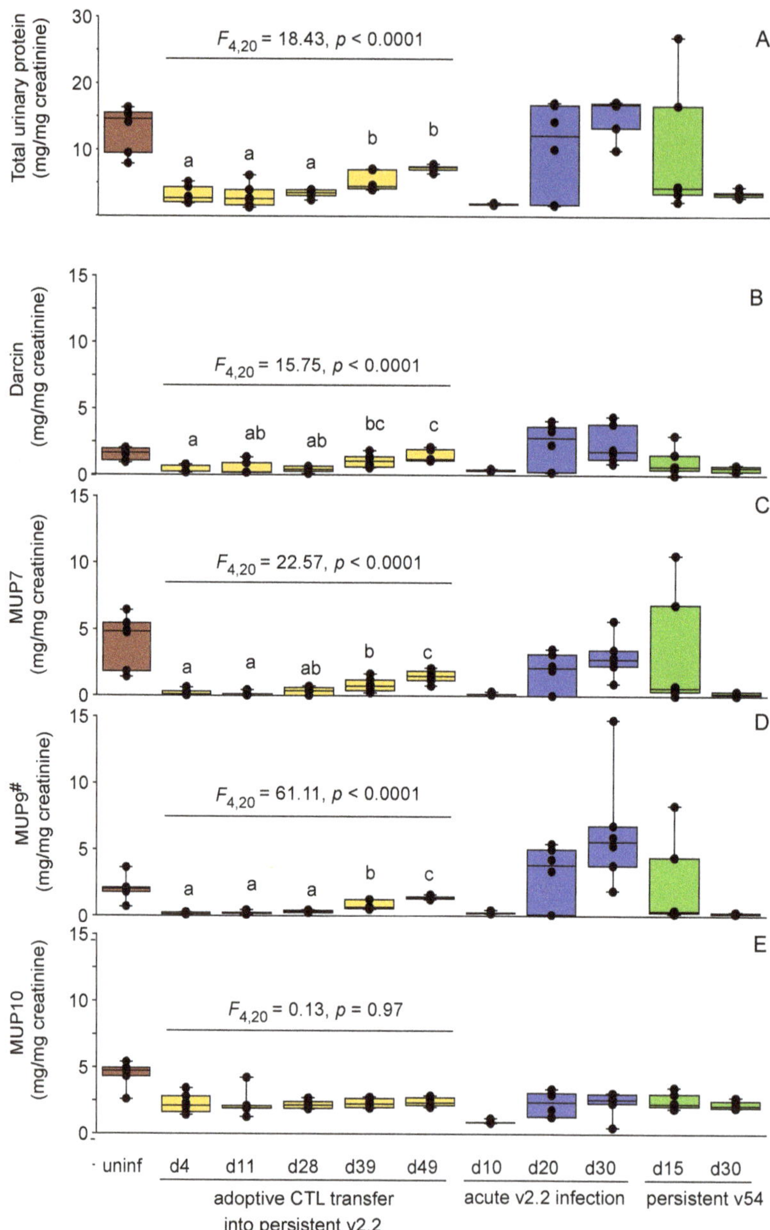

Figure 4. Changes in MUP output during and after CTL clearance of persistent or acute LCMV infection. Males with persistent LCMV v2.2 infection from birth (yellow) were tracked through viral clearance by adoptive CTL transfer (N = 7, see Figure 3 for viral titers). This was compared with males of the same age undergoing acute v2.2 infection and natural clearance (blue), inoculated with a persistent v54 infection (green) or uninfected (red) (N = 6 males per time point). (**A**): total urinary protein corrected for urine dilution. (**B–E**): amounts of each major mass peak (darcin, MUP7, MUP9$^{\#}$ and MUP10, respectively), corrected for urine dilution. Different letters above each column for adoptive CTL transfer experiment indicate time points that differ significantly using Bonferroni post-hoc comparisons ($p < 0.05$). MUP9$^{\#}$ is used to refer to the group of central MUPs of very similar or identical mass (see Figure 1).

The pattern of central MUPs expressed by males during persistent v2.2 or v54 infection, or during an acute adult v2.2 infection (day 10), was similar to that of adolescent males that have an immature MUP pattern; the profile is dominated by MUP10 (non sex-specific) with very little expression of androgen-dependent central MUPs normally expressed by mature adult males (Figure 5). However, while adolescent males generally express very little darcin (Figure 1C,D), adult infected males retain darcin expression during infection, even when their total MUP output is very low (Figures 4B and 5). Following adoptive CTL transfer into males with persistent v2.2 infection, the amount of MUP10 produced remained constant over all five time points sampled ($F_{4,20} = 0.13$, $p = 0.97$; Figure 4E), but expression of mature male central MUPs (MUP7, MUP9#) and darcin slowly increased (Figure 4B–D). By day 49 post-transfer, males expressed a profile of MUPs typical of healthy adult males (Figure 5).

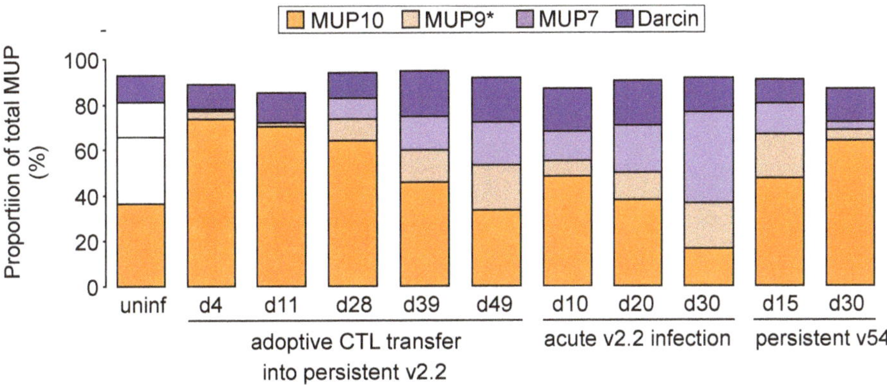

Figure 5. Proportional changes in MUP profile during LCMV infection. The MUP profiles from Figure 4 were expressed in relative terms, with each MUP mass displayed as a percentage of the contribution to total MUP output. Data are means for each treatment group and time point, sample sizes as in Figure 4. Uninfected males were of equivalent age and singly housed.

Thus, persistent LCMV WE infection led to a dramatic reduction in MUP output that affected expression of all MUPs, but the strongest impact was on central MUPs that have strongly male-biased expression in healthy adult mice (MUP7, MUP9#; Figure 5). Output of central MUP10, which is not sex-specific in C57BL/6 mice, decreased during acute v2.2 viraemia, but otherwise was maintained at quite consistent levels of expression (Figure 4E). The proportion of darcin in each male's MUP profile showed relatively limited variation across infection and recovery, with the amount expressed largely tracking infection-related changes in total MUP output, although the proportion of darcin was significantly elevated on recovery from infection and was lower in younger adult males.

4. Discussion

We had previously shown suppressed MUP output among adult C57BL/6 or FVB/N male mice that were unable to generate a virus-specific CD8 T cell response to LCMV WE v54 infection and were unable to clear the virus infection. By contrast, generation of a biologically active CTL response that cleared v2.2 infection led to a substantial increase in MUP investment which was greater than that of matched singly housed control males [31]. Here we report that such changes are not equal across all MUP isoforms, with infection differentially affecting the expression of the male pheromone darcin, central MUPs that have strongly male-biased expression and MUP10 that does not show sex-biased expression.

MUPs reflect a significant proportion of the protein synthesis in mouse liver and MUP mRNA may constitute as much as 5% of the total mRNA pool in male laboratory mice [45]. The irreversible loss of protein amounts to several 10s of milligrams a day for mature male mice, which imposes a significant energetic demand on the liver. Any changes that substantially influence the ability of this tissue to translate and secrete proteins are likely to impact on MUP output, such that MUP signals in mouse urine may reflect shifts in hepatic metabolic status. For example, calorie restriction can substantially diminish MUP production [46–48], probably mediated through growth hormone. Metabolic dysfunction in the diabetic obese mouse can also diminish production of specific MUPs (MUP1, [49]). Experimental models of murine schistosomiasis, which causes liver dysfunction and the disruption of major metabolic pathways, substantially diminish overall MUP production according to the severity of disease [50,51]. Male MUP output also declines with senescence, correlating with a decline in epididymal sperm counts and with reduced attractiveness of male urine signals to females [52]. Lopes and Konig [10] show that administering an immune challenge by injection of lipopolysaccharide reduces darcin production among captive bred wild males, as well as reducing male activity and ultrasonic signaling, although the production of other MUPs was not assessed. However, darcin output was much less diminished in mice parasitized with *Aspiculuris tetraptera* [53]. The overall picture is that there may be multiple levels of control on MUP output, reflecting overall energetic demand of their synthesis, mediation through altered endocrine signals, or effects that are specific to particular isoforms to elicit specific responses. However, very few studies have addressed how this differential control alters male MUP signals and the impact that this has on their functions in competitive scent signaling and female mate selection.

Infection of adult males with LCMV WE v2.2 led to an early and substantial drop in male MUP production during acute infection, but males recovered expression of a normal mature male MUP pattern with consistently strong output by 30 days after the initial infection. This contrasted with a number of control sibs that showed a substantial decline in MUP investment after 30 days of single housing, including a major reduction in expression of androgen-dependent MUPs, such that their output resembled that of immature mice. This much reduced expression among some control males may have been a response to prolonged social isolation, which can induce a range of endocrine, brain and behavioural changes in mice, including an increased physiological response to stress, chronic activation of the HPA stress pathway and increased anxiety and depression-like behaviour [54]. However, in our wild house mouse colony, singly housed adult males that are regularly exposed to conspecific scents typically retain a normal mature male output of 10–30 mg MUP/mg creatinine [52,55]. The experience of competitive breeding conditions in enclosures stimulates males to increase their investment in MUPs even further [11,14,55], with outputs in some males rising to over 100 mg MUP/mg creatinine in semi-natural populations, although the level of investment varies widely between individuals (range of 10–113 mg/mg creatinine, [6]), with socially dominant males having the highest MUP outputs [11,14]. MUP output shows a similar wide range among males captured from the wild [8]. Under laboratory conditions, uninfected healthy males of most laboratory strains including C57BL/6 typically express around 10 mg MUP/mg creatinine [3,6,41], very similar to MUP levels expressed by uninfected mature males in the colony used in this study. After recovery from v2.2 infection, all males fully recovered a high level of MUP investment (mean of 13.8, range of 9.8–17.8 mg/mg creatinine), including relatively high levels of darcin and male-biased central MUPs. This suggests that mounting an immune response and clearance of v2.2 challenge enhanced subsequent outlay in MUPs, in contrast to the decline in MUP output among singly housed control males that were socially isolated for the same length of time.

Both darcin and central MUPs function together to provide an important male competitive sexual signal in mice. The effects of LCMV infection on these signals may help to explain why the prevalence of LCMV is relatively low in wild mouse populations despite both horizontal and vertical transmission of this virus. In nature, LCMV can be transmitted horizontally to conspecifics through contact with contaminated excretions (urine, faeces, saliva, tears, semen, milk). This usually leads to acute transient infection in naïve animals, which is cleared primarily by a virus-specific $CD8^+$ cytotoxic T-cell response. However, most natural transmission is likely to be vertical, by in-utero transfer from mother to offspring, which leads to lifelong infection in the offspring [33,56]. Some LCMV WE variants, such as v54, can also induce persistent infection in adults by aborting the virus-specific CD8 T response [31,33]. As offspring of infected mothers become persistent infective carriers and naïve females can easily become infected via horizontal transmission through copious virus in excreta, we should expect the prevalence of LCMV in mouse populations to increase rapidly and reach an equilibrium close to 100%. However, surveys indicate that LCMV prevalence typically is relatively low among wild-caught mice. Approximately 9% of house mice were seropositive for LCMV in urban sites around Baltimore [57], 9.5% in a survey of farm and zoo populations in UK [58], 7% around the port of Yokohama [59] and 3.6% in a survey of house mice in Germany (cited in [57]). A number of mechanisms might contribute to reducing LCMV prevalence in mouse populations. For example, infected animals may have reduced survival under more challenging conditions outside of the laboratory. However, our findings suggest that LCMV infection is likely to have a substantial impact on female mate choice via effects on competitive male sexual signals. If this is the case, sexual signaling and mate choice could play a major role in limiting the spread of this pathogen in natural mouse populations.

Our hypothesis is represented in Figure 6. Adult male house mice advertise their competitive ability to females through competitive scent signaling. This involves close interaction between the darcin sex pheromone (expressed by all mature males) and a male's individual signature which is encoded by a set of polymorphic central MUPs that bind and slowly release a male's airborne odour signature [12,38,60]. Males invest heavily in scent-marking their territories with urine that contains a strong darcin signal together with much more substantial investment in their individual-specific central MUP signature; with frequent refreshment, these scent marks continually release the male's airborne signature. Females preferentially approach and investigate the strongest scent signals in the vicinity [38,61]. When females detect darcin on nasal contact with a male's scent, the pheromone induces rapid learned attraction to the male's associated odour signature, shaped by the male's MUP profile [12,18]. They also learn attraction to spatial cues associated with the darcin pheromone location [15,20]. Subsequently, females evince a remembered attraction to that male's odour and his location, which both focuses and amplifies female attraction to a male that can dominate and scent mark the local territory (Figure 6A). In the presence of females and competitors, males increase the strength and duration of their scent signals by increasing both their MUP output and rate of scent refreshment (e.g., [55]).

However, MUP output is compromised in males with a persistent congenital LCMV infection gained from an infected mother. These males produce less darcin, although still at a level that could induce a positive female response [12,16]. Of greater significance, their total MUP output is substantially reduced, largely due to suppression of their central MUP signature (Figure 6A). This is likely to have several consequences for sexual signaling and for the infection risk that males pose to susceptible females. First, low MUP output will reduce the strength and duration of airborne odours released from urine marks that stimulate females to approach and contact a male's scent [52]. This direct scent contact is essential for females to detect the non-volatile darcin pheromone and for females to learn any attraction to a male's odour or location [18,20,62,63]. Secondly, even if females do contact and detect darcin in the male's scent, suppression of the male's polymorphic individual signature suggests that females will be unable to learn attraction to a specific individual odour associated with that male [12,18]. Thus, the changes in MUP production in persistently infected males predict that their scent signals will not stimulate normal sexual attraction from females. Male MUP signals are similarly compromised if susceptible uninfected males become infected after the perinatal period but are unable to generate an effective CTL response against a particular LCMV variant, again leading to persistent LCMV infection (Figure 6B). However, adult males that are able to mount an effective virus-specific CTL response and clear the virus subsequently produce strong MUP signals.

After recovery from infection, males not only produce a relatively large amount of darcin, but they also invest particularly heavily in the central MUP signature that broadcasts their individual odour through scent marks. This will increase the likelihood that females will detect, investigate and learn attraction to the male's individual scent signature and location (Figure 6B).

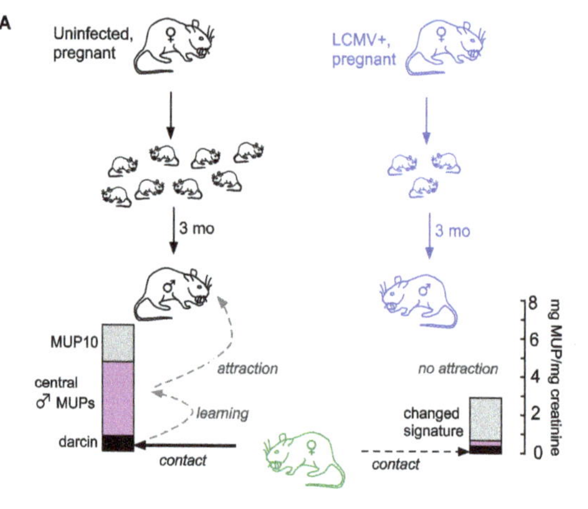

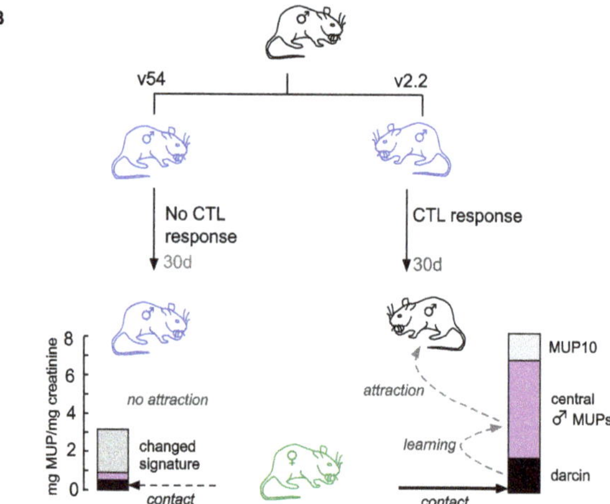

Figure 6. Effects of LCMV infection on male MUP expression and expected influence on female attraction. The cartoon depicts MUP signals produced by males that are infected with LCMV (blue mice) or uninfected/recovered from LCMV infection (black mice), with the predicted response of an adult female prospecting for a mate (green mice). Bars indicate the amount of darcin (black), male-biased central MUPs (purple) and central MUP10 (grey) produced by males of each type (data are mean mg MUP/mg creatine produced by singly housed C57BL/6 males in this study). (**A**) Fewer offspring survive from infected mothers and carry persistent infection with very weak male MUP signals when adult. Uninfected males produce much stronger MUP signals, including a central MUP signature and darcin, which stimulates females to contact scent and learn attraction to the individual owner. (**B**) Males that are able to clear LCMV infection gained in adulthood produce much stronger MUP signals including an individual signature and darcin, stimulating females to contact and learn attraction to the individual owner. See text for further details.

From a female's perspective, an ability to mediate attraction to potential mates by using male sexual signals that are highly sensitive to LCMV infection status would reduce the likelihood that susceptible females will acquire infection from a mate. Mating and gestation are likely stages of vulnerability for female fitness. Laboratory studies suggest that maternal LCMV infection during early gestation, particularly in the first trimester, leads to high rates of abortion, foetal resorption and increased pup death through starvation [32,64], compromising the immediate survival and growth of pups. Our findings suggest that persistent infection among surviving pups also could greatly compromise the offspring's future reproductive success. However, while our study has demonstrated clear effects of LCMV infection on male sexual signals in the laboratory, it should be acknowledged that the impact on reproductive success in natural populations is not yet known. In addition to the immediate infection risks for females seeking mates (a risk only for susceptible females), males vary in their ability to overcome horizontal LCMV infection and gain immunity according to both their genetic background and the specific LCMV variant [31,65]. To avoid passing genes to their offspring that may be ineffective in overcoming local LCMV variants, females should avoid infective mates that have not yet gained immunity, even if the female herself is not susceptible to infection transmitted by the male. This raises the intriguing possibility that both MUP and MHC responses to viral infection may interact in influencing female responses to male scent signals at this critical time. An MHC H-2 complex is essential for the generation of a virus-specific CD8 CTL response to acute viral infection to clear the virus. Between 15 and 20 days after CTL generation, the majority of CTLs are decreased to form virus-specific memory CTLs [66]. These memory CTLs respond more quickly, strongly and effectively in preventing reinfection with the same or a cross-reacting virus. Thus, once immune, males will be resistant to virus reinfection and provide females with little risk of infection through mating or through subsequent contact with the male or his excreta. However, when susceptible adult males become infected, viral loads increase before male MUP output is suppressed (see Figure 3, d3 post v2.2 infection, and Figure 4, d15 post v54 infection). Not only does this put susceptible females at risk if they have been attracted by scent from an apparently healthy male, but there will be no proof that the selected mate has the ability to overcome infection with that particular variant.

Here, we speculate whether changes in a male's scent that are indicative of LCMV infection may have the potential to induce pregnancy failure, known as the Bruce effect, in recently mated females [67]. This would allow females to avoid investing in pups from sires that may not be able to develop LCMV immunity, as well as avoiding gestation if they are susceptible to LCMV infection. In laboratory experiments, females form a memory of the chemosignals of their mate specifically in response to vaginocervical stimulation during mating [68,69]. Over the next few days, exposure to chemosignals from an unfamiliar male in the absence of their familiar mate's imprinted chemosignals inhibits prolactin surges in newly inseminated females that are needed to maintain luteal function during early pregnancy, resulting in pre-implantation pregnancy failure [70,71]. However, the presence of remembered chemosignals, either from the familiar mate, or from genetically identical males of the same strain, prevents the unfamiliar male scent from blocking the female's pregnancy. The functional significance and evolutionary advantage of this response for females under natural conditions has been widely debated but remains enigmatic [70,72,73]. Newly mated females are sensitive to a variety of chemosignal differences compared to the imprinted mate's scent, each of which can stimulate pregnancy block when manipulated separately on the mate's background scent, including differences in MHC peptides in male urine [74], volatile odour signatures [75,76], low molecular weight ligands bound to MUPs [77], the level of the peptide ESP1 excreted in mouse tears [78] and polymorphic nonformylated NADH dehydrogenase peptides synthesised in mitochondria [79]. Studies of pregnancy block in mice so far have focused on chemosignal differences between different males, with the proposed explanation that females might block pregnancy if their mate disappears and is replaced by a foreign male, potentially allowing the female to remate with a better male or to avoid potential infanticide from the new male before making gestational investment [80–82]. However, some of these chemosignal components may also change within sires that undergo an acute LCMV infection during the critical period after a female has mated. Given the potentially very high costs to female reproductive success if mothers become infected with LCMV in early gestation, pregnancy block could be a highly effective strategy for mothers to delay investment in offspring until they have developed immunity. Further, as the ability of susceptible offspring to develop immunity to LCMV infection postnatally will depend on their inherited genotype, pregnancy block could prevent investment when it is unclear if sires can clear an acute infection. The suppression of a male's individual MUP signature is very likely to change his profile of low molecular weight MUP ligands, such that females no longer recognise the male's airborne odour [12]. As low molecular weight MUP ligands from an unfamiliar laboratory strain donor are highly effective in blocking embryo implantation [77], we speculate that this change in a male's own individual signature may be sufficient to induce pregnancy block (Figure 7). Recently mated females are also highly sensitive to the presence of foreign MHC peptides in male urine [74,83], so it would be intriguing to test whether females can directly detect the presence of viral peptides in a male's urine. However, other components of male chemosignals might also be particularly sensitive to LCMV infection as LCMV replicates not only in the liver, where MUPs are produced, but in many other tissues, including salivary and lacrimal glands [32,33,84]. Exocrine gland secreting peptide 1 (ESP1) is an androgen-dependent peptide that plays a key role in enhancing female receptive behaviour to allow successful copulation once females have selected a mate [85]. The peptide is produced in the extraorbital lacrimal glands by most wild-derived mouse strains and is released in male tears, although (similar to darcin) many domesticated laboratory strains do not express this peptide at functional levels. While we can only speculate that LCMV infection in lacrimal glands will alter expression of this key peptide, it is notable that pregnancy block occurs when newly mated females are exposed to males that excrete a different level of ESP1 to the male they mated with, in the absence of the mating male [78].

It remains to be tested whether females block pregnancy in response to the changing scent of a mate that is infective with LCMV. A number of factors make this an intriguing hypothesis. First, LCMV infection potentially has high short- and long-term costs for offspring (particularly for males) which could provide strong selection for preimplantation termination of pregnancy when females are at strong risk of infection in early gestation, or when a male's infection becomes evident only after mating. Second, timing of the sensitive period for pregnancy block after mating (within the first 4 days) is appropriate to the likely development of scent changes in recently infected males. Third, males undergo substantial changes in scent that we might expect to stimulate pregnancy block based on experimental manipulations of scent components that have stimulated preimplantation failure in laboratory studies using healthy males. Lastly, pregnancy block does not occur if scent from the mating male remains present when females are exposed to scent from an unfamiliar male. This continued presence of an unchanged mate's scent would allow females to recognise that the unfamiliar scent comes from a different male, whereas an unfamiliar scent that replaces the mate's scent could signify a change in the familiar mate's scent.

In conclusion, our findings suggest that MUP signalling in male mice is highly sensitive to LCMV WE infection and to the development of immunity through the generation of a virus-specific CD8 CTL response. Such T cells are generated after virus infects antigen presenting cells (primary dendritic cells). Dendritic cells process viral proteins into peptides that bind to the host's MHC and are transported to the infected cell's surface [33,86–88]. Given that MUP signals play key roles in female mate choice and attraction, it is tempting to speculate that the negative impact of LCMV on female reproductive success and that of their offspring may have contributed to the evolution of a sexual signaling system that is sensitive to a male's infective status and ability to develop immunity to local variants of this virus. There has been a long-standing interest in the role of MHC in generating pheromones that influence mate choice ever since Boyse and colleagues used inbred mice and their H-2 congenic strains to demonstrate the ability of mice to discriminate between urine scents from strains that differ only in MHC type [89–93]. However, it has been more challenging to isolate MHC effects on a variable genetic background [94–97]. The roles that MHC-associated chemosignals play in mate choice have also proven controversial. For example, when differences in both MHC and MUP genotypes among wild mice are taken into account, females use the polymorphic profile of MUPs in male scent marks to recognise males but not their MHC-associated chemosignals [38], while mice mating freely in semi-natural enclosures avoid mates that share MUP genotype but not MHC [98]. By contrast, pregnancy block among laboratory mice is stimulated by MHC differences [74,76], by ESP1 expression [78] (encoded by a gene tightly linked to MHC on mouse chromosome 17, [5]) and by strain differences in low molecular weight ligands bound to MUPs [77]. We propose that further study of impact of LCMV and its variants on sexual signaling in mice might provide a test system to explain the relationship between MUP and MHC haplotypes, potentially reconciling one of the most notable longstanding debates in mouse semiochemistry.

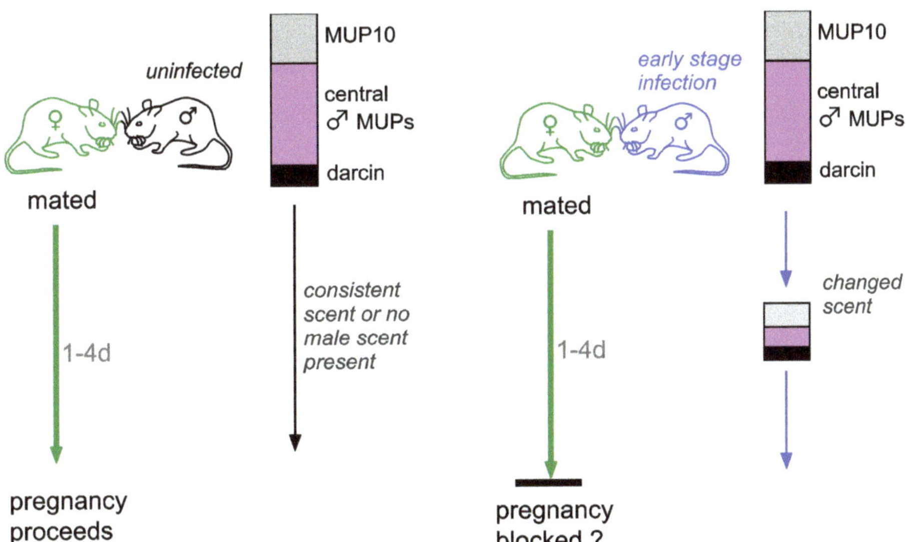

Figure 7. A putative model for the potential of LCMV to elicit pregnancy block (the Bruce effect). LCMV can be transmitted via semen and other excreta from infected mice. Females learn the scent signature of their mate in response to vaginocervical stimulation during mating. Males that are not infected with LCMV (black mouse) produce a strong and consistent signature of MUPs and associated ligands. Detection of consistent mate's scent (or no male scent) over first 3 days post mating protects against implantation failure and pregnancy proceeds. Males at an early stage of LCMV infection during viral replication (blue mouse) initially produce strong MUP signals that rapidly become suppressed (also changing the profile of bound ligands). Exposure to changed scent of mate over first 4 days post mating in the absence of the remembered mate's scent is predicted to inhibit prolactin release and leads to pregnancy failure. Bars indicate the amount of darcin (black), male-biased central MUPs (purple) and central MUP10 (grey) produced by males (data are mean mg MUP / mg creatine produced by singly housed C57BL/6 males in this study). See text for further details.

Author Contributions: Conceptualization, M.B.A.O., J.L.H. and R.J.B.; methodology, M.B.A.O., B.C.W., J.L.H., A.D., M.C.P. and R.J.B.; software, B.C.W., J.L.H. and R.J.B.; validation, M.B.A.O., B.C.W., J.L.H. and R.J.B.; formal analysis, M.B.A.O., J.L.H. and R.J.B.; investigation, B.C.W., J.L.H. and R.J.B.; resources, M.B.A.O., J.L.H. and R.J.B.; data curation, B.C.W., J.L.H. and R.J.B.; writing—original draft preparation, M.B.A.O., J.L.H., R.J.B.; writing—review and editing, M.B.A.O., B.C.W., J.L.H.

and R.J.B.; visualization, B.C.W., J.L.H. and R.J.B.; supervision, M.B.A.O., J.L.H. and R.J.B.; project administration, M.B.A.O., J.L.H. and R.J.B.; funding acquisition, M.B.A.O., J.L.H. and R.J.B. All authors have read and agreed to the published version of the manuscript.

Funding: This research was funded by the National Institutes of Health grant number AI009484 and Biotechnology and Biological Sciences Research Council grant number BB/J002631/1.

Institutional Review Board Statement: The study was conducted according to the guidelines of the Association for Assessment and Accreditation of Laboratory Animal Care (AAALAC) and approved by the Institutional Animal Care and Use Committee (IACUC) of The Scripps Research Institute (Protocol Number 09-0098, approved 5 May 2018).

Informed Consent Statement: Not applicable.

Data Availability Statement: Data supporting reported results can be found in this manuscript.

Acknowledgments: We are grateful to Philip Brownridge for exceptional instrument support. This is Publication Number 30070 from Department of Immunology and Microbiology, The Scripps Research Institute, La Jolla, CA.

Conflicts of Interest: The authors declare no conflict of interest.

References

1. Finlayson, J.S.; Asofsky, R.; Potter, M.; Runner, C.C. Major urinary protein complex of normal mice: Origin. *Science* **1965**, *149*, 981–982. [CrossRef] [PubMed]
2. Beynon, R.J.; Hurst, J.L. Multiple roles of major urinary proteins in the house mouse, *Mus domesticus*. *Biochem. Soc. Trans.* **2003**, *31*, 142–146. [CrossRef]
3. Beynon, R.J.; Hurst, J.L. Urinary proteins and the modulation of chemical scents in mice and rats. *Peptides* **2004**, *25*, 1553–1563. [CrossRef] [PubMed]
4. Beynon, R.J.; Armstrong, S.D.; Gómez-Baena, G.; Lee, V.; Simpson, D.; Unsworth, J.; Hurst, J.L. The complexity of protein semiochemistry in mammals. *Biochem. Soc. Trans.* **2014**, *42*, 837–845. [CrossRef]
5. Green, J.P.; Holmes, A.M.; Davidson, A.J.; Paterson, S.; Stockley, P.; Beynon, R.J.; Hurst, J.L. The Genetic Basis of Kin Recognition in a Cooperatively Breeding Mammal. *Curr. Biol.* **2015**, *25*, 2631–2641. [CrossRef]
6. Hurst, J.L.; Beynon, R.J. Rodent urinary proteins: Genetic identity signals and pheromones. In *Chemical Signals in Vertebrates 12*; Springer: Berlin/Heidelberg, Germany, 2013; pp. 117–133.
7. Hurst, J.L.; Payne, C.E.; Nevison, C.M.; Marie, A.D.; Humphries, R.E.; Robertson, D.H.; Cavaggioni, A.; Beynon, R.J. Individual recognition in mice mediated by major urinary proteins. *Nature* **2001**, *414*, 631–634. [CrossRef]
8. Hurst, J.L.; Beynon, R.J.; Armstrong, S.D.; Davidson, A.J.; Roberts, S.A.; Gómez-Baena, G.; Smadja, C.M.; Ganem, G. Molecular heterogeneity in major urinary proteins of *Mus musculus* subspecies: Potential candidates involved in speciation. *Sci. Rep.* **2017**, *7*, 44992. [CrossRef]
9. Kaur, A.W.; Ackels, T.; Kuo, T.H.; Cichy, A.; Dey, S.; Hays, C.; Kateri, M.; Logan, D.W.; Marton, T.F.; Spehr, M.; et al. Murine pheromone proteins constitute a context-dependent combinatorial code governing multiple social behaviors. *Cell* **2014**, *157*, 676–688. [CrossRef]
10. Lopes, P.C.; König, B. Choosing a healthy mate: Sexually attractive traits as reliable indicators of current disease status in house mice. *Anim. Behav.* **2016**, *111*, 119–126. [CrossRef]
11. Nelson, A.C.; Cunningham, C.B.; Ruff, J.S.; Potts, W.K. Protein pheromone expression levels predict and respond to the formation of social dominance networks. *J. Evol. Biol.* **2015**, *28*, 1213–1224. [CrossRef]
12. Roberts, S.A.; Prescott, M.C.; Davidson, A.J.; Mclean, L.; Beynon, R.J.; Hurst, J.L. Individual odour signatures that mice learn are shaped by involatile major urinary proteins (MUPs). *BMC Biol.* **2018**, *16*, 48. [CrossRef]
13. Stopka, P.; Janotova, K.; Heyrovsky, D. The advertisement role of major urinary proteins in mice. *Physiol. Behav.* **2007**, *91*, 667–670. [CrossRef] [PubMed]
14. Thoß, M.; Luzynski, K.C.; Enk, V.M.; Razzazi-Fazeli, E.; Kwak, J.; Ortner, I.; Penn, D.J. Regulation of volatile and non-volatile pheromone attractants depends upon male social status. *Sci. Rep.* **2019**, *9*, 489. [CrossRef] [PubMed]
15. Demir, E.; Li, K.; Bobrowski-Khoury, N.; Sanders, J.I.; Beynon, R.J.; Hurst, J.L.; Kepecs, A.; Axel, R. The pheromone darcin drives a circuit for innate and reinforced behaviours. *Nature* **2020**, *578*, 137–141. [CrossRef] [PubMed]
16. Hoffman, E.; Pickavance, L.; Thippeswamy, T.; Beynon, R.J.; Hurst, J.L. The male sex pheromone darcin stimulates hippocampal neurogenesis and cell proliferation in the subventricular zone in female mice. *Front. Behav. Neurosci.* **2015**, *9*, 106. [CrossRef]
17. Phelan, M.M.; Mclean, L.; Simpson, D.M.; Hurst, J.L.; Beynon, R.J.; Lian, L.Y. 1H, 15N and 13C resonance assignment of darcin, a mouse major urinary protein. *Biomol. NMR Assign.* **2010**, *4*, 239–241. [CrossRef] [PubMed]
18. Roberts, S.A.; Simpson, D.M.; Armstrong, S.D.; Davidson, A.J.; Robertson, D.H.; Mclean, L.; Beynon, R.J.; Hurst, J.L. Darcin: A male pheromone that stimulates female memory and sexual attraction to an individual male's odour. *BMC Biol.* **2010**, *8*, 75. [CrossRef] [PubMed]

19. Austen, J. *Pride and Prejudice*; T. Egerton: Whitehall, UK, 1813.
20. Roberts, S.A.; Davidson, A.J.; Mclean, L.; Beynon, R.J.; Hurst, J.L. Pheromonal induction of spatial learning in mice. *Science* **2012**, *338*, 1462–1465. [CrossRef]
21. Sheehan, M.J.; Lee, V.; Corbett-Detig, R.; Bi, K.; Beynon, R.J.; Hurst, J.L.; Nachman, M.W. Selection on Coding and Regulatory Variation Maintains Individuality in Major Urinary Protein Scent Marks in Wild Mice. *PLoS Genet.* **2016**, *12*, e1005891. [CrossRef]
22. Beynon, R.J.; Armstron, S.D.; Claydon, A.J.; Davidson, A.; Eyers, C.E.; Langridge, J.I.; Gómez-Baena, G.; Harman, V.M.; Hurst, J.L.; Victoria, L.; et al. Mass spectrometry for structural analysis and quantification of the Major Urinary Proteins of the house mouse. *Int. J. Mass Spectrom.* **2015**, *391*, 146–156. [CrossRef]
23. Mudge, J.M.; Armstrong, S.D.; Mclaren, K.; Beynon, R.J.; Hurst, J.L.; Nicholson, C.; Robertson, D.H.; Wilming, L.G.; Harrow, J.L. Dynamic instability of the major urinary protein gene family revealed by genomic and phenotypic comparisons between C57 and 129 strain mice. *Genome Biol.* **2008**, *9*, R91. [CrossRef] [PubMed]
24. Janotova, K.; Stopka, P. The level of major urinary proteins is socially regulated in wild *Mus musculus* musculus. *J. Chem. Ecol.* **2011**, *37*, 647–656. [CrossRef]
25. Stockley, P.; Bottell, L.; Hurst, J.L. Wake up and smell the conflict: Odour signals in female competition. *Philos. Trans. R. Soc. Lond B Biol. Sci.* **2013**, *368*, 20130082. [CrossRef] [PubMed]
26. Nelson, A.C.; Cauceglia, J.W.; Merkley, S.D.; Youngson, N.A.; Oler, A.J.; Nelson, R.J.; Cairns, B.R.; Whitelaw, E.; Potts, W.K. Reintroducing domesticated wild mice to sociality induces adaptive transgenerational effects on MUP expression. *Proc. Natl. Acad. Sci. USA* **2013**, *110*, 19848–19853. [CrossRef] [PubMed]
27. Andersson, M. *Sexual Selection*; Princeton University Press: Princeton, NJ, USA, 1994.
28. Wyatt, T.D. *Pheromones and Animal Behavior*; Cambridge University Press: Cambridge, UK, 2014.
29. Teng, M.N.; Borrow, P.; Oldstone, M.B.; De La Torre, J.C. A single amino acid change in the glycoprotein of lymphocytic choriomeningitis virus is associated with the ability to cause growth hormone deficiency syndrome. *J. Virol.* **1996**, *70*, 8438–8443. [CrossRef] [PubMed]
30. Teng, M.N.; Oldstone, M.B.; De La Torre, J.C. Suppression of lymphocytic choriomeningitis virus—Induced growth hormone deficiency syndrome by disease-negative virus variants. *Virology* **1996**, *223*, 113–119. [CrossRef] [PubMed]
31. Ware, B.C.; Sullivan, B.M.; Lavergne, S.; Marro, B.S.; Egashira, T.; Campbell, K.P.; Elder, J.; Oldstone, M.B.A. A unique variant of lymphocytic choriomeningitis virus that induces pheromone binding protein MUP: Critical role for CTL. *Proc. Natl. Acad. Sci. USA* **2019**, *116*, 18001–18008. [CrossRef] [PubMed]
32. Lipkin, W.I.; Villarreal, L.P.; Oldstone, M.B. Whole animal section in situ hybridization and protein blotting: New tools in molecular analysis of animal models for human disease. *Curr. Top. Microbiol. Immunol.* **1989**, *143*, 33–54. [CrossRef] [PubMed]
33. Oldstone, M.B. Biology and pathogenesis of lymphocytic choriomeningitis virus infection. *Curr. Top. Microbiol. Immunol.* **2002**, *263*, 83–117. [CrossRef]
34. Roberts, S.C.; Little, A.C.; Burriss, R.P.; Cobey, K.D.; Klapilová, K.; Havlíček, J.; Jones, B.C.; Debruine, L.; Petrie, M. Partner choice, relationship satisfaction, and oral contraception: The congruency hypothesis. *Psychol. Sci.* **2014**, *25*, 1497–1503. [CrossRef]
35. Ahmed, R.; Oldstone, M.B. Organ-specific selection of viral variants during chronic infection. *J. Exp. Med.* **1988**, *167*, 1719–1724. [CrossRef]
36. Berger, D.P.; Homann, D.; Oldstone, M.B. Defining parameters for successful immunocytotherapy of persistent viral infection. *Virology* **2000**, *266*, 257–263. [CrossRef]
37. Kunz, S.; Sevilla, N.; Mcgavern, D.B.; Campbell, K.P.; Oldstone, M.B. Molecular analysis of the interaction of LCMV with its cellular receptor [alpha]-dystroglycan. *J. Cell Biol.* **2001**, *155*, 301–310. [CrossRef]
38. Cheetham, S.A.; Thom, M.D.; Jury, F.; Ollier, W.E.; Beynon, R.J.; Hurst, J.L. The genetic basis of individual-recognition signals in the mouse. *Curr. Biol.* **2007**, *17*, 1771–1777. [CrossRef]
39. Evershed, R.P.; Robertson, D.H.; Beynon, R.J.; Green, B.N. Application of electrospray ionization mass spectrometry with maximum-entropy analysis to allelic 'fingerprinting' of major urinary proteins. *Rapid Commun. Mass Spectrom.* **1993**, *7*, 882–886. [CrossRef]
40. Robertson, D.H.; Hurst, J.L.; Searle, J.B.; Gündüz, I.; Beynon, R.J. Characterization and comparison of major urinary proteins from the house mouse, *Mus musculus domesticus*, and the aboriginal mouse, *Mus macedonicus*. *J. Chem. Ecol.* **2007**, *33*, 613–630. [CrossRef] [PubMed]
41. Cheetham, S.A.; Smith, A.L.; Armstrong, S.D.; Beynon, R.J.; Hurst, J.L. Limited variation in the major urinary proteins of laboratory mice. *Physiol. Behav.* **2009**, *96*, 253–261. [CrossRef] [PubMed]
42. Clissold, P.M.; Bishop, J.O. Variation in mouse major urinary protein (MUP) genes and the MUP gene products within and between inbred lines. *Gene* **1982**, *18*, 211–220. [CrossRef]
43. Armstrong, S.D.; Robertson, D.H.; Cheetham, S.A.; Hurst, J.L.; Beynon, R.J. Structural and functional differences in isoforms of mouse major urinary proteins: A male-specific protein that preferentially binds a male pheromone. *Biochem. J.* **2005**, *391*, 343–350. [CrossRef] [PubMed]
44. Lee, W.; Khan, A.; Curley, J.P. Major urinary protein levels are associated with social status and context in mouse social hierarchies. *Proc. Biol. Sci.* **2017**, 284. [CrossRef]
45. Berger, F.G.; Szoka, P. Biosynthesis of the major urinary proteins in mouse liver: A biochemical genetic study. *Biochem. Genet.* **1981**, *19*, 1261–1273. [CrossRef]

46. Derous, D.; Mitchell, S.E.; Wang, L.; Green, C.L.; Wang, Y.; Chen, L.; Han, J.J.; Promislow, D.E.L.; Lusseau, D.; Douglas, A.; et al. The effects of graded levels of calorie restriction: XI. Evaluation of the main hypotheses underpinning the life extension effects of CR using the hepatic transcriptome. *Aging* **2017**, *9*, 1770–1824. [CrossRef]
47. Mitchell, S.J.; Madrigal-Matute, J.; Scheibye-Knudsen, M.; Fang, E.; Aon, M.; González-Reyes, J.A.; Cortassa, S.; Kaushik, S.; Gonzalez-Freire, M.; Patel, B.; et al. Effects of Sex, Strain, and Energy Intake on Hallmarks of Aging in Mice. *Cell Metab.* **2016**, *23*, 1093–1112. [CrossRef] [PubMed]
48. Pallauf, K.; Günther, I.; Chin, D.; Rimbach, G. In Contrast to Dietary Restriction, Application of Resveratrol in Mice Does not Alter Mouse Major Urinary Protein Expression. *Nutrients* **2020**, *12*, 815. [CrossRef]
49. Hui, X.; Zhu, W.; Wang, Y.; Lam, K.S.; Zhang, J.; Wu, D.; Kraegen, E.W.; Li, Y.; Xu, A. Major urinary protein-1 increases energy expenditure and improves glucose intolerance through enhancing mitochondrial function in skeletal muscle of diabetic mice. *J. Biol. Chem.* **2009**, *284*, 14050–14057. [CrossRef] [PubMed]
50. Isseroff, H.; Sylvester, P.W.; Held, W.A. Effects of Schistosoma mansoni on androgen regulated gene expression in the mouse. *Mol. Biochem. Parasitol.* **1986**, *18*, 401–412. [CrossRef]
51. Manivannan, B.; Rawson, P.; Jordan, T.W.; Secor, W.E.; La Flamme, A.C. Differential patterns of liver proteins in experimental murine hepatosplenic schistosomiasis. *Infect. Immun.* **2010**, *78*, 618–628. [CrossRef] [PubMed]
52. Garratt, M.; Stockley, P.; Armstrong, S.D.; Beynon, R.J.; Hurst, J.L. The scent of senescence: Sexual signalling and female preference in house mice. *J. Evol. Biol.* **2011**, *24*, 2398–2409. [CrossRef] [PubMed]
53. Lanuza, E.; Martín-Sánchez, A.; Marco-Manclús, P.; Cádiz-Moretti, B.; Fortes-Marco, L.; Hernández-Martínez, A.; Mclean, L.; Beynon, R.J.; Hurst, J.L.; Martínez-García, F. Sex pheromones are not always attractive: Changes induced by learning and illness in mice. *Anim. Behav.* **2014**, *97*, 265–272. [CrossRef]
54. Berry, A.; Bellisario, V.; Capoccia, S.; Tirassa, P.; Calza, A.; Alleva, E.; Cirulli, F. Social deprivation stress is a triggering factor for the emergence of anxiety- and depression-like behaviours and leads to reduced brain BDNF levels in C57BL/6J mice. *Psychoneuroendocrinology* **2012**, *37*, 762–772. [CrossRef]
55. Garratt, M.; Mcardle, F.; Stockley, P.; Vasilaki, A.; Beynon, R.J.; Jackson, M.J.; Hurst, J.L. Tissue-dependent changes in oxidative damage with male reproductive effort in house mice. *Funct. Ecol.* **2012**, *26*, 423–433. [CrossRef]
56. Rivière, Y.; Gresser, I.; Guillon, J.C.; Bandu, M.T.; Ronco, P.; Morel-Maroger, L.; Verroust, P. Severity of lymphocytic choriomeningitis virus disease in different strains of suckling mice correlates with increasing amounts of endogenous interferon. *J. Exp. Med.* **1980**, *152*, 633–640. [CrossRef] [PubMed]
57. Childs, J.E.; Glass, G.E.; Korch, G.W.; Ksiazek, T.G.; Leduc, J.W. Lymphocytic choriomeningitis virus infection and house mouse (Mus musculus) distribution in urban Baltimore. *Am. J. Trop. Med. Hyg.* **1992**, *47*, 27–34. [CrossRef]
58. Blasdell, K.R.; Becker, S.D.; Hurst, J.; Begon, M.; Bennett, M. Host range and genetic diversity of arenaviruses in rodents, United Kingdom. *Emerg. Infect. Dis.* **2008**, *14*, 1455–1458. [CrossRef] [PubMed]
59. Morita, C.; Matsuura, Y.; Kawashima, E.; Takahashi, S.; Kawaguchi, J.; Iida, S.; Yamanaka, T.; Jitsukawa, W. Seroepidemiological survey of lymphocytic choriomeningitis virus in wild house mouse (*Mus musculus*) in Yokohama Port, Japan. *J. Vet. Med. Sci.* **1991**, *53*, 219–222. [CrossRef] [PubMed]
60. Kwak, J.; Grigsby, C.C.; Rizki, M.M.; Preti, G.; Köksal, M.; Josue, J.; Yamazaki, K.; Beauchamp, G.K. Differential binding between volatile ligands and major urinary proteins due to genetic variation in mice. *Physiol. Behav.* **2012**, *107*, 112–120. [CrossRef]
61. Rich, T.; Hurst, J.L. The competing countermarks hypothesis: Reliable assessment of competitive ability by potential mates. *Anim. Behav.* **1999**, *58*, 1027–1037. [CrossRef]
62. Martínez-García, F.; Martínez-Ricós, J.; Agustín-Pavón, C.; Martínez-Hernández, J.; Novejarque, A.; Lanuza, E. Refining the dual olfactory hypothesis: Pheromone reward and odour experience. *Behav. Brain Res.* **2009**, *200*, 277–286. [CrossRef] [PubMed]
63. Ramm, S.A.; Cheetham, S.A.; Hurst, J.L. Encoding choosiness: Female attraction requires prior physical contact with individual male scents in mice. *Proc. Biol. Sci.* **2008**, *275*, 1727–1735. [CrossRef]
64. Kreshover, S.J.; Hancock, J.A. The effect of lymphocytic choriomeningitis on pregnancy and dental tissues in mice. *J. Dent. Res.* **1956**, *35*, 467–478. [CrossRef]
65. Oldstone, M.B.A.; Ware, B.C.; Horton, L.E.; Welch, M.J.; Aiolfi, R.; Zarpellon, A.; Ruggeri, Z.M.; Sullivan, B.M. Lymphocytic choriomeningitis virus Clone 13 infection causes either persistence or acute death dependent on IFN-1, cytotoxic T lymphocytes (CTLs), and host genetics. *Proc. Natl. Acad. Sci. USA* **2018**, *115*, E7814–E7823. [CrossRef] [PubMed]
66. Asano, M.S.; Ahmed, R. CD8 T cell memory in B cell-deficient mice. *J. Exp. Med.* **1996**, *183*, 2165–2174. [CrossRef] [PubMed]
67. Bruce, H.M. An exteroceptive block to pregnancy in the mouse. *Nature* **1959**, *184*, 105. [CrossRef] [PubMed]
68. Brennan, P.; Kaba, H.; Keverne, E.B. Olfactory recognition: A simple memory system. *Science* **1990**, *250*, 1223–1226. [CrossRef] [PubMed]
69. Kaba, H.; Rosser, A.; Keverne, B. Neural basis of olfactory memory in the context of pregnancy block. *Neuroscience* **1989**, *32*, 657–662. [CrossRef]
70. Becker, S.D.; Hurst, J.L. Female behaviour plays a critical role in controlling murine pregnancy block. *Proc. Biol. Sci.* **2009**, *276*, 1723–1729. [CrossRef]
71. Brennan, P.A. Outstanding issues surrounding vomeronasal mechanisms of pregnancy block and individual recognition in mice. *Behav. Brain Res.* **2009**, *200*, 287–294. [CrossRef]

72. Brennan, P.A. The nose knows who's who: Chemosensory individuality and mate recognition in mice. *Horm. Behav.* **2004**, *46*, 231–240. [CrossRef] [PubMed]
73. Stokes, R.H.; Sandel, A.A. Data quality and the comparative method: The case of pregnancy failure in rodents. *J. Mammal.* **2019**, *100*, 1436–1446. [CrossRef]
74. Leinders-Zufall, T.; Brennan, P.; Widmayer, P.; Chandramani, S.P.; Maul-Pavicic, A.; Jäger, M.; Li, X.H.; Breer, H.; Zufall, F.; Boehm, T. MHC class I peptides as chemosensory signals in the vomeronasal organ. *Science* **2004**, *306*, 1033–1037. [CrossRef]
75. Thomas, K.J.; Preeji, K.P.; Ranjith, S. Imprinting of a Nonpheromonal Cue and Its Protective Effect on Alien Male-Induced Implantation Failure in Mice. *Chem. Sens.* **2018**, *43*, 523–527. [CrossRef] [PubMed]
76. Yamazaki, K.; Beauchamp, G.K.; Wysocki, C.J.; Bard, J.; Thomas, L.; Boyse, E.A. Recognition of H-2 types in relation to the blocking of pregnancy in mice. *Science* **1983**, *221*, 186–188. [CrossRef] [PubMed]
77. Peele, P.; Salazar, I.; Mimmack, M.; Keverne, E.B.; Brennan, P.A. Low molecular weight constituents of male mouse urine mediate the pregnancy block effect and convey information about the identity of the mating male. *Eur. J. Neurosci.* **2003**, *18*, 622–628. [CrossRef] [PubMed]
78. Hattori, T.; Osakada, T.; Masaoka, T.; Ooyama, R.; Horio, N.; Mogi, K.; Nagasawa, M.; Haga-Yamanaka, S.; Touhara, K.; Kikusui, T. Exocrine Gland-Secreting Peptide 1 Is a Key Chemosensory Signal Responsible for the Bruce Effect in Mice. *Curr. Biol.* **2017**, *27*, 3197–3201.e3. [CrossRef] [PubMed]
79. Kaba, H.; Fujita, H.; Agatsuma, T.; Matsunami, H. Maternally inherited peptides as strain-specific chemosignals. *Proc. Natl. Acad. Sci. USA* **2020**, *117*, 30738–30743. [CrossRef] [PubMed]
80. Coopersmith, C.B.; Lenington, S. Pregnancy block in house mice (Mus domesticus) as a function of t-complex genotype: Examination of the mate choice and male infanticide hypotheses. *J. Comp. Psychol.* **1998**, *112*, 82–91. [CrossRef]
81. Labov, J.B. Pregnancy blocking in rodents: Adaptive advantages for females. *Am. Nat.* **1981**, *118*, 361–371. [CrossRef]
82. Rülicke, T.; Guncz, N.; Wedekind, C. Early maternal investment in mice: No evidence for compatible-genes sexual selection despite hybrid vigor. *J. Evolut. Biol.* **2006**, *19*, 922–928. [CrossRef]
83. Sturm, T.; Leinders-Zufall, T.; Maček, B.; Walzer, M.; Jung, S.; Pömmerl, B.; Stevanović, S.; Zufall, F.; Overath, P.; Rammensee, H.G. Mouse urinary peptides provide a molecular basis for genotype discrimination by nasal sensory neurons. *Nat. Commun.* **2013**, *4*, 1616. [CrossRef]
84. Fazakerley, J.K.; Southern, P.; Bloom, F.; Buchmeier, M.J. High resolution in situ hybridization to determine the cellular distribution of lymphocytic choriomeningitis virus RNA in the tissues of persistently infected mice: Relevance to arenavirus disease and mechanisms of viral persistence. *J. Gen. Virol.* **1991**, *72*, 1611–1625. [CrossRef]
85. Haga, S.; Hattori, T.; Sato, T.; Sato, K.; Matsuda, S.; Kobayakawa, R.; Sakano, H.; Yoshihara, Y.; Kikusui, T.; Touhara, K. The male mouse pheromone ESP1 enhances female sexual receptive behaviour through a specific vomeronasal receptor. *Nature* **2010**, *466*, 118–122. [CrossRef]
86. Ng, C.T.; Sullivan, B.M.; Oldstone, M.B. The role of dendritic cells in viral persistence. *Curr. Opin. Virol.* **2011**, *1*, 160–166. [CrossRef]
87. Zinkernagel, R.M.; Doherty, P.C. Restriction of in vitro T cell-mediated cytotoxicity in lymphocytic choriomeningitis within a syngeneic or semiallogeneic system. *Nature* **1974**, *248*, 701–702. [CrossRef]
88. Zinkernagel, R.M.; Oldstone, M.B. Cells that express viral antigens but lack H-2 determinants are not lysed by immune thymus-derived lymphocytes but are lysed by other antiviral immune attack mechanisms. *Proc. Natl. Acad. Sci. USA* **1976**, *73*, 3666–3670. [CrossRef]
89. Singer, A.G.; Tsuchiya, H.; Wellington, J.L.; Beauchamp, G.K.; Yamazaki, K. Chemistry of odortypes in mice: Fractionation and bioassay. *J. Chem. Ecol.* **1993**, *19*, 569–579. [CrossRef]
90. Yamaguchi, M.; Yamazaki, K.; Beauchamp, G.K.; Bard, J.; Thomas, L.; Boyse, E.A. Distinctive urinary odors governed by the major histocompatibility locus of the mouse. *Proc. Natl. Acad. Sci. USA* **1981**, *78*, 5817–5820. [CrossRef] [PubMed]
91. Yamazaki, K.; Beauchamp, G.K.; Kupniewski, D.; Bard, J.; Thomas, L.; Boyse, E.A. Familial imprinting determines H-2 selective mating preferences. *Science* **1988**, *240*, 1331–1332. [CrossRef] [PubMed]
92. Yamazaki, K.; Singer, A.; Beauchamp, G.K. Origin, functions and chemistry of H-2 regulated odorants. *Genetica* **1999**, *104*, 235–240. [CrossRef] [PubMed]
93. Yamazaki, K.; Beauchamp, G.K.; Curran, M.; Bard, J.; Boyse, E.A. Parent-progeny recognition as a function of MHC odortype identity. *Proc. Natl. Acad. Sci. USA* **2000**, *97*, 10500–10502. [CrossRef] [PubMed]
94. Carroll, L.S.; Penn, D.J.; Potts, W.K. Discrimination of MHC-derived odors by untrained mice is consistent with divergence in peptide-binding region residues. *Proc. Natl. Acad. Sci. USA* **2002**, *99*, 2187–2192. [CrossRef] [PubMed]
95. Kwak, J.; Willse, A.; Preti, G.; Yamazaki, K.; Beauchamp, G.K. In search of the chemical basis for MHC odourtypes. *Proc. Biol. Sci.* **2010**, *277*, 2417–2425. [CrossRef] [PubMed]
96. Overath, P.; Sturm, T.; Rammensee, H.G. Of volatiles and peptides: In search for MHC-dependent olfactory signals in social communication. *Cell Mol. Life Sci.* **2014**, *71*, 2429–2442. [CrossRef] [PubMed]
97. Willse, A.; Kwak, J.; Yamazaki, K.; Preti, G.; Wahl, J.H.; Beauchamp, G.K. Individual odortypes: Interaction of MHC and background genes. *Immunogenetics* **2006**, *58*, 967–982. [CrossRef] [PubMed]
98. Sherborne, A.L.; Thom, M.D.; Paterson, S.; Jury, F.; Ollier, W.E.; Stockley, P.; Beynon, R.J.; Hurst, J.L. The genetic basis of inbreeding avoidance in house mice. *Curr. Biol.* **2007**, *17*, 2061–2066. [CrossRef]

MDPI
St. Alban-Anlage 66
4052 Basel
Switzerland
Tel. +41 61 683 77 34
Fax +41 61 302 89 18
www.mdpi.com

Viruses Editorial Office
E-mail: viruses@mdpi.com
www.mdpi.com/journal/viruses

www.ingramcontent.com/pod-product-compliance
Lightning Source LLC
LaVergne TN
LVHW070951170726
843515LV00005B/968